Privatisation and Liberalisation in European Telecommunications

The last three decades have witnessed a significant reorganisation of established European telecommunications administrations and the transformation of monopolistic markets to more open and regulated forms of competition. Prompted by structural advancements such as technological innovation, increasing demand for competitive and specialised services, together with the globalisation of business and European harmonisation and integration, these recent developments have marked a fundamental shift away from the traditional organisation of this rapidly expanding industry as a national, state controlled affair.

Privatisation and Liberalisation in European Telecommunications combines a detailed, sector-specific study of comparative telecommunications regimes set in the context of the EC, with an extensive historical and empirical analysis of individual policy management and change as experienced by three diverse regulatory cultures, namely Britain, the Netherlands and France. By adopting a comprehensive analytical framework based on far-reaching literature, the author explores a wide range of theories, addressing key issues at the forefront of contemporary political and academic debate. Do nation states matter in the globalising telecommunications industry? Does the common challenge of techno-global telecommunications restructuring elicit different national responses? What is the significance of a single-speed or multi-speed Europe in implementing telecommunications governance regimes?

This book will be of interest to people studying in the fields of comparative and European business, economics and politics.

Willem Hulsink is Assistant Professor at the Faculty of Business Administration, Erasmus University, Rotterdam. Previously he was a Research Fellow at the Science Policy Research Unit (SPRU) of the University of Sussex at Brighton, and a Lecturer in Policy and Organisation Studies at Tilburg University. He has written several articles in the field of European telecommunications policy and management.

Routledge Studies in International Business and the World Economy

1 **States and Firms**
Multinational enterprises in institutional competition
Razeen Sally

2 **Multinational Restructuring, Internationalization and Small Economies**
The Swedish case
Thomas Andersson, Torbjörn Fredriksson and Roger Svensson

3 **Foreign Direct Investment and Governments**
Catalysts for economic restructuring
Edited by John H. Dunning and Rajneesh Narula

4 **Multinational Investment and Economic Structure**
Globalization and competitiveness
Rajneesh Narula

5 **Entrepreneurship in a Global Context**
Edited by Sue Birley and Ian Macmillan

6 **The Global Structure of Financial Markets**
An overview
Edited by Dilip K. Ghosh and Edgar Ortiz

7 **Alliance Capitalism and Global Business**
John H. Dunning

8 **Multinational Enterprises from the Netherlands**
Edited by Roger van Hoesel and Rajneesh Narula

9 **Competition, Growth Strategies and the Globalization of Services**
Real estate advisory services in Japan, Europe and the United States
Terrence LaPier

10 **European Integration and Foreign Direct Investment in the EU**
The case of the Korean consumer electronics industry
Sang Hyup Shin

11 **New Multinational Enterprises from Korea and Taiwan**
Roger van Hoesel

12 **Competitive Industrial Development in the Age of Information**
The role of co-operation in the technology sector
Edited by Richard J. Braudo and Jeffrey G. Macintosh

13 **The Global Restructuring of the Steel Industry**
Innovations, institutions and industrial change
Anthony P. D'Costa

14 **Privatisation and Liberalisation in European Telecommunications**
Comparing Britain, the Netherlands and France
Willem Hulsink

Privatisation and Liberalisation in European Telecommunications

Comparing Britain, the Netherlands and France

Willem Hulsink

ROUTLEDGE

London and New York

First published 1999
by Routledge
11 New Fetter Lane, London EC4P 4EE

Simultaneously published in the USA and Canada
by Routledge
29 West 35th Street, New York, NY 10001

Typeset in Baskerville by
M Rules
Printed and bound in Great Britain by
Mackays of Chatham PLC, Chatham, Kent

British Library Cataloguing in Publication Data
A catalogue record for this book is available from the British Library

Library of Congress Cataloguing in Publication Data
Hulsink, Willem
Privatisation and liberalisation in European telecommunications: comparing Britain, the Netherlands, and France/Willem Hulsink.
p. cm.
Includes bibliographical references and index.
ISBN 0-415-18003-1 (hb)
1. Telecommunication–Deregulation–Great Britain.
2. Telecommunication–Deregulation–Netherlands.
3. Telecommunication–Deregulation–France
4. Privatization–Europe–Case studies. I. Title.
HE8094.H85 1998
384′.041–dc21 98-21492
CIP

ISBN 0 415 18003 1

I remember that we, Bavink and I, once arrived at the sea, when the half-sun was on the horizon, big, cold and red. Bavink struck his forehead with his fist and cursed: "God, God, I will never paint that, I never could." (. . .) It's as if that sea wants something from me. In it, God is too. God is calling. It is truly no picnic. He is everywhere. And everywhere He is calling Bavink. You go mad from your own name, when it is called out that often. And then Bavink has to paint. Then God has to be put on a piece of canvas with paint. Then Bavink is calling "God". And thus they keep calling one another. To God, it's a game, He is infinite and everywhere. He just keeps calling. But Bavink only has one silly head and one silly right hand and he can only work on one little painting at a time. And whenever he thinks he's caught God, then he has canvas and paint. Then God is everywhere, except where Bavink wants Him. (. . .) Do you know what I wish? That I could make railway timetables. Such a man God leaves in peace, he is not worth the trouble.

(Nescio (1914), *Titaantjes (Little Titans)*. Nijgh & Van Ditmar (1983:45–58))

Contents

List of figures and tables x
Acknowledgements xi

1 Introduction 1

1.1 Introduction 1
1.2 The restructuring of public service industries in Western Europe 2
1.3 Traditional telecommunications regime: domestic monopoly/international cartel 5
1.4 Structural forces setting the national telecommunications agenda 10
1.5 Comparative analysis of telecommunications restructuring in Europe: global forces and domestic responses 15
1.6 Conceptual framework 22
1.7 Explaining governance transformation: an institutional perspective 27
1.8 Comparing telecommunications governance regimes across nations and across time: a research framework 32

2 International restructuring and national strategies and contingencies 37

2.1 Introduction 37
2.2 Comparative public policy studies 38
2.3 Comparative organisation studies 43
2.4 Comparative political economy 48
2.5 Gourevitch's integrative approach: five national variables 53
2.6 Summary 73

3 Structural forces challenging the European national telecommunications regime 75

3.1 Introduction 75
3.2 The ancien régime of telecommunications: public monopoly, PTT and rent-seeking coalitions 76

3.3 Technological (r)evolution 84
3.4 Globalisation and differentiation of communication markets 88
3.5 International deregulation 93
3.6 European integration 99
3.7 Summary and concluding remarks 108

4 The liberalisation, privatisation and regulatory reform of telecommunications in the UK: in care of the market? 111

4.1 Introduction 111
4.2 The wider political-economic setting of UK telecommunications 112
4.3 The politics of economic adjustment in the UK 119
4.4 Demonopolisation of telecommunications in the UK 126
4.5 The liberalisation of the UK telecommunications market 145
4.6 The privatisation of British Telecom 151
4.7 Regulatory reform of UK telecommunications policy 163
4.8 Summary and concluding remarks 168

5 The liberalisation, privatisation and regulatory reform of telecommunications in the Netherlands: in pursuit of a new consensus? 170

5.1 Introduction 170
5.2 The wider political-economic setting of Dutch telecommunications 171
5.3 The politics of economic adjustment in the Netherlands 176
5.4 Demonopolisation of telecommunications in the Netherlands 180
5.5 The liberalisation of the Dutch telecommunications market 199
5.6 The privatisation of KPN/PTT Telecom 207
5.7 Regulatory reform of Dutch telecommunications policy 216
5.8 Summary and concluding remarks 222

6 The liberalisation, privatisation and regulatory reform of French telecommunications: still in care of the state? 225

6.1 Introduction 225
6.2 The wider political-economic setting of French telecommunications 226
6.3 The politics of economic adjustment in France 236
6.4 Demonopolisation of telecommunications in France 241
6.5 The liberalisation of the French telecommunications market 258
6.6 The privatisation of France Télécom 262
6.7 Regulatory reform of French telecommunications policy 270
6.8 Summary and concluding remarks 275

7 A comparative institutional analysis of British, Dutch and French telecommunications policy 279

7.1 Introduction 279

7.2 Liberalisation, privatisation and regulatory reform of Dutch, French and UK telecommunications 280

7.3 Explaining the strategic responses of the UK, the Netherlands and France: domestic ruling coalitions and institutional constraints 291

7.4 Concluding remarks 302

8 Convergence and/or divergence in European telecommunications restructuring? 307

8.1 Introduction 307

8.2 Structural forces forging convergence and the persistence of national diversity 308

8.3 Suggestions for further research 315

References 322

Index 343

Figures and tables

Figures

1.1	Ways of comparing political and economic structures and processes	33
1.2	Research framework	35
2.1	Different policy styles	39
2.2	The complete national industrial system	57
3.1	The internal organisation of PTT: functional and discretional areas	78
3.2	The rent-seeking coalition around the PTT monopoly	80
3.3	The disentanglement of the PTT functions and responsibilities	83
3.4	The changing industrial organisation of telecommunications	83
7.1	Governance transformation in European telecommunications, 1980–98	281
8.1	Extended research framework	320

Tables

2.1	Business systems in East Asia	47
4.1	British Telecom: key data	156
4.2	British Telecom: major partnerships	158
5.1	PTT Telecom Netherlands: key data	209
5.2	PTT Telecom Netherlands: major partnerships	215
6.1	France Télécom: key data	263
6.2	France Télécom: strategic linkages	266
7.1	Market, ownership and regulatory changes in British, Dutch and French telecommunications, 1980–98	282

Acknowledgements

The similarities between academic research and the art of painting are striking. Both the academic and the painter feel an almost obsessive need to transform a vague and implicit set of ideas into an artefact that claims a certain degree of originality among the sources it uses and by which it is inspired, in terms of the method and approach, and the distinct argument or perspective it offers. In addition, there are parallels concerning the intangibility of time, which make planning these endeavours extremely hard, and the difficulty of striking a balance between hard work and skill, and between concentration and inspiration. Furthermore, both researcher and artist will experience mixed feelings when the process nears completion, ranging from relief because the job is almost done, to vexation because the work is never really finished, and further details can always be found and nuances added, to the sheer pragmatic necessity of putting in the final touches. Finally, both are painfully aware that the chances of their work ever receiving a wider audience are slim to say the least.

Perhaps the differences between the two disciplines are as striking as the similarities. In academia, given the cumulative and evolutionary development of knowledge, one has to acknowledge one's sources explicitly, and indicate where the ideas and data come from upon which one has elaborated. The artist, on the other hand, has a far greater degree of freedom: he can stay within an established and recognised art form, or combine artistic elements into a new eclectic style, or overhaul established routines by ignoring the past and making his individual contribution on canvas. Second, the academic is less of an individualist than the painter: he has to depend much more on his colleagues and supportive institutions.

In light of the above, I am indebted to a large number of people and organisations, without whose support this study would never have been accomplished. Three people stand out for their involvement in shaping this study throughout. From the early beginning until the final stages, Frans van den Bosch, Giandomenico Majone and Arthur Wassenberg have contributed in a variety of ways to the completion of this job. Their intellectual support and analytical focus have certainly made the argument and structure of the book more transparent and rigorous. They never tired of reading yet another rough draft and coming up with useful suggestions, while at the same time reminding me that all things must eventually come to an end. For this, I am extremely grateful.

While conducting this study, I was fortunate enough to be able to work at the following institutions. All have contributed in their particular way to its completion: Erasmus University Rotterdam (Netherlands): Department of Strategic Management and Business Environment; European University Institute, Florence (Italy): Department of Political and Social Sciences; Tilburg University (The Netherlands): Department of Policy and Organisation Sciences; University of Sussex, Brighton (United Kingdom): Science Policy Research Unit. I would also like to express my gratitude to the Rathenau Institute (The Hague) in the Netherlands for their financial support to carry out this study. In this respect, I especially want to thank my collaborators on the New Telecommunications in the Netherlands research project, namely Andrew Davies, Robin Mansell, Ronald Schapendonk and Anja de Wit, all of whom wholeheartedly contributed to its finalisation.

For the field work of this study I relied on the input and support of officials and experts in the three countries investigated. In France, I want to mention the generosity of J. Arlandis (IDATE), J-P. Chamoux (Droit et Informatique & Université de Marne-la-Vallée), J-P. Nioche (HEC), and T. Védel (CEVIPOF/CNRS) who were patient and willing in replying to my questions and sharing their ideas on French telecommunications policy with me. In the Netherlands, H. Albeda (Consumentenbond), J. Arnbak (Technical University Delft), P. de Graaf (VNO/NCW), M. Kopijn (PA Consulting Group), T. de Liefde (then Ministry of Transport & Public, now BTG), H.J. de Ru (Vrije Universiteit Amsterdam) and N. Siljee (PTT Telecom) took the time to discuss all the 'ins and outs' of restructuring telecommunications in the Netherlands. In the United Kingdom, B. Hogwood (University of Strathclyde), P. Strickland (BT), D. Souter (National Communications Union), J. Taylor (University of Strathclyde) and W. Wigglesworth (Oftel) showed me the idiosyncrasies of UK telecommunications.

Leigh Hancher and Jan Kooiman went through an earlier version of the manuscript; their comments are gratefully acknowledged. Andrew Davies, Frank Boons, Geert-Jan Kemme, Harry Bouwman, Rob van Tulder and Henk Volberda saw sections of the book and sharpened its argument. The contribution of certain people in the final shaping of the book should not be underestimated. Gert Stronkhorst improved the English, and Wil Geurtsen did the final formatting of the manuscript.

This study is dedicated to my family, friends and colleagues, all of whom showed their interest in the progress of the research, either by encouraging me to finish quickly, or by putting things into perspective by saying that it is only a book.

Willem Hulsink

1 Introduction

1.1 Introduction

Since the late 1970s, serious doubts have been cast upon the established telecommunications regime as a public monopoly in all industrial democracies. This is a result of structural developments such as technological innovation, increasing demand for competitive and specialised services, globalisation of business, international deregulation and European integration. Together with the recent moves towards harmonisation and integration of national policies into a larger European Community, the question 'Does the common challenge of techno-global telecommunications restructuring elicit different national responses?' has moved to the forefront of political and academic debate. This book provides a comparative study of three different European responses to the common stimulus of structural changes in the telecommunications industry.

The process of adjustment in European telecommunications manifested itself in the early 1980s, when national policy makers started to realise that the challenge posed by structural developments in technology, markets and international regulation, together with Europe's lack of competitiveness in high-tech industries *vis-à-vis* the USA and Japan, required an appropriate policy response. Recent political debate in the three chosen Western European countries has centred on whether and to what extent the telecommunications monopoly should be liberalised and state control ought to be privatised (and, if yes, at what speed) and how an appropriate new framework for market regulation should be set up.

Instead of a cross-national convergence towards open markets and a level playing field, the outcome of this study reflects a variety of policy measures being implemented, such as non-decision making in one country, incremental adjustments in another or policy innovations in a third. These differences in policy response might be explained by examining the domestic priorities as defined and advocated by key actors and ruling coalitions in the policy arena and the larger national setting, made up of persistent rules, institutions and economic conditions (for example, the scale and the degree of openness of the economy), which shape the process of telecommunications decision making in a particular way. In this study we will examine how the United Kingdom, the Netherlands, France and (to some extent) the European Community as a whole, have adjusted to the new technological, economic, regulatory and international conditions of modern telecommunications. Do the aforementioned states have any manœuvrability in the formulation of a national strategy with respect to telecommunications, reflecting domestic values and priorities? And if so, what factors explain the strategic choice made within a given country?

The aim of this research is to analyse and compare the British, Dutch and French strategies with regard to the restructuring of telecommunications. Section 1.2 discusses the putting into question and the subsequent restructuring of the traditional ownership and control structures of the public service industries in Western Europe. Section 1.3 briefly introduces the traditional telecommunications regime based on domestic public monopoly and international cartel arrangements. Section 1.4 discusses four exogenous variables challenging the sovereignty of national telecommunications regimes in Western Europe, namely the structural forces of technological progress, changing demand and supply conditions, international deregulation, and European integration. Section 1.5 examines the search for an adequate policy response to international techno-economic restructuring. The formation of industrial adjustment strategies will be discussed against the background of the academic debate in comparative policy and business studies on the validity of cross-national convergence versus national diversity. The theoretical concepts throughout the study, deriving from organisation theory, strategic management and political economy, are introduced in Section 1.6. The organisational and institutional perspective adopted in this comparative study of the changes in telecommunications policy in the three countries, is offered in Section 1.7. The research model on the basis of which the study is carried out, is spelt out in Section 1.8.

1.2 The restructuring of public service industries in Western Europe

The focus of this research is on the process of monopolisation and demonopolisation in the West European telecommunications industry. The collectivisation of the provision of telephony and telegraphy that started at the end of the nineteenth century and lasted until the 1980s, became institutionalised through statutory monopolies, authorised by the central government and operated by state-owned PTT administrations. In Europe's mixed economies the state has traditionally played a leading role in key sectors such as railways, telecommunications and energy by acting as entrepreneur, direct supplier of services, lender of last resort and planner for the industry in question or the economy as a whole (Ambrosius 1984; Jänicke 1990). The European governments relied upon interventionist instruments, such as indicative planning, nationalisations and state monopolies, extensive public works and procurement programmes, and demand creation/income redistribution policies. The increase of state intervention at the sector level was paralleled by the establishment of the welfare state at the macro-level. In his *In Care of the State,* De Swaan (1988) has referred to the gradual rise of nationwide, collective and compulsory arrangements to structure the provision of public services, social security and welfare systems. In the formation and implementation of these Keynesian macro-economic policies, the state apparatus was supported by the active participation of centralised labour and employers, organisations, jointly combating inflation and unemployment, and securing substantial levels of economic growth (Shonfield 1965). European governments and their recognised social partners established tripartite arrangements to settle wage/price levels, working conditions and labour market issues. At the end of the 1970s, European public sectors were relatively large and characterised by an extensive bureaucracy, a substantial degree of public ownership and public resource allocation, and ambitious redistribution and employment programmes. Consequently, civil servants and their labour unions had substantial power: public sector employees had a solid legal and political position, based on the civil servants' statute, favourable working conditions and extensive participation mechanisms. Ministerial intervention in the day-to-day operations of the nationalised industries and interference with the pricing, personnel and investment decisions of public managers was frequent and all-pervasive. Majone (1994a, b) has identified a failure of nationalised industries, based on the following elements: the capture of public managers by politicians and trade unions, a general overmanning of state monopolies, ambiguous and inconsistent corporate objectives and portfolio, poor co-ordination among the various public enterprises, and the absence of effective control over public enterprises by Parliament, the courts or the sponsoring minister.

As a consequence of the collapse of the Bretton Woods system of fixed exchange rates and the two oil crises in the 1970s, the post-war period of prolonged economic growth, full employment, low inflation and organised capitalism came to an end (Lash & Urry 1987). European economies were hit hard by the subsequent economic recession and increasing competition from Japan and the newly industrialising countries, resulting eventually in de-industrialisation, long-term unemployment and cutbacks in public spending. From the late 1970s onwards, governments became concerned with how to adjust their stagnating economies to the new techno-economic and international conditions. In order to strengthen Europe's position in a more turbulent and volatile world economy, the established national frameworks, based on extensive welfare state provisions, state-owned infrastructure provision and industrial policies aimed at strengthening the countries' key sectors and national champions, needed rethinking and restructuring. The big challenge for national policy makers in the 1980s became to prepare domestic firms, strategic sectors and the national economy as a whole for international competition, and to develop an appropriate mix of strategic responses (for instance deregulation, privatisation, innovation policy, selective liberalisation).

At the end of the 1970s, politicians, businessmen and intellectuals rediscovered economic liberalism: they promoted the de-collectivisation of the economy by giving clear priority to market mechanisms over government intervention. Littlechild (1978) raised serious doubts about large-scale and far-reaching state intervention in the national economy. He argued that private enterprise (*laissez faire)* and the smooth functioning of markets and competition would not only generate overall efficiency gains, but stimulate entrepreneurship and innovation. As illustrated by Swann (1988) in *The Retreat of the State*, before radical deregulation and privatisation programmes were implemented in the US and the UK, the various activities carried out by the governments in the two economies were critically reviewed and discussed in Parliament. On the European continent, conservative and Christian Democratic parties followed the Anglo-Saxon path by giving priority to scaling down welfare provisions, contracting out public services and privatising state-owned enterprises. By the mid-1980s, the Social Democratic parties were also converted to economic neo-liberalism and they too actively supported pro-market adjustment policies. In their responses to economic stagnation, European governments laid emphasis on curbing public expenditure, on administrative reform and deregulation, on the decentralisation of collective bargaining, on shifting resources from sunset to sunrise sectors and on modernising traditional industries.

Western European governments are now in the process of gradually withdrawing from their active role in the provision of public services and are transferring assets from the public to the private sector. They have granted substantial managerial autonomy to former state enterprises and have separated operational activities from regulatory controls. Deregulation has been described as the process of reducing state control over an industry or activity so as to make it structurally more responsive to market forces (Baldwin & McCrudden 1987). In practice, however, deregulation often means less restrictive or rigid regulation, rather than no regulation at all. Public policy objectives, like universal service, price controls, innovation and fair competition, are now achieved through different means, for instance through market co-ordination, administrative regulation and private law techniques. The former two instruments safeguarding the public interest refer to the control of market entry and market behaviour and to the scrutiny of quality of service provision and overall industrial performance. The latter includes bilateral contracts between the government and the franchisee and specific provisions in the articles of association to prevent hostile take-overs or restrict foreign ownership. Furthermore, in Western Europe, deregulation at the national level has often been followed by reregulation at the EC level. As a consequence, public utilities in Western Europe have lost their pre-existing statutory immunity from the Community's competition legislation and their business activities have become subject to the anti-trust provisions of the Treaty of Rome.

These drastic policy changes implied a shift towards the (managed) liberalisation of market segments, the corporatisation and (intended) privatisation of the former monopolies and a reform of the institutional structure, in order to avoid the conflicting interests of the administration as market player and referee at the same time. The process of liberalising the public utilities refers to the removal of statutory restrictions on market access, henceforward encouraging new entry for private sector participants. The traditional state monopolies have often been replaced by a mixture of (more) market-oriented arrangements, such as new entry, the increase of competitive tendering and franchising, the introduction of yardstick competition and competition between different technological modes and capital market competition. Privatisation refers to the idea and practice of the transfer of productive assets from public ownership and control into private ownership through either public offering, management buy-out or traded sale (Veljanovski 1989). In addition to the fiscal objective to use the receipts from the sale of public assets to reduce government ownership, the decision to privatise has often been motivated by a desire to improve the management of public corporations by reducing state intervention and stimulating efficiency, innovation and customer responsiveness. It often implies a change in the legal form (from public to civil law), a replacement of senior management and a drastic corporate reorganisation. The utilities are transformed from departmental administrations to state-controlled or state-regulated corporations and granted access to capital markets and allowed to develop more flexible personnel policies. Although put at arm's length, the privatised company still has direct links with the government, which more often than not remains the key shareholder in the new situation.

1.3 Traditional telecommunications regime: domestic monopoly/international cartel

The telecommunications industry until recently was characterised by a stable and predictable environment, in which technology, institutional patterns and decision making were relatively stable and well organised. There existed a broad consensus about the organisation of the industry as a national public monopoly, relying upon the collective good of the telecommunications network, the large sunk costs of the infrastructure and a basic orientation towards the domestic market. The telecommunications network was operated by a government-controlled monopoly provider with the exclusive responsibility to serve everyone without discrimination. Three versions of such a hierarchical governance regime were in use throughout the telecommunications domain: government department, public enterprise or regulated private monopoly. In most West European countries the operation of the telecommunications system was exclusively assigned to a government department or a public enterprise, generally known as the PTT, with responsibility for the postal, telegraph and telephone monopolies, and sometimes also for the public money services. These PTTs belonged to the public administration and as such were subject to strong government interference: revenues went into the treasury and decisions regarding tariffs, investments, capital and labour conditions needed legislative acceptance. Special cases were Belgium and Sweden, where separate public monopolies for telecommunications and mail were established. The second mode refers to the public enterprise regime: in Japan and the UK one public corporation, separated (more or less) from the public administration, was in charge of operating the telephone system. The third mode is the private regulated monopoly, in which private corporations were given an exclusive licence to manage the network and provide the postal, telephone and telegraph services. Illustrations can be found in the US and Canada where private firms such as AT&T were given the exclusive licence to function as telephone operators.

The postal and telecommunications sector has traditionally been characterised by a high degree of government intervention. Besides the concern about strategic industries crucial to national security and safety, state involvement existed for two reasons: the natural monopoly attributes of the postal and telecommunications network, and the fact that a nationwide coverage of these networks would contribute to a country's socio-economic cohesion. The economies of scale and the network characteristics of the telecommunications industry generate positive network externalities, in that the value of the telephone service increases as the number of interconnections increases. Furthermore, the telecommunications and mail services were regarded as essential public utilities with an overriding importance for the national economy; they provided socio-economic infrastructures that, when implemented nationwide, reached, interconnected and integrated every citizen into society. National governments promoted the objective of universal public service for telecommunications by an ingenious system of cross-subsidisation, long-distance telephone services sponsoring local calls and international calls financing domestic services (more generally, business use subsidising residential use). The legitimate use of cross-subsidisation in price setting on the one hand served to spread the overall costs of enlarging the network as much as possible and, on the other hand, defended the national monopoly by referring to the need to prevent cream skimming in lucrative market segments. In other words, the public telecommunications network was not only a techno-economic system, but also a socio-political system of burden sharing that promoted the transfer of benefits towards the economically weak in the form of redistribution.

The governance regime of traditional telecommunication, characterised by the provision of a public infrastructure, universal service at the lowest cost, standard price setting and redistribution in charges, has become known as the public monopoly model. To typify the long-standing and protectionist arrangements between domestic public and private actors around the PTT monopoly, Noam coined the term 'postal-industrial complex', referring to a closed system in which government, parliament, the telecommunications administration, the domestic suppliers of telecommunications equipment, the domestic newspaper industry, residential users and the labour unions were involved in a 'rent-seeking' coalition of mutual benefits (Noam 1987, 1992). The exclusive position of the PTT was acknowledged on the condition that all the other major stakeholders in the PTT system were allowed to share in the revenues from its services. The cartel-like system was profitable for insiders and, furthermore, its inefficiencies were hidden by the general downward trend in the cost of electronic technology (Noam 1987:32). Within this regime PTT was the monopolistic network operator and the domestic equipment manufacturer(s) served as its monopsonistic supplier(s). These telecommunications equipment firms produced mainly for the domestic market to meet (public) demand for switching and terminal equipment; they became the PTT's preferred suppliers. The interests of equipment suppliers were protected by 'buy domestic' policies and idiosyncratic standards set by the PTT in narrow collaboration with its established supplier(s) to the exclusion of domestic and foreign outsiders. For governments the PTT monopoly was a major source of revenue for the national treasury; in many countries the PTT was seen as the goose with the golden egg that helped to settle the government's budget deficit.

The PTT configuration also received strong support within the political system from the social-democratic factions in government and parliament, and from trade unions, whose material and ideological interests lay in policy objectives such as the redistribution of charges, universal service, employment and job security, and industrial and regional policy. The interests of households were also incorporated in this postal-industrial complex; the revenues of long-distance telecommunications were redistributed and used to subsidise the local telephone costs of residential users.

As the monopolistic provider of the telecommunications infrastructure, services and terminal equipment, the PTT had to meet two ambiguous goals: it had to perform reasonably well economically and technically, and at the same time offer a universal public service. The performance of the PTT was furthermore hampered by the difficulty of managing internal dependencies between the growing and capital-intensive telecommunications and the saturating and labour-intensive postal subsidiaries. It was common practice within the PTT administration for the flourishing telecommunications subsidiary to sponsor the loss-making mail subsidiary. Externally, the PTTs were confronted with national governments using the telecommunications revenues to balance the budget, while on the other hand refusing the PTT the necessary investment funds for upgrading its networks. The long waiting lists for telephone connections that existed in the 1970s in Western Europe were partly due to the inefficiencies within the then overtly bureaucratic PTT administrations and the difficulties PTTs had in carrying out their daily activities, due to government interference. The PTT administration was a single-service organisation with a strong technocratic and engineering orientation. Both the corporate strategy of the PTT organisation and the national telecommunications policy were devised by technical experts rather than politicians, and largely escaped public scrutiny (Noam 1987; Mansell 1993). Traditionally, the PTT's orientation towards its customers could be characterised as paternalistic and household-oriented. Its view on telecommunications management was essentially that of an engineer, with heavy emphasis on technical performance, long-term planning, economies of scale and scope, system integration and centrally co-ordinated end-to-end services. Basically, the PTT offered standard homogeneous services (telephone and mail) to the general public, whereby any special demands on the part of the business community were largely ignored. The somewhat rigid and inflexible attitude towards business users changed in the late 1970s when the markets for telephone services became saturated and demand for dataprocessing services from professional and business users started to grow rapidly. After the objective of universal availability of telephony was achieved, the PTT developed a diversification strategy by extending its range of services, to include data-communications services, videotex, fax, video conferencing, etc.

Apart from its pivotal position in the governance regime, much of the PTT's influence came from allowing other groups within society to share in its monopoly revenues. The huge procurement contracts of equipment by the PTT provided a large market for the domestic equipment industry in which maximum economies of scale could be achieved. The relationship between the PTT and the equipment industry in designing and manufacturing network facilities developed into long-standing and close arrangements. Transaction-specific investments in network engineering and interoperability, and standard setting, reinforced by restrictive government procurement policies, produced a monopsonistic market with the PTT as the primary buyer and the domestic equipment industry as its sole supplier. The highly concentrated market for public switching equipment was protected from foreign competition by a variety of protective measures that safeguarded the position of the domestic supplier(s): national preferences, joint research and development and the setting of common standards. The consequence of such a politicised market environment and the tight quasi-vertically integrated links between the PTT and one or two manufacturers was that the price of equipment was too high and the quality too low (Dang-N'Guyen 1986). Examples of such quasi-vertical relationships between national PTTs and their domestic equipment manufacturers were: Philips and the Dutch PTT, GEC & Plessey and British Telecom, CIT-Alcatel and France Télécom, and Siemens and Deutsche Bundespost.

Other stakeholders in the postal-industrial complex were the labour unions, left-wing cabinets and government in general. The PTT administration and the equipment firm(s) ranked among the largest national employers with relatively high levels of union membership. On the basis of such a strong representational basis the labour unions established a close relationship with the PTT administration and the domestic equipment industry. As a consequence, the PTT employees and the employees of the equipment manufacturers enjoyed relatively high levels of job security and salaries. To left-wing cabinets the PTT, as a public corporation in a key industry, was relevant in the formation and implementation of technological and industrial policies. With its vast procurement budgets and huge labour forces, the PTT was one of the largest investors and employers in the country. Traditionally, government involvement in telecommunications has been extensive. The reasons for this high degree of state intervention were manifold. Besides the natural monopoly character of the telecommunications network, governments regarded the telecommunications infrastructure as strategically important for both the national economy and national defence/security. Sometimes the PTT was used by government to achieve other objectives than providing universal telecommunications services: competition policies, innovation policies, macro-economic policies, employment and regional policies. The centralised and hierarchical telecommunications network enjoyed broad public and political approval because it supported the concept of public service: it was universal in reach, common carrier-based in access, price-controlled by necessity, and redistributive with regard to its charges (Noam 1987). In exchange for their monopoly status, the telephone administrations were obliged to extend their service to all. Therefore consumers, small firms, the self-employed and rural communities supported the PTT system because they feared that a liberalised regime would threaten the subsidy of their service. Consumer organisations, the chambers of commerce, employer associations and labour unions were directly represented in an integrated institutional framework of various advisory and consultative councils for PTT affairs. The poor and the elderly played a mere token role in the policy-making process, as symbols of the universal service character of the telecommunications network.

The office equipment manufacturers and dataprocessing industry have remained somewhat outside the postal-industrial complex, at least in the past. In the last two decades the PTTs have been able to draw some of them into their dominant coalition of organisations by giving them a key role in the introduction of office automation and the upgrading of dataprocessing facilities in the PTT administration. For the computer industry, PTT has been both a highly valued customer with a solid financial backbone, and also a (potential) competitor. As we will see later on, the computing industry maintained a position on the periphery of the postal-industrial complex that proved strategically important. They served another big group of customers who were left out of the postal-industrial coalition: big business users such as banks, news agencies, publishers, insurance companies, airline companies, and other big multinationals with subsidiaries all over the world, to name a few. These businesses demanded cheap international calls, customised (data)communications facilities and flexible regulations, demands that at that time could hardly be met by the PTTs. To these large users telecommunications has simply grown too important, in terms of corporate strategy and costs, to be regulated in the traditional way. Rather than be made subject to the vagaries of monopolistic pricing by a public telephone company that offered mainly conventional telephone services, large business users slowly started to develop their own telecommunications links and operated their own private networks, or they leased them at low and stable rates from new service providers such as IBM and General Electric Information Services (GEIS).

The national network operators jointly controlled the market for international telecommunications services. The point-to-point nature of international telephone, telex and telegraph services fostered a bilateral governance mode, based on the conception of international telecommunications as a jointly provided service. The provision of cross-border communications services was regarded as the result of a shared investment by the carriers of origin, transit (or through-traffic) and ultimate destination. Besides being exclusively responsible for the provision and operation of the domestic network, the national PTTs were joint monopolists in the operation and management of global telecommunications services. The rates for international calls and the division of revenues were negotiated by the two national network operators involved in establishing the link. The established network operators furthermore agreed on setting tariffs for international telecommunications high above cost in order to subsidise domestic telecommunications. As well as through bilateral agreements, international telecommunications was also organised through multilateral arrangements between public network operators, for instance negotiated within the International Telecommunications Union (ITU). The ITU is the key intergovernmental organisation to monitor and ensure the interconnection and interoperability of national systems in international communications on a technical and administrative basis. As an agency of the United Nations, the ITU is intended to maintain and extend economic co-operation between its more than 150 member states in all fields of telecommunications, such as developing recommendations about standards and tariffs, allocating frequencies, and coordinating the development of telecommunications infrastructures (Aronson & Cowhey 1988; Cowhey 1990; Renaud 1990). Officially, the ITU negotiations were led by representatives of the nation states, but in practice the influence of PTT administrations was always substantial; PTT engineers tended to dominate the discussions on technical and regulatory matters. In this exclusive regime, entry of new service providers was restricted at both the national and international level.

In short, the national postal-industrial complex consisted of the major stakeholders in the telecommunications industry, all sharing in the revenues generated by the domestic monopoly of the PTT. The international telecommunications system consisted of a system of national monopolies embedded in an international market/price cartel centred around the ITU. To identify this traditional governance regime of telecommunications, we will put forward the term *national public monopoly/international cartel.* However, the traditional framework of telecommunications was seriously questioned when the foundations on which the monopoly/cartel regime was built started to erode as a result of a conjunction of technological, economic, international and institutional developments, which we will discuss in the following section.

1.4 Structural forces setting the national telecommunications agenda

The shifts in the telecommunications polity that took place in the 1970s and 1980s can be explained by referring to technological, economic, international, European, and domestic-political factors. The technological factors included the diffusion of new transmission alternatives, challenging the natural monopoly characteristics of the network, and the convergence of technologies, blurring the traditional boundaries between telecommunications and sectors like broadcasting, dataprocessing, and electronic publishing (Pool 1983, 1990). The economic factors were the non-sustainability and inefficiencies of the established market structure, leading to the introduction of more competition and the consequent promotion of innovation, quality of service and customer responsiveness by suppliers (Noam 1992; Mansell 1993). The international factors were the internationalisation of business and politics, furthered by transnational users and the early-deregulating governments of the USA, the UK and Japan, which had a substantial impact on mercantilist policies in Western Europe (Noam 1992; Mansell 1993; Davies 1994). The European factors were the ongoing process of European integration and the implementation of supranational telecommunications regulations, constraining the policy process in the constituting member states even further (Schneider & Werle 1990). Domestic-political explanations focus on the strength and interplay of stakeholders in the policy process and the legal-institutional setting that constrains their interplay in the formulation of an appropriate reaction to the ongoing restructuring in telecommunications. The actual strategic response of particular countries in terms of degree and timing of implementing changes, as assumed in this study, will be shaped partly by the impact these structural trends have upon the ruling coalition and institutions at a national level. Telecommunications policy and decision making also reflects the distribution of power among domestic stakeholders and their strategic interactions intended to shape appropriate institutional structures.

The first development that seriously questioned the state-controlled European telecommunications industry could be labelled the *technological revolution*; computerisation and digitalisation have eroded the boundaries between technologies and industrial sectors and have allowed for a sharp rise in both quality and quantity of communications facilities. Innovations like micro-electronics, computers, radiowave, satellites, cable and videotechnologies and their functional linkages have enabled a convergence between print, dataprocessing, broadcasting and telecommunications by integrating these formerly separated sectors in one integrated information/communications market. These developments have enabled a flexible and customised use of the telecommunications network by allowing to unbundle the distribution of information through infrastructure provision, bearer services and enhanced services. Information and communications technologies proved to be what Pool (1983, 1990) has called 'technologies of freedom' and 'technologies without boundaries', eroding the existing national monopolies and making telecommunications almost insensitive to geographical distance. As a consequence the position of incumbents has eroded and new entrants have been encouraged to penetrate various segments of the telecommunications market. The diffusion of cable, satellite and radiowave technologies has offered new opportunities for outsiders to provide alternative networks and competitive products and services.

One also has to take into account changes in the economic structure of the telecommunications industry, like the development of the information society and the service economy, the emergence of new markets, and the evolution of multimedia conglomerates. These developments have led to a proliferation of new value-added services and to the entry of private service providers, challenging the PTT monopoly. This diffusion of new technologies was stimulated by an increasing demand from large businesses for cheap and professional services, encouraging even more competition among equipment and service providers and the traditional operators. Together, these two developments have resulted in a considerable downfall in the costs of information and communications. Large business users and private service providers have fiercely lobbied against the cross-subsidisation of tariffs, criticising poor performance of some of the public operators in favour of a more cost-related structure and greater freedom to operate private (inter)national communications systems. The established administrative structure of mutual dependencies between PTT, government, domestic equipment industry, labour unions, and households was not seriously adapted to meet business demands for specialised, international and cheap communications services. These mutually beneficial relationships had led to major economic inefficiencies, such as overpriced equipment, low quality levels, idiosyncratic standards, barriers to innovation, expensive long-distance tariffs, and low levels of international trade.

Telecommunications, once a stable and protected industry, gradually became an innovative high-tech industry, where rising R&D costs and shortening production life cycles were demanding a more international outlook. The markets for equipment and services, once subject to national protection and market fragmentation, evolved into a global and oligopolistic affair, dominated by multinational corporations like AT&T, Siemens, Alcatel, IBM, MCI, as well as the strategic alliances formed by the various European PTTs within their own ranks. This globalisation and differentiation of communications markets also made a reconsideration of the existing national policies necessary.

Another development that demanded such a fundamental reappraisal was the process of *international deregulation*, which in the US had already started in the early 1960s. In the early 1980s, these American deregulatory policies were emulated by ambitious privatisation and liberalisation programmes in the United Kingdom and Japan. The promotion of international deregulation also spilled over to discussions in the GATT framework, where the early liberalisers sought to liberalise free trade in telecommunications equipment and services. The regulatory reforms of US telecommunications implied a gradual relaxation of the entry conditions for new businesses in specific market segments of the domestic industry (peripheral equipment, cellular radio and long distance telephony) and an overall enlargement of discretion to market co-ordination and a curtailing of the regulated private monopoly. A landmark in American telecommunications deregulation was the Modified Final Judgement of 1984, leading to the divestiture of AT&T as a nationally regulated monopoly. AT&T was broken up into seven separate local/regional companies, the so-called Regional Bell Operating Companies (RBOCs) to be regulated by the government, and the slimmed-down AT&T was freed from any regulatory impediments to enter new markets. As a consequence, competition was fostered between the telecommunications and the dataprocessing industry and their major protagonists AT&T and IBM.

The deregulatory package was part of a deliberate American strategy to strengthen its global hegemony in information technology and telecommunications (Schiller 1982; Hills 1986; Pool 1990). Given their strong dependence on US technology and the large size of its market, European and Japanese policy makers had to take the American move to deregulate its domestic telecommunications and computing industries into account. First, the American programme served as an impetus and model for other countries also wanting to restructure their domestic telecommunications monopoly. Second, the global shift of AT&T and IBM exerted a strong pressure on the European and Japanese governments to open up – at least to some extent – their highly protectionist telecommunications markets. One of the consequences of US deregulation was that firms like AT&T, IBM, the RBOCs and MCI sought market expansion abroad. When confronted with the US demand for market access, Japanese and European policy makers had to take into account the actual strength and the (potential) entry of and competition from large American firms in formulating an appropriate policy response to the new techno-economic developments.

The strategic move of the US government to deregulate its telecommunications industry was quickly emulated by the Japanese and British governments, who decided to privatise their public operators and to a large extent liberalise their home markets between 1980 and 1994. European states were more or less forced to respond and adjust their domestic public monopoly regime to the new conditions. The position of European countries, however, has remained relatively weak: their national markets are too small to compete with the US and Japan, nationally oriented operators and manufacturers are behind in transnationalisation, and the Common Telecommunications Market is as yet not fully accomplished.

Another major challenge facing Western European states is the process of European integration, illustrated by the growing power of EC institutions in the design of Community-wide telecommunications and Internal Market policies. Although the European Commission could have intervened since 1962 when the rules of free movement came into effect, there existed a consensus within the European Community to keep the telecommunications sector subject to the member states' control and organised through a state-controlled monopoly, with hardly any concern for domestic and international competition. The launch of the Internal Market programme in the mid-1980s was inspired by the idea that market integration, industrial restructuring and the upgrading of the Community's technological base would prove vital in catching up with Europe's major competitors, the US and Japan. The supremacy of Community Law over national legislation and the ongoing process of European integration has reduced the freedom the Community's member states have to pursue public policy objectives (Volcansek 1992b). As a consequence, protective arrangements between national government and their domestic industry were increasingly overruled by Community regulations to ensure the operation of free markets and encourage 'national policy competition' between the member states (Jacquemin & Wright 1993; Sun & Pelkmans 1995). The growing importance of the European level of decision making is not only the result of the increase of responsibilities for Community institutions. The 'Brusselisation' of policy making has also fostered the formation of influential supranational interest groups and their participation in the policy process and the development of cross-national strategic alliances in the European market place.

For a long time the European Community hardly put any restrictions on national telecommunications policies. There was a change, however, in the early 1980s, when telecommunications deregulation in the UK forced the Commission to intervene to deal with attempts by British Telecom to abuse its monopoly power. With the publication of the Green Paper for European Telecommunications and subsequent legislation in the late 1980s, the leeway to develop specific domestic policies, reflecting particular ownership structures, domestic priorities and national legacies, was seriously reduced. For instance, the Commission successfully attacked protective procurement policies and telecommunications lost its immunity from the anti-trust provisions of the EEC Treaty. The role of the European Commission has changed from being a mere international and collective regime, relying upon the consent of its constituencies, to that of a corporate actor in its own right, actively engaged in a comprehensive R&D framework, collective standardisation and the creation of a European telecommunications market (Schneider & Werle 1990;

Hulsink 1994). Nothwithstanding the Community's efforts and relative successes in harmonising the various domestic policies and creating a Common Market for terminal equipment, value-added services, satellite and cellular communications, and access to public networks by 1993, the existing public monopolies on infrastructure provision and basic services ('voice telephony') remained unaffected. Only recently, the European Council of Ministers decided that from 1998 the entire telecommunications sector would be open to competitive entry, including the aforementioned key markets.

So far the Commission has been relatively unsuccessful in forging a Community-wide consensus on the aims and scope of an adequate common strategy. The political responses to the economic recession and transnational pressures have remained biased in favour of the member states and domestic stakeholders. The reason for this lack of support can be traced back to a decision-making structure in which the interests of the member states still prevail through the body of the Council of Ministers. So far little discretion has been given to the supranational bodies of the Commission and Parliament. EC-wide telecommunications policies cannot be imposed upon member states, but are jointly decided by the intergovernmental Council of Ministers. This body, in which national governments directly participate, offers the possibility either to further or block national interests. Basically, the policy outcome of this joint decision making in the European telecommunications domain is a political compromise between the Scylla of uncontrolled liberalisation leading to destructive competition and cuts in public service provision, and the Charybdis of too much state intervention, effectively blocking institutional change and innovation.

Structural developments in technology, markets and regulation, however, do not automatically produce new institutional arrangements; a new industrial order needs to be negotiated, agreed upon and implemented by national and sectoral actors. As argued by Blankart and Knieps (1989:594): 'technical progress and efficiency arguments are a necessary but not sufficient reason to change the status quo towards deregulation.' Besides being constrained by techno-global developments, the formulation of appropriate national strategies is to some extent shaped by factors and actors at the macro-level of the national economy, for instance a country's legal-administrative tradition, the scale and degree of openness of the economy, particular arrangements linking the state with business and labour, and the prevailing economic ideology. Blankart and Knieps (1989) have referred to diverging institutions of political decision making to explain different paths of telecommunication deregulation. In comparing the policies of the US and Germany in the 1980s, they have found differences in interest group structure and access to the political system. While the US exhibits a more decentralised and pluralist structure of decision making at both the federal and state level, German decision making is centralised at the federal level. According to Blankart and Knieps, these institutional differences between the US and Germany explain why the former has followed the more radical strategy. In addition to these national contingencies, policy responses are also influenced by constraints at the sectoral level, referring to the intrinsic features of the domestic telecommunications industry, like the power distribution between the incumbents and their political allies and the attempts by (potential) new entrants seeking access to the market place and the policy arena.

1.5 Comparative analysis of telecommunications restructuring in Europe: global forces and domestic responses

In strategic management, organisation theory and political economy, there is a lively debate about whether 'nations matter' in an internationalising political economy with respect to the public policy choices and private strategies pursued by leading domestic actors (e.g. Zysman 1983, 1994; Katzenstein 1985; Gourevitch 1986; Porter 1990; Whitley 1992a, b; Ruigrok & Van Tulder 1995). Researchers disagree as to the extent to which policy processes and outcomes are shaped by factors unique to a particular policy field or whether they are a function of the political and social characteristics of the nation in which they are developed. Two different approaches to the comparison of public policies across nations and across sectors have been put forward (Freeman 1985; Feick 1992). The *cross-national convergence* or policy sector thesis stipulates a differentiation within individual countries across sectors and a convergence across nations within sectors, whereas the *national diversity thesis* stipulates a differentiation within individual sectors across nations and convergence across sectors. In this study, we will apply the two theses to analyse two opposing trends in the high-tech industry of telecommunications, that seemingly pose a dilemma both for governments and (multi)national corporations, namely the choice between *techno-globalism* and *techno-nationalism* (Stevens 1990; Ostry & Nelson 1995; Sally 1995).

The convergence thesis assumes the appearance of similar problem definitions, policy goals, instruments, styles, and outcomes across nations, ultimately leading to increasing similarities between nations' policies (Bennett 1991). These corresponding policies have proven or promise to be the most efficient, effective and/or legitimate, and hence it is assumed these 'best practices' ought to be widely diffused and universally adopted. With the globalisation of particular industries and European harmonisation and integration, domestic politics have become more and more intertwined with international matters, forging a convergence of national responses to structural economic developments. The intensification of transnational economic and political linkages, manifested by the globalisation of firms and industries and the internationalisation of policy communities and regimes, have restricted the room for manœuvre of small and medium-sized states. However, there still seems to be significant room for diversity in the strategies pursued by individual European states to cope with an increasingly complex and global environment (Dyson & Humphreys 1986, 1990; Humphreys 1990b; Grande & Schneider 1991; Unger & van Waarden 1995; Berger & Dore 1996). Despite the gradual process of globalisation, there are arguments to support the claim that the national level, where both public and (multinational) private actors are deeply embedded in and rely on supportive domestic networks, is still important. The national diversity (or techno-nationalism) thesis claims that domestic politics and policies still matter, and that, as a consequence, there continues to exist a number of diverging approaches. Each country has its unique mix of economic, political, social, and historical elements. Consequently, each has its own particular inheritance in terms of economic organisation, interest intermediation and decision making.

The convergence hypothesis

In the post-war period the level of economic and political interdependence between the economies of Japan, the USA and Western Europe has increased sharply. Today's companies, triggered by the search for economies of scale and scope, rising R&D costs and a quickening pace of technological change and worldwide convergence of consumer preferences, are competing for world market shares and technological leadership. Although, admittedly, the notion of an open world economy is to a degree wishful thinking and one should not forget that the internationalisation of economic activities is still limited, the rise of powerful multinational companies, the increasing worldwide flow of goods and money, and the formation of supranational trading zones and economic blocs all make it clear that things have changed (Ruigrok & Van Tulder 1995; Hirst & Thompson 1996). For instance, the globalisation of the automobiles, electronics, computing and aerospace sectors took place through the formation of foreign direct investment strategies and international strategic alliances: if one wants to operate a global business, one needs to co-operate with international partners to serve customers in a global environment. As argued forcefully by Ohmae (1985, 1990), every corporation that wants to have a strong position in the 'interlinked' world economy needs to create a direct presence in each of the key markets. Transnational firms must be able to penetrate and strengthen their position in the three regional markets of Japan, the USA and Europe simultaneously. A foothold in the regional market one wants to conquer can be created by establishing joint ventures or mergers with enterprises from that region. Once an 'insider' in the host market, protectionist trade barriers and a foreign business environment are successfully overcome: European firms went transnational and have become insiders in Japan and the USA, and Japanese and American corporations in their turn have moved into Europe and have also become insiders in that part of the world market.

The international economic environment is not only shaped by increasing competition between multinational firms from the three leading regions and a consequent concentration of production, trade and finance at world level, but also by the interactions between governments and (multi)national corporations concerning investments at home or abroad (Porter 1986; Gilpin 1987; Stopford & Strange 1991; Strange 1992). The process of world market integration has reduced the ability of national governments to control and manage their domestic industries. In a global economy states can no longer follow costly and risky mercantilist strategies, exemplified by support for national champions, protective trade barriers and massive government subsidies. Constrained by cuts in public spending and free trade provisions enacted through the EC and the GATT/WTO frameworks, governments keep an eye on the competitive position of their domestic industries, welcoming foreign investments in key economic sectors that help to strengthen their domestic techno-industrial base and hoping to find an attractive international partner for their national champions to enable access to foreign markets. For instance, in heavily regulated markets such as telecommunications, financial services and high technologies, it is almost impossible for foreign firms to enter those markets without joint ventures with domestic partners: market entry has to be negotiated. A clear free trade policy, whereby foreign investors are welcomed without any local content provisions, is not a viable alternative for local governments seeking to achieve durable economic wealth. In line with Lindblom's claim with regard to the privileged position of business, multinational companies always have the (potential) ability to invest and operate abroad to avoid high taxes and a burdensome regulatory environment (Lindblom 1977). Therefore governments have to find a balance between the restructuring of national champions and domestic industries and the promotion of free trade and market access for its domestic firms abroad.

This strategy of liberalising markets sequentially in order to give domestic firms sufficient time to adjust to the new conditions in international markets, is followed by a majority of nation states. The emerging global economy seems to be characterised by a system of free trade within the regions and managed trade between the regions ('quasi-trading blocks'), in which national governments or regional administrations are involved in setting the terms and conditions (Ruigrok & Van Tulder 1995; Hirst & Thompson 1996). The present relationships between states and firms in the international political economy could be interpreted as a mixed-motive game in which 'rival states and rival firms' attempt to strengthen their economic position in the worldwide competition by relying upon a mixture of competitive and collaborative strategies (Stopford & Strange 1991; Strange 1992). This new game of competing for world market share forces firms to bargain with host governments and/or other companies seeking access to new technologies and markets (Ostry 1990; Cowhey & Aronson 1993). National governments also have to be alert and responsive, and try to attract foreign investments in key economic areas as well as support the domestic activities and the presence abroad of its 'home' multinational companies.

Although it has lost substantial market share in some industries, like car manufacturing, steel and electronics as well as losing monetary controls (Keohane 1984; Hart 1992), the United States is still the unquestioned hegemon of the world system (Strange 1987, 1988). The control of US government and US business over key areas such as security, economic production, finance, technology, transport and communications systems and energy supply (i.e. the Pax Americana), has allowed for substantial bargaining power in determining the terms of individual international exchanges and the trade negotiations within the GATT framework in the post-war period (Vernon 1971; Gilpin 1975). The US strengthened its strategic position and created a competitive edge in the airline, finance and telecommunications sectors by being the first to deregulate and to demand a free trade regime in the 1970s and 1980s, based on the principles of market access, reciprocity, non-discrimination and the most-favoured nation principle applied multilaterally. This clear relationship between a country's position in the international division of labour (i.e. degree of international competitiveness) and particular trade policy preferences (i.e. degree of free trade) has been described by Gourevitch (1986:56):

> Economic actors at the cutting edge of international competition are likely to support policies that promote open trading, not only for their own products, but more generally for intensified specialisation in international trade. Actors less well placed in international competition are more likely to support protection or modes of shelter or subsidy.

The increasing growth of economic interdependence at world level has been the outcome of major transnational corporations establishing subsidiaries, agreements and alliances all over the world. Especially big American firms like AT&T and IBM, relying upon a vast home market and tight links with the American defence industry through large procurement contracts, have become major players in global telecommunications. After the deregulation of their home market, these two leading American firms have acquired minority stakes in a number of European firms engaged in the communications and electronics industry. Compared to Japanese and especially European companies, there is a big difference in size of the companies, technology and market leadership, and a high level of transnationalisation. The American companies, encouraged and supported by their government's pro-market programme, were at the forefront of the globalisation process in telecommunications. The US government was the first to question the monopoly regime as well as the first to turn it into a market-oriented industry by loosening and lifting government regulations in order to stimulate innovation, competition and international trade. It was believed that such a deregulated regime would be beneficial not just for the American telecommunications industry, but for American business in general, operating all over the world. Given US world leadership in computing and telecommunications in the early 1980s, Japan and the EC had to react to this challenge. At that time the European Community was made up of a number of small, fragmented and sheltered markets, which had become a handicap to innovation, growth and international competitiveness.

Although it is true that globalisation has to a large extent been accomplished by international firms, business was supported in its effort to transnationalise by international economic and political institutions operating at the regional or world level, such as the EC, and the multilateral arrangements established in the OECD and the GATT/WTO frameworks. These international institutions are highly involved in the making and shaping of policies concerning transnational competition, trade and innovation issues in either gathering data about these issues (OECD), enforcing a common European market (Commission) or facilitating worldwide trade (the GATT/WTO). EC law has an impact both on member states who are directly bound by it and on companies who are constrained by the competition provisions. The GATT/WTO framework is a forum for negotiation that is becoming increasingly important for international telecommunications. For a long time cross-border traffic was organised in a cartel-like way through various agreements between the national PTTs and regulated by the ITU. As it is losing its public and monopolistic character, telecommunications is increasingly becoming an industry like any other, and as a consequence it faces the rise of falling under GATT jurisdiction. In the Uruguay Round of GATT negotiations a broader coverage of trade liberalisation was suggested, one of the areas targeted being barriers to international trade in services and market access (for example standardisation and public procurement). Discussions are taking place to subject the international trade in telecommunications services to the free trade rules in the foreseeable future.

The national diversity hypothesis

The institutional environment of a particular country provides both constraints on and incentives for discretionary action in the formation and implementation of industrial policy and corporate behaviour. The term 'institutional contingencies' refers to whether (and how) countries organise and regulate their domestic economy in a particular way, reflecting a unique mix of social, economic, political and historical elements, such as a specific production profile, distinctive domestic priorities and industrial culture, and the particular role(s) of the state and its interaction patterns with private actors (such as firms, financial institutions, trade associations, and labour unions) in the policy process. When a nation state consistently shows a distinctive way of acting, structuring and performing in terms of policy making and/or implementation processes, influencing the development of public policies across time and across sectors, one could argue that a specific national policy style exists (Kelman 1981; Richardson *et al.* 1982; Vogel 1986; Feick 1992). The concept of style has originally been defined by Gombrich (1979:352) as 'any distinctive and therefore recognisable way in which an act is performed or an artefact made, or ought to be performed and made'. In this research we will investigate whether there are diverging responses, based on particular national policy styles, to the techno-economic challenges posed by international telecommunications. When comparing public policies, we must include both the macro-economic and the legal-political environments, as they, too, constrain industrial behaviour.

A first difference is a country's comparative advantage based on natural resources, factor endowments (the level or amount of available skills, technology, capital), particular relationships among the nation's competitive industries, and various macro-economic variables (exchange rates, aggregate level of demand, the level of savings, interest rates and investment) (Shonfield 1965; Porter 1990). A second difference is the size and openness of the domestic economy (Katzenstein 1978, 1985): the first element refers to the magnitude of the domestic market in terms of economies of scale and scope and of R&D efforts, and the second refers to the relative weight of transnational actors, including multinational companies and international institutions, on national policy formation. Some countries have a small and open economy (e.g. the Netherlands), other countries have a medium-sized economy with a relatively open (e.g. United Kingdom) or with a rather closed character (e.g. France). Others still have a large economy which could be qualified as relatively open (USA) or closed (Japan). Compared with big economies, small industrialised countries have to rely upon niche strategies and exports to gain the benefits of economies of scale and recoup R&D costs. Because of their openness and relatively small size, states like the Netherlands and Belgium are very sensitive to trends in the world economy. In response to meet this high international sensitivity, the small European countries have developed a stable system of political accommodation at home, in which the major parties in economic policy (business, labour, government) work very closely together in the formation and implementation of economic policies (Katzenstein 1985). Their innovation policies are oriented towards the diffusion and widespread application of advanced technologies, seeking to combine domestic capabilities with inward investment by foreign multinational firms (Freeman & Lundvall 1988). While smaller countries will conceive a structural industrial crisis as a cyclical occurrence generated outside their economy over which they have hardly any control, bigger countries have the ability to initiate mission-oriented innovation programmes on their own, aiming at technological leadership (space, semiconductors, aircraft manufacturing, etc.), which may to some extent combat the crisis (Ergas 1984). While a large majority of the industrialised countries in the early 1980s were considering or already implementing privatisation and deregulation measures, the French socialist government, for instance, opted in favour of nationalising the advanced sectors of its economy.

A third difference between national economies has to do with the relative strength and the autonomy of the state *vis-à-vis* society as a whole. In explaining differences in international competitiveness between sectors and between countries, Hart (1992) has referred to the varying patterns in the creation and diffusion of new technologies, that are in turn affected by particular state–society arrangements. The distribution of power between state, business, and organised labour and their specific interaction patterns helps to explain changes in international competitiveness, because these state–society arrangements can accelerate or impede the development and diffusion of technological innovations and breakthroughs crucial to a nation's competitiveness. Particular state–society relationships manifest themselves not only at the macro level of economic policy making, but at micro and meso level as well. Individual businesses and industries are embedded in a complex network of distinctive relationships with central and local governments, national financial institutions and trade unions etc.; even multinational companies are strongly entrenched in specific networks of suppliers and supportive institutions at home (Porter 1990; Grabher 1993; Ruigrok & Van Tulder 1995; Hirst & Thompson 1996).

Zysman (1983, 1994) and Katzenstein (1985) have referred to the different ways in which the costs and benefits of industrial change can be distributed throughout society and the different means by which political settlements as to who benefits and suffers from growth can be reached. The Anglo-Saxon countries are characterised by a liberal tradition with business operating at arm's length from the state.

The principal role for government is to provide the appropriate economic conditions for domestic companies and markets to operate effectively. In these market-led or company-led countries the political settlement with regard to the distribution of the costs and gains of industrial growth is left to the market mechanism and competitive rivalry. The German, Swedish and Dutch states are examples of so-called negotiation-led or neo-corporatist systems with moderate government intervention, where the state and its 'social partners' (organised business, labour, the banking community and agriculture) collaborate intensively to secure domestic peace and together they develop a joint trade policy. In these countries industrial adjustment is achieved through explicit bargaining and consensus building to make arrangements that will benefit all the domestic stakeholders as well as the national economy. France and Japan can be characterised by the leading role played by central government in guiding and orchestrating the national economy. In these *dirigiste* or state-led countries, the distribution of the costs and benefits of industrial change is determined by political manipulation of the market. Through indicative planning, public ownership, investment subsidies and large public procurement programmes, industrial targeting policies, control over the financial system, and superior expert knowledge, the state bureaucracy is able to develop and effectively control the crucial sectors and key technologies of the national economy.

The participation of industrialised countries in world markets and international economic rivalry constrains domestic decision making with regards to the formulation of appropriate industrial adjustment strategies and macro-economic policies. International developments, however, do not necessarily have a direct impact on national economies, but are reflected or refracted through policy choices, developed and negotiated by political actors and dominant organised interests at the national level. In his comparative study on the impact of international economic crises on domestic economic policy making, Gourevitch (1986) has distinguished five options to adapt domestic economic policies to the new techno-economic and political conditions in the world system:

1 the neo-liberal market orientation;
2 socialisation of planning and production;
3 protectionism;
4 demand management;
5 mercantilism.

These policy alternatives may prove useful to our study if we place them between free trade and mercantilism/nationalisation: the first assumes a subordination of politics to a dynamic market rationality generating economic efficiency and industrial innovation, the second assumes the supremacy of state intervention over market and political *grand design* strategies.

The answer to the question: Is it convergence or national diversity that matters most? may be that they both play a significant role, but that they do so at different stages in the policy process. The convergence dictum is relevant because all countries feel the impact of international restructuring in telecommunications: every nation state participating in the world economy is confronted with the problem of how to adjust its domestic industries, policies and institutional structures to drastic changes in technology, markets, competition and international deregulation. National diversity, however, may prove significant in the implementation stage, when national governments, firms, labour unions, and other actors decide upon the actual enforcement of adjustment programmes. Domestic decision making and policy making could be relevant because variations across nations may occur in the degree and the pace of responding and adjusting to international restructuring.

1.6 Conceptual framework

Strategic formation and public policy making will be analysed from a power perspective, characterised by a focus on the resource dependencies among stakeholders and the strategic interactions and negotiations between public and private actors in the policy arena (Crozier & Friedberg 1977; Olsen 1981; Wassenberg 1985). The strategies of the various actors and the interactions between them as they (re)shape the negotiated order, are the central elements in such an approach. Negotiation behaviour is aimed at the scope of the mutual dependence of the actors involved, by recognising the ambitions, plans and interests of allies and adversaries, and the support or opposition from the various stakeholders with regard to a change in public policy. The negotiation perspective attempts to make clear that the availability of blueprints is no guarantee for the effective development of policies, but that, basically, policy and strategy are the outcome of bargaining within and between mutually dependent organisations. In that respect, the discretionary power the players involved have at their disposal is crucial, and dictates whether they are able to change the range of options open to others without apparently putting direct pressure on them to make one decision rather than another. As made clear by Schattschneider (1975), one should not only pay attention to the final outcomes of decision making, but also take into account how subjects and proposals emerge in the first place and eventually get sufficient political attention and support. His emphasis is on the variation and selection mechanisms inherent in the political process: out of the various conflicts that arise, only a few are filtered out by the interplay of dominant stakeholders, and finally generalised to the broader public. Some (potential) issues, values and participants are not considered seriously, or even excluded, and kept out of the public domain. So the scope for conflict and available alternatives is continuously being redefined and rearranged by the organised interests involved in the policy process:

> All forms of political organisation have a bias in favour of the exploitation of some kinds of conflict and the suppression of others because organisation is the mobilisation of bias. Some issues are organised into politics while others are organised out.
>
> (Schattschneider 1975:69)

Changes in policy making and the duration and pace of implementation are the result of power shifts in the policy arena and/or changes in the composition of the political agenda, prioritising some issues and excluding others.

Actors seek to strengthen their position and enhance their effectiveness in the political system and the market place. It is in those places where the struggle for control over the terms of resource exchanges and the pursuit of dominance take place. Each actor seeks to reduce environmental uncertainty and control critical resources and contingencies by aiming at governing, directly or indirectly, the other actors involved and thus escape the constraints that others would impose on him. In their effort to control interdependencies and overall uncertainty, organisations can follow several strategies to achieve a relatively steady state of social closure, varying between collaboration, conflict seeking, competition, secretive manipulation and overt integration (Murphy 1988). Cawson *et al.* (1990) have suggested that organisations can follow at least three strategies to safeguard a solid power base and achieve some state of closure in the market place and/or the policy arena: a market-orientated strategy aimed at strengthening its competitive position *vis-à-vis* other competitors, cartel-oriented strategies promoting associative action to guarantee monopolised advantages for the members of the élite, and state-orientated strategies demanding protection and shelter from the state.

Furthermore, one should pay attention to both actual and potential issues, as well as any observable, repressed and latent conflicts between stakeholders in the policy process. Lukes (1979) has discerned three dimensions of power: the pluralist notion of power covering the behavioural analysis of overt and observable decision making by political actors (i.e. the execution of force), the revisionist view defining power in terms of control over the political agenda and the mobilisation of bias in the political system (i.e. the manipulation of decisions and non-decisions), and the radical perspective conceiving power as working through the suppression of latent conflicts and the shaping of consensual authority (i.e. power or domination as a legitimate system property). Anyone wanting to analyse institutional change from a power perspective must also pay close attention to overt decision making and hidden control over the political agenda, taking into account both the subjective and concealed interests of the parties involved.

The negotiation/power perspective provides a lucid process-oriented outlook on the strategic interactions among public and private actors. The approach disregards, however, the persistent influence of institutional arrangements that may constrain the bargaining process. The term 'institutions' is used in a rather general sense in this study, referring to a relatively persistent set of rules, norms and values that structure the economic and political exchanges between stakeholders. In analysing the acquisition and exchange of critical resources and the mutual search for external control by organisations, we have to ask in what way these transaction and interaction patterns are governed. As made clear by Hage and Clignet (1982) and Williamson (1993), economic production and distribution processes are contingent upon an institutional environment, made up of relatively stable and persistent state–society relations and socio-legal ground rules. Because different types of business have different requirements with regards to governing transactions and interaction patterns, there does not exist a single and optimal way to organise particular industries or public programmes and to maximise their economic performance. The concept of *governance structure* has been put forward to refer to alternative institutional arrangements that control the processes of production, resource allocation and decision making in a specific industry. Or stated differently, governance structures are ideal-typical organisational forms that co-ordinate and control economic (or political) activity and transactions among interdependent organisations within a particular domain.

Five institutional forms can be distinguished (Williamson 1975, 1985, 1991; Streeck & Schmitter 1985; Powell 1990; Lindberg *et al.* 1991; Hollingsworth *et al.* 1994):

1. market forms based on bilateral exchange, classical contracting, price co-ordination and incentive-intensity;
2. hierarchy: based on administrative command, authority relations and legitimate powers (e.g. horizontal and/or vertical integration, state intervention);
3. obligational networks: based on resource dependencies, partnering, relational contracting and contractual bonds;
4. promotional networks: based on group loyalty, socialisation, joint interests, mutual trust and confidence relations;
5. association: based on membership agreements, recognition by the state and intermediation.

Markets are self-regulating exchange systems in which buying and selling takes place in a competitive bidding process between atomised utility maximisers. The market process is governed by the price mechanism that reveals more or less adequate information in the allocation of goods between buyers and sellers. Market forms co-ordinate economic transactions between private actors, holding separate property rights in the different resources to be bartered, actors that are voluntarily engaged in the contractual exchange of goods and services. Hierarchies are administrative systems that govern economic activities by fiat, centralised controls, and formal procedures. In this institutional form the property rights are vested in one integrated formal organisation to which both supervisor and subordinates belong. The exercise of managerial prerogatives by one category of actors in wielding authority over others through hierarchy is often explained by the uncertainty and frequency of transactions and the need to address the hazards of opportunism, free riding problems and secure task-specific knowledge and skills.

The network form can be located in between bilateral market or contractual resolutions and administrative hierarchical control as a kind of intermediate or hybrid form. In this horizontal and multilateral governance structure economic activities are conducted by reciprocal communications and exchange patterns and a strong value consensus. The network lacks a formal authority structure and decision making relies upon a mixture of mutual adjustment, negotiation and arbitration (or mediated enforcement). Obligational networks maintain control through corporate interlocks between a core firm and its satellites or between two firms of equal size through equity-sharing arrangements, joint ventures and quasi-vertical structures, such as long-term subcontracting, franchise arrangements and corporate transfer pricing. Promotional networks are temporary coalitions or clans, established to accomplish a common purpose, such as R&D alliances and bidding consortia. Associations promote the establishment of a representative monopoly among specific categories of actors in identical, similar or adjacent market positions or policy domains and the organised concertation of individual actors in the policy process. These collective organisations mobilise sectoral or collective interests and organise and enforce co-operative behaviour among their members in order to define and further public or categorical goods (e.g. sector-wide training, standardisation). Besides sectoral inclusion and internal coherence, associations derive their political and economic powers from recognition by the state as the single interlocutor, acting on behalf of one particular sector or category of interests in the policy process.

A governance regime is the totality of institutional arrangements, including rules as well as rule-making agents, that regulate economic activities and transactions inside and across the boundaries of an economic system at certain moments in time (Lindberg *et al.* 1991; Hollingsworth *et al.* 1994). Economic action and co-ordination is shaped, not just by markets, corporate hierarchies and control over the means of production, but also by a wide range of non-economic institutions that create and maintain the minimal conditions without which economic exchange and organisation would not be possible. Economic institutions are embedded in and modified by a social and political institutional context, including shared cultural beliefs, interdependent networks, systems of interest representation and the degree and mode of state intervention, that create pressures for economic change or preserve the existing situation (Granovetter 1985; Zukin & DiMaggio 1990). While private property rights, free market exchanges and the overall efficiency imperatives to integrate economic activities within managerial hierarchies make up the common elements of capitalism as an economic system, social and political institutions contribute to the differences among capitalist economies.

The state plays a special role in shaping the actions and transaction of private actors and the stipulation and enforcement of governance structures. Governments are both actors (or a configuration of actors) pursuing goals of their own, a political arena where deliberation and struggles over power take place, and institutional structures in which economic transactions are embedded (Lindberg & Campbell 1991; Hollingsworth *et al.* 1994). The state is a political and economic player acting in its own right, relying upon hierarchical control and authoritarian command on the basis of its unique coercive capacity to provide, enforce and adjust the laws and traditions in society. The state is exclusively authorised to define and manipulate the rules that determine the conditions of ownership and control over the means of production in society, to enforce contracts, and to set general rules of competition associated with the co-ordination of economic activities. The state is prominent in the organisation of political participation at both the sectoral and national levels by providing diverse economic actors and various societal groups differentiated access to the economic policy-making process. Last but not least, the state also provides the legal-institutional framework by shaping the strategic choices and relative opportunities of economic actors and their exchange patterns. Governments legitimise particular industrial structures by not only defining the relationships of economic actors to property, but also settling the wider institutional basis of economic production, allocation, and accumulation within the national economy.

In its broadest sense, regulation refers to government intervention and control over specific markets and industries, channelling and directing economic activities generally regarded as desirable to society (Reagan 1987). The purpose of overall regulation is to facilitate and accomplish the realisation of socio-economic goals such as allocative efficiency and economic growth, price stability, employment, and the provision of public goods by shaping the structural characteristics of an industry. The justification for administrative regulation has to do with market imperfections that produce sub-optimal outcomes in terms of efficiency (e.g. natural monopolies, externalities, excessive competition, and information asymmetry) and – to some extent – equity (e.g. universal service provision). Anti-trust regulation aims at encouraging competition and curtailing the influence of monopolies, cartels and other restrictive business practices that might disturb the proper functioning of markets. Economic regulation refers to the imposition of controls over prices, entry, exit, output, services rendered, markets served and profitability in particular industries. Social regulation deals with the protection of the environment, consumer protection, occupational health and safety etc. Regulation can be interpreted as an informal system of rule making, which operates though negotiation and bargaining between politicians, civil servants, industry, consumers and administrative bodies in the shadow of the law (Hancher & Moran 1989; Veljanovski 1991a). Such a regulatory framework of organisations includes vertical relationships (e.g. goal setting, licensing, performance control, mandatory reporting, sanctions) and horizontal relationships (e.g. competition, co-operation, conflict, accommodation).

Regulation may be organised in a variety of ways: from self-regulation through self-imposed codes of conduct for a specific industry or profession to externally imposed regulation whereby public organisations have legitimate coercive responsibilities in regulatory policy making. Two variations of externally imposed regulation may be discerned: public monopoly and public regulation. The first one works through direct public intervention whereby a state enterprise has the exclusive responsibility for the provision of particular services. Here the government controls the operational and strategic activities of the state enterprise, such as capital allocations, investment plans, pricing and personnel policy. Public ownership is a means to deal with the natural monopoly conditions described above and the (actual or potential) abuse of private monopoly power. It could also be conceived as a means of dealing with socio-economic problems that cause the market to function badly or not at all: national security, externalities, universal service provision, industrial adjustment and restructuring, regional development and/or ideological concerns. The second form works through administrative agencies that enjoy a considerable degree of independence from the industry (and sometimes from government as well) in regulating a particular market (Selznick

1985). The role of a regulatory agency is restricted to that of referee in charge of general oversight and legal enforcement, reflecting a more judicial relationship between the state and the private sector (Volcansek 1992a; Majone 1994a, b). Public regulation refers to a package of legislative and administrative controls designed to structure and alter politically the operation of particular markets of basic importance to the national economy through an independent regulatory authority, while leaving private property intact. Reagan (1987) has characterised public regulation as a half-way-house between government intervention and *laissez faire*, with markets being organised by the stipulation of entry and ownership conditions and control of market behaviour through rules on price setting, fair competition, and universal service requirements.

The strategic choice between different governance modes of industry can be seen as a strategic game in which interests and values of the involved parties play a vital part and the political-economic organisation of society is at stake. Regulated firms and industries, public agencies and other actors operate within the political and administrative process in exactly the same way as they operate in the market (Owen & Braeutigam 1978; Vietor 1989, 1994). Public or private companies that want to operate effectively in a regulatory environment must be active both in the market place and in the political arena, and complement their business plans with political action. The larger business environment provides opportunities for strategic and tactical behaviour: the realisation of economic objectives and innovation could be used as an offensive instrument, whereas the gathering of information, co-optation of experts, lobbying, cross-subsidisation and litigation are illustrations of more defensive means to strengthen one's position. Established parties and prospective entrants alike make extensive strategic use of information in the administrative process, taking virtually every opportunity to make detailed economic and technical information available and understandable in ways which emphasise the strong points in their cases and obscure or at least diminish the weaker points. The affected parties in regulatory policy making can also respond by resorting to delaying tactics: for instance, by flooding the regulatory agency with more information than it can absorb.

1.7 Explaining governance transformation: an institutional perspective

From the 1970s onwards, the academic interest in the study of economic, political and legal institutions has increased. New Institutional Economics and New Institutionalism in sociology and political science have started to conceptualise and investigate institutions as organisational forms, socially constructed and routinely reproduced rule systems, and particular state–society arrangements that structure politics, respectively. According to the New Institutional Economics (NIE), institutions can be seen as being incentive and governance structures designed by rational actors to respond to economic needs by reducing transaction costs and information uncertainty (Williamson 1975, 1985, 1991). The emergence, functioning and transformation of institutions is explained with the help of micro-economics, with the focus on the rational choice for the most efficient outcome and the incentives associated with a particular complex of interdependent rules and informal constraints, which in total determine economic performance (North 1990). For instance, the transaction costs associated with hierarchical forms tends to be lower than the costs of market co-ordination under conditions of high frequency, high asset specificity and high uncertainty. The underlying assumptions of NIE: the primacy of economic goals, the rational actor in pursuit of self-interest, and the emergence of institutions as efficient responses to external requirements, have been strongly criticised by the institutionalist approaches, as developed in sociology and political science.

According to the New Economic Sociology, institutions are shared cognitions and socially constructed rule systems by agents whose actions are embedded in concrete networks of social relations (Granovetter 1985, 1992; Powell & DiMaggio 1991). Besides pursuing the goal of economic efficiency, actors strive for non-economic objectives as well, such as power, dominance, status and public recognition. In analysing rule-bound organisational behaviour, the sociologist J. Thompson (1967:24) has made a useful distinction between constraints and contingencies: the former are fixed institutional conditions to which organisations must adapt, and the latter are not necessarily fixed conditions, but they are not subject to arbitrary control by the organisation. The development and transformation of both political and economic institutions is contingent upon ongoing interaction patterns, lasting cognitive frames and shared norms as well as the constraints imposed by the historical legacy of (in)formal rules and procedures and the overall structure of society (Scott 1995). In their more or less deliberate search for appropriate institutional solutions to their problems, actors creatively select, (re)combine and modify already existing rules, belief systems and scripts, thereby creating new institutions which are different but still resemble the old ones. Institutional arrangements are both a dependent variable at time t and an independent variable at time $t + 1$ (Krasner 1988). In interpreting the enabling and constraining effects of institutions, Douglas (1986) has characterised these ongoing processes of puzzling, shaping and (re)integrating various institutional elements in a variety of new ways as *bricolage*.

In political science, institutions refer to 'formal rules, compliance procedures, and standard operating practices that structure the relationships in various units of the polity and economy' (Hall 1986:19). Central in this respect is the notion of the institution as a set of rules affecting the structure of particular situations at various levels and in particular ways by requiring particular actions, prohibiting certain other outcomes and/or allowing others (Ostrom 1986). In the political version of institutionalism, it is assumed that the socio-economic organisation of a country both conditions and reflects the distribution of power among the key stakeholders affected by particular state–society arrangements and policy outcomes (March & Olsen 1989). Specific institutional arrangements shape particular kinds of politics by affecting the way an actor defines his own interests as well as the degree of power that any set of agents has over policy outcomes. These institutions have evolved out of previous situations and rule settings in which human actors found themselves and that have caused policy makers to take certain courses of action.

Given that each alternative course of action tends to favour the interests of particular stakeholders over others, the parties involved will bargain about certain institutions and policy outcomes to further their interests. So agents will actively attempt to further their interests and particular policy outcomes while at the same time being subject to institutional structures that set limitations to their operations and interactions.

Governance transformations are the outcome of the dynamic interplay between actors, private and public, either in favour of change or eager to preserve the *status quo*, and the larger technological, economic, cultural and political context, constraining their strategic behaviour. In the explanation of why governance regimes emerge, persist or change, Campbell and Lindberg (1991) have suggested five possible factors:

1 the search for economic efficiency;
2 innovation and technological developments;
3 the quest for power and control;
4 the impact of cultural beliefs;
5 the role of the state in promoting or foreclosing particular courses of action.

The first factor pressing for governance transformation, economic efficiency, relies upon rational actors trying to find more efficient and/or effective ways of doing business. Consequently, economic actors will probably select those governance structures they believe will reduce transaction costs and improve economies of scale and scope. Technology provides new opportunities for governance transformation by expanding the range of choices to innovate production, restructure product markets and facilitate new entry for outsiders.

Besides the continuous search for economic efficiency and innovation, one has to include the importance of power struggles and the pursuit of control over technology, critical dependencies and market competition in maintaining or challenging established governance structures. Actors are concerned with safeguarding their private interests, balancing the power relations with their stakeholders and controlling the terms under which these transactions and economic activities take place. Culture also plays a role in the occurrence and transformation of economic governance, although its explanatory power is limited. Cultural and ideological factors constrain the range of available governance mechanisms from which actors might choose by inhibiting or facilitating particular co-ordination mechanisms that fit within the encompassing system of beliefs, norms and values of a particular sector. The role of the state in structuring and transforming industries is more complex. State actors shape the selection of governance regimes by defining and enforcing the property rights within the national economy and they determine the conditions under which each of the governance arrangements may exist. The government devises policies that prohibit certain courses of action, promote institutional change and certain kinds of behaviour through incentives, and legitimise and enforce the (newly established) governance regime.

In explaining policy change, shifts in power dependencies between stakeholders in the policy network and the larger governance regime play a dominant role. However, one should not underestimate the significance of new persuasive ideas, intellectual debates and prior experiences with related policies in the political process. Social scientists are seemingly more interested in the 'political' aspects of the policy process, referring to elements as power struggles, bargaining and conflict resolution, and neglecting such cognitive and argumentative aspects as problem solving, advocacy, communication, deliberation, persuasion and learning. Ideas, being a propelling force in the search and implementation of new solutions to collective societal problems, will be seen as an equally relevant explanation for the development of public policy (Hall 1989, 1993; Majone 1989a, 1992). Hall (1993) has distinguished between three forms of change in policy content: first-order change, caused by incremental and routinised decision making regarding socio-economic problems, second-order change, caused by altering the instruments without challenging the overarching hierarchy of objectives behind the policy, and third-order change referring to a coherent set of new ideas that redefine the persistent problem and its possible solutions. Hall considers first- and second-order change as manifestations of normal incremental problem solving and third-order change as a paradigm shift, provoked by an accumulation of policy failures that trigger the promotion and discussion of new conceptions and institutional structures.

The extent to which these alternatives become widely accepted and implemented depends on their techno-economic, political and administrative viability and the support they receive from communities of experts, leading socio-economic groups, politicians, and administrators (Hall 1986, 1989; Rose 1993). March (1989) and Kingdon (1984) have referred to the linkage of four largely independent streams, namely the flows of energy from participants, problem recognition, available alternatives and choice opportunities, that makes policy breakthroughs possible. Derthick and Quirk (1985) have described the drastic changes in US regulatory policy on transport, airlines and telecommunications, that were the result of a shift in the dominant interests and a surge of conceptual innovation generating new arguments, evidence and proposals. Originally, the alternative of regulatory reform was developed by economic experts advocating deregulation as a way to cut back often unnecessary forms of regulation and hence reduce social costs. These intellectual suggestions for pro-competition policies were picked up and supported by leading politicians responsive to public concerns about high levels of inflation and government intervention and/or inclined to free market ideologies. Although hindered by a relatively late and ineffective reaction from the affected industries to protect their vested interests and slow down far-reaching reform measures, entrepreneurial office holders, commissions and advisory working groups further advanced the ideas for deregulation and eventually implemented them.

We will follow an evolutionary theory of institutional change that emphasises both rational variation and (re)adaptation, and the constrained selection of appropriate forms and the elimination of all the others (Aldrich 1979; Nelson & Winter

1982; DiMaggio & Powell 1983; Campbell & Lindberg 1991). The mechanism of variation refers to the more or less intentional motivations and cumulative search efforts of actors to improve their capabilities and strategic position. The strategic responses of organisations to transform governance regimes more favourable to their interests, are triggered by changes in their techno-economic or political environment. The selection mechanism refers to the filtering process of how better-fit forms are picked out by the environment and less-fit forms are replaced or rejected. The mechanism of structural adaptation refers to the incremental choices made by policy makers aimed at reforming and adapting institutional structures in response to environmental changes, threats and opportunities. Those particular forms are selected out, that are effectively pursued by powerful actors and are best suited to the changing environmental conditions. The shape organisational forms eventually take is the outcome of a step-by-step process of trial-and-error learning, adaptation and persuasion.

The more or less intentional variation process is seriously constrained by strong inertial forces like sunk costs in plant, equipment and personnel, established routines and experiences, regulations, and barriers to entry and exit, all of which may set strong limits to organisational adaptability (Hannan & Freeman 1984). Organisations confronted with high levels of uncertainty, resource scarcity and constraints often follow 'best practice patterns' and 'survival paths' by borrowing or emulating structural forms of relevant other organisations. DiMaggio and Powell (1983) have emphasised that organisational fields are continuously being (re)shaped by environmental developments and ultimately stabilise through the emergence of institutional forms that are the most compatible with their environmental characteristics and requirements. Organisations and institutions may, deliberately or necessarily, model themselves on similar organisations and institutions that they perceive as more successful or legitimate. They may adopt a form that has been shown to be efficacious in another context, or be forced to emulate a powerful organisational form in a neighbouring field. The mechanism of imitation refers to the availability of blueprints that could function in cases of high uncertainty and ambiguity as exemplars of problem solutions in the reform of organisational and institutional forms at hand. Majone (1991) and Rose (1993) have pointed out that policy makers, operating under tight resource and time constraints, often rely on pre-existing models from politically and economically powerful countries and make incremental adjustments only at the margin, instead of developing new frameworks. Given the inevitable differences in cultural, economic and social settings, foreign institutional forms are never exactly and directly implemented, but only imitated partially with such minor adaptations and adjustments as fit with local circumstances.

The thrust of the evolutionary approach is that environments select only those policy ideas and forms that fit within a predefined configuration of interests, and filter out those that are maladapted to meet the demands of the political and economic environment. The selection process within the environment is conditioned by historical contingencies and constraints, including the legacy of past policy choices, locked-in phenomena, and long-term commitments. As argued by North

(1990), the continuity of history and institutional persistence often acts as a constraint on the ability of organisations to respond to changes in their environment:

> History matters. It matters not just because we can learn from the past, but because the present and the future are connected to the past by the continuity of a society's institutions. Today's and tomorrow's choices are shaped by the past. And the past can only be made intelligible as a story of institutional evolution.
>
> (ibid. 1990:vii)

So policy responds less directly to the current social and economic conditions than it does to the consequences of previous initiatives: 'Policy makers are inheritors before they are choosers' (Rose 1993:78).

This path-dependent behaviour is inspired by the fact that an alteration of rules and practices usually involves high switching costs and impaired learning effects and adaptive expectations; instead yesterday's experiences will provide the opportunity set for today's actors and organisations (David 1986; Arthur 1988, 1996). Institutional settings, however, may temporarily shield an organisation or an industry from selection pressures. Organisations will then seek to buffer their technical core and interests from environmental disturbances through the pursuit of various closure strategies, aimed at reducing or suppressing competition and technical innovation. Random initial institutional choices and small historical events which preclude future options, including those that would have been more efficient and effective, may be locked in early, and increasingly impede the future adoption of alternative policies and institutional arrangements. While the room for future strategic choices may eventually become limited over the long run, those particular routines that were adopted initially may become more efficient over time, due to the increasing return mechanisms that reinforce the direction once on a given path; in the words of Krasner (1988:83):

> Institutions (. . .) persist because they follow path-dependent patterns of development. Path-dependent patterns are characterised by self-reinforcing positive feedback. Initial choices, often small and random, may determine future historical trajectories. Once a particular path is chosen, it precludes other paths, even if these alternatives might, in the long run, have proven to be more efficient or adaptive.

1.8 Comparing telecommunications governance regimes across nations and across time: a research framework

In this institutional analysis of the restructuring in European telecommunications, we investigate the occurrence of similar and different policy responses over time and across countries. The underlying assumption is that nation-specific adjustment trajectories are shaped through the search strategies of domestic actors, enduring government–business patterns and external institutional contingencies that narrow the range of policy choices (Zysman 1994). A comparative historical approach is, according to Chandler (1990), essential for the evaluation of industrial and institutional transformation across nations, sectors and over time:

> To be valid, historical analyses must be comparative. They must compare the histories of enterprises within the same industry, and then they must compare the collective history of the enterprises within that particular industry with that of other industries in the same nation and also with that of the same industry in other nations. Only such broad-based data can provide the comparisons that indicate common patterns of institutional growth and reveal the impact of cultural, economic, and historical differences on institutional evolution. Such comparisons, in turn, provide the underpinnings for a systematic analysis of the dynamics of modern industrial capitalism.
>
> (ibid. 1990:10)

For the comparative analysis of big structures and large processes, Tilly (1984) has introduced a 2 × 2 framework based on two dimensions: the *multiplicity of forms*, i.e. the number of different kinds of forms studied, and the *share of all instances*, ranging from the study of an example to all known examples of the form (see Figure 1.1). *Individualising comparisons* clarify the characteristics of the case at hand by means of contrast with another well-documented case(s). The goal of the comparison is to specify the uniqueness of a case and to establish exactly what is particular about a particular historical experience. *Encompassing* cross-national and cross-sectoral studies put all the different instances at various locations within the same system with the objective of explaining their characteristics as a function of their varying relationships to the system as a whole. *Variation-finding* comparisons

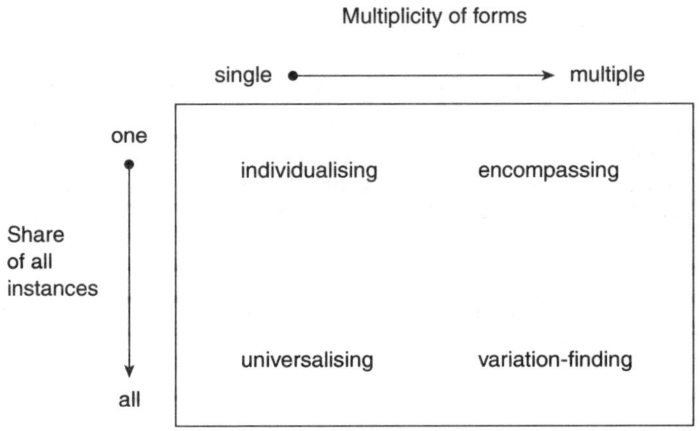

Figure 1.1 Ways of comparing political and economic structures and processes
Source: Tilly 1984

establish a principle of variation in the character or intensity of a phenomenon by examining systematic differences among instances. No claim is made to account for the full range of experiences, but instead the goal is to make sense of particular social structures and processes and eventually to extend principles and properties of one case readily to new ones. *Universalising* studies identify common properties among all instances of a phenomenon and aim to prove the point that every instance of a phenomenon follows essentially the same stages, rules and standard sequences and transitions. This comparative study on telecommunications restructuring in France, the Netherlands and the UK can be located right in the centre of this matrix: the multiplicity of forms is restricted to three, and the share of all instances is set within a time span of almost twenty years, namely the 1980s and 1990s, when the telecommunications industry in those countries was substantially transformed.

The aim of this study is to compare and explain telecommunications restructuring as it took place in three countries over a period of time. The period of study includes 1980–98 and the countries under investigation are the United Kingdom, the Netherlands and France. These states were selected on their distinct patterns of government action to stimulate the pace of technological and economic development in telecommunications. In previous research, analysing the process of states adjusting their domestic industrial structure to the changes in world economy, three national strategies have been distinguished: company- or market-led adjustment, state-led adjustment, and negotiation-led (or corporatist) adjustment (Zysman 1983, 1994; Katzenstein 1985). The UK approach in restructuring domestic telecommunications could be considered an example of the first strategy, as it relies on the extended role of market forces in the domestic and international economy, with a modest regulatory role for central government. The French approach is an example of the second strategy, emphasising far-reaching state intervention and political manipulation of the key markets of the national economy. The Dutch approach can be regarded as a manifestation of the third strategy, exhibiting explicit bargaining and consensus seeking among all the stakeholders involved, as to which adjustment policy to follow.

In this study we will examine whether there exist clear differences in the style, the degree and the timing of strategic responses of national governments and other stakeholders to adjust their domestic industry to the new techno-economic and international conditions in telecommunications. To investigate the various strategic responses to persistent structural forces, we distinguish three independent variables at different levels of analysis:

1 *(X 1) an extra-national variable*: includes the aforementioned four structural forces that effectively challenge the established telecommunications regime in the countries under investigation in this study. (i) Technological revolution, (ii) globalisation and differentiation of markets, (iii) international deregulation, and (iv) European integration are treated as exogenous variables. These structural forces serve as an impetus for the restructuring of telecommunications and set the agenda for the debate in the three political arenas to be analysed;

2 *(X 2) the sectoral regime variable*: refers to the distinctive distribution of power between the incumbents and their political allies, and (potential) new entrants, seeking access to the market place and the policy process, as well as the prevailing governance regime of the telecommunications sector of one particular country;

3 *(X 3) the national context variable*: determines whether states exhibit a unique mix of socio-political, economic and historical elements, making up a distinctive framework of incentives and constraints for policy making. Three national institutional settings are distinguished: a liberal regime (UK), a negotiated or corporatist regime (Netherlands) and a statist regime (France). This national context variable will be elaborated in Chapter 2, where an extended framework is presented. On the basis of Gourevitch (1986), a further distinction will be made between the particular production profile, the role of the state in the economy, the system of interest intermediation, the dominant economic approach (or ideology), and the positioning of the country in the international political economy. Together, these national variables make up the extended national context variable (X 3).

The *dependent variable* (Y) in this research is the pace and timing of change in British, French and Dutch telecommunications policy (see Figure 1.2). The time frame of this study runs between the early 1980s, when state monopolies were still (largely) intact in France, the Netherlands, and the UK, and the late 1990s, when all three countries had restructured their telecommunications sectors. The yardstick for comparison will be the timing and degree of implementation of the new governance regime which succeeds the traditional public monopoly model. The negotiation process and policy development over time are also investigated: were there any radical breakthroughs or incremental adjustments of the existing policies

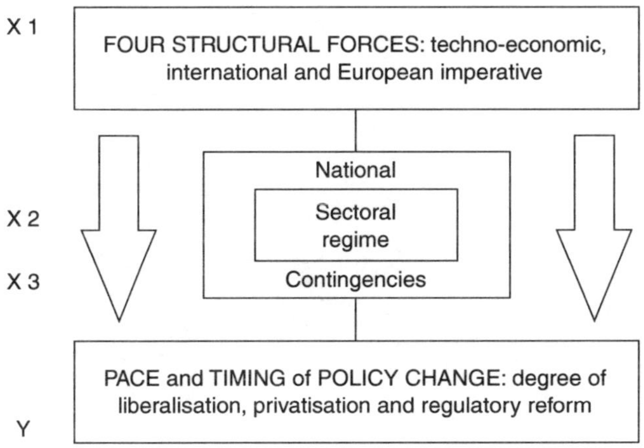

Figure 1.2 Research framework

and how did these major and/or minor changes take place? Interviews with officials, consultation of experts, official policy documents, position papers of (private) stakeholders, annual reports, background studies, etc., provide the data and insights for this study. Pace and timing refer to the implementation of decisions concerning the liberalisation of the telecommunications market, the privatisation of the public operator, and regulatory reform to ensure public service provision and fair competition.

Following this introductory chapter, we provide an elaborate discussion of the various approaches in the field of comparative government–industry relations and economic policy making. Chapter 3 will give an overview of the techno-economic and international forces in the telecommunications industry changing it from a national public monopoly into one that fosters new entry and innovation, the privatisation of PTTs, regulatory reform, and the creation of international strategic alliances between telecommunications companies. Chapters 4, 5 and 6 contain case studies on the restructuring of the British, Dutch and French telecommunications industries and investigate the deliberations, actual developments and experiences with demonopolisation in these countries. The degree and timing of the implementation of liberalisation, privatisation and regulatory reform measures in the United Kingdom, the Netherlands and France is discussed in Chapter 7. After this comparative institutional analysis of the decisive actors, factors and institutional contingencies in the three national settings being investigated, some final concluding remarks are given in Chapter 8.

2 International restructuring and national strategies and contingencies

2.1 Introduction

The aim of this study is to point out in what way three Western European countries have responded to changes in technologies, demand, regulation, and the international political economy of telecommunications, and whether the adjustment strategies of France, the Netherlands and the United Kingdom in the 1980s and 1990s have varied or converged. Comparative studies contribute to the debate on national and sectoral diversity versus international convergence in analysing industrial restructuring and explaining policy development. The question to be addressed is whether public policies vary across nations, across policy sectors and even over time or whether, eventually, they become more similar. This chapter provides an overview of various theories and concepts that deal with the convergence–divergence debate in comparative strategy, organisation and policy studies. In section 2.2 we will discuss the 'policy styles' approach, referring to distinctive forms of decision making stemming from national or sectoral settings, as investigated in comparative public policy studies. In section 2.3 we will introduce the business system approach and related concepts that refer to management and economic organisation being culture-bound and societally-embedded, as advocated in comparative organisation studies. In section 2.4 we will discuss the contributions from comparative political economy. Here, the emphasis is on the international context within which national institutions are embedded and on the different structural adjustment strategies countries pursue. Finally, in section 2.5, finally, we will elaborate on the research model, as mentioned in Chapter 1, by introducing and discussing the five national variables, originally introduced by Gourevitch (1986), that will allow for a comparison between telecommunications restructuring in France, the Netherlands, and the United Kingdom. The contingencies and constraints that shape and constrain a country's adjustment strategy are the production profile, the role of the state, the system of interest intermediation, the economic ideology and culture, and the nation's position in the international political economy.

2.2 Comparative public policy studies

Comparative public policy scientists investigate in which way different government agencies, interest associations of organised business and labour, individual firms and other actors, such as banks and consumer associations, act and interact in the formation and implementation of public policies. The comparative public policy approach has been defined by Heidenheimer *et al.* (1990:3) as 'the study of how, why, and to what effect different governments pursue particular courses of action and inaction'. The aim of this approach is to find answers to the three questions posed in this definition. The question of how public policies differ (or not) focuses on how different governments deal with certain problems, within which institutional framework they operate and with which configuration of actors (e.g. regulatory agencies, interest groups, individual companies etc.) they interact when trying to solve these problems. The question of why public policies differ (or not) pays attention to the constraints and determinants in the political process: decisions can be shaped by the ideas policy makers put forward, power dependencies in the political system and intensive lobbying by interest groups. The question of what the consequences are of government (in)action concentrates on the evaluation of the chosen policies and the assessment of a government's capacity to cope with problems. This question deals with the pay-off of certain policies. Sometimes the objectives of governments are achieved, or when the results are disappointing, political debate starts anew in order to adapt objectives to be more successful, or the chosen policies give way to externalities and side effects. The question of to what effect different governments pursue particular courses of action is related to implementation and evaluation analysis, referring to the appraisal by the major stakeholders, the media and the general public as to the effects policies have in practice.

The primary focus of comparative public policy is on the way states diverge in the manner in which their policies are founded politically. The comparative public policy approach has its origins in the discipline of comparative politics with its focus on the role and organisation of distinctive political and administrative institutions across nations. Here one can think of the study of cabinets, parliaments, political parties, public bureaucracies, independent regulatory agencies, courts of law (e.g. Lane & Ersson 1991). All these organisations may – to a varying extent – shape the policy outcomes and institutional development of particular countries. An illustration of cross-national research conducted within this tradition is Lijphart's comparative study of the political systems of the major long-standing democracies in the world. Lijphart (1984) systematically compares the (degree of) concentration of the executive power (cabinet), the composition of the legislature (parliament) and the relations between them, the party and electoral system, and the organisation of central–local relations. He divides the various democratic regimes into two basic models: majoritarian democracies such as the UK and New Zealand (and to some extent France), correlating with homogeneous societies, and consensus democracies such as Belgium, Switzerland and the Netherlands, correlating with pluralist societies.

A further theoretical contribution was made by Richardson *et al.* in the 'policy styles' approach with its focus on the central characteristics of the way societies formulate their public policies. Richardson *et al.* (1982:13) have defined national policy styles as 'the interaction between the government's approach to problem solving and the relationship between government and the other actors in the policy process.' In a series of national case studies, the researchers have made an attempt to classify European countries according to their standard operating procedures for reaching decisions (Richardson 1982). Although they admit that in every country a variety of styles can be found, certain dominant national styles can be distinguished. Richardson *et al.* (1982) have identified two primary factors in their typology: the nature of the government's approach to problem solving and the nature of societal governance. The combination of these two factors has made possible a categorisation of societies into four basic policy styles (see Figure 2.1). The first deals with a government's approach to policy change in terms of either an anticipatory/active attitude towards problem solving, favouring radical change, or an essentially reactive approach to problem solving, promoting incremental change. The anticipatory approach can be qualified as active and assertive: there is a central authority with a high coercion ability, clear objectives for the future, limited conflict over values, a wide search for the options and clear possibilities for radical change. The reactive approach can be qualified as incrementalist, and is characterised by a conflict over values, emphasis on consultation and mutual adjustment, successive limited comparisons, low coercion and managerial change. The second factor is related to the relationship of a government with other actors in the policy process. The relevant question here is what the relationship is between governments and interest groups: is the government eager to reach a consensus with organised groups by way of accommodation, or is it more inclined to impose decisions, notwithstanding opposition from groups?

Germany and the Netherlands can be found in category I, in that they emphasise consensus and show a reactive attitude to problem solving by closely following international developments. Sweden and Japan can be located in category II: they

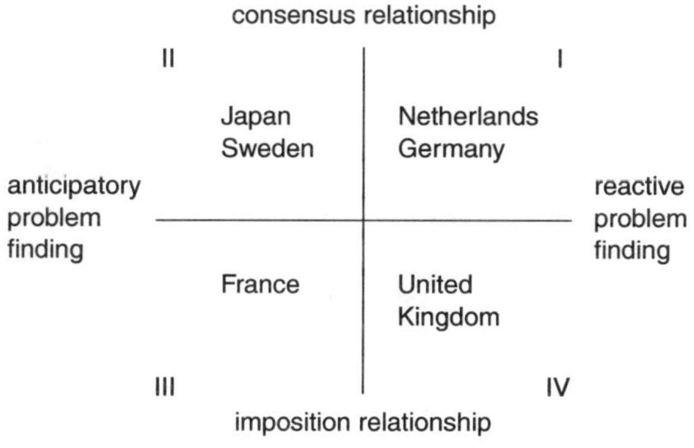

Figure 2.1 Different policy styles
Source: Richardson *et al.* 1982:13

have also stressed the need to reach consensus, but – inspired by normative values – seek to combine this with a more anticipatory or active approach to problem solving. France belongs to category III: it is less concerned with consensus, but instead exhibits a policy style in which the state often takes the lead in carrying out reform programmes. Category IV is where governments are increasingly reactive in their approach to problem solving and, yet, if any policy change is to be achieved, it has to be enforced against the resistance of at least some organised groups. The United Kingdom may be characterised as having a distinctive policy style, placing great emphasis on confrontation in the context of a very reactive approach to problem solving.

In his cross-national study on occupational health and safety policy in Sweden and the United States, Kelman (1981) has pointed out that differences in national styles left a much sharper imprint on rule implementation than on rule making. Despite major differences between the relative strength of business and labour in the US and Sweden and the presence of diverging political cultures (Sweden: social-democratic; the USA: liberal-conservative), Kelman found that the goals and content of regulatory control were to a large degree the same in the two countries. The decisions taken by the Swedish agency for occupational health and safety (ASV) and its American counterpart (OSHA) were almost alike; only a few differences existed in the stringency of the regulations. Kelman (1981) discovered, however, that there were major differences in the ability of government agencies, labour and business in the two countries to reach joint agreements regarding the content of regulations and their enforcement. The assumptions, operating styles and organisational environment of the ASV agency reflected the values and institutions of the Swedish *overhet* tradition (i.e. the Continental state–society conception). The OSHA agency revealed the American liberal tradition (the stateless society conception) with its more adversarial and legalistic system of government–industry relations.

The Swedish enforcement system was characterised by the presence of dominant normative inducements, informality, searching for compromises and mutual adjustment. This facilitated the development of political institutions that fostered accommodation among the contending groups, bringing the representatives of the conflicting parties together into a small forum for negotiations. In this limited and informal setting, mutual adjustment between the stakeholders is encouraged and joint agreements may be worked out. In the US, the implementation of regulations for occupational health and safety takes place in a system characterised by formality, antagonism, judicial review, and the use of court trials and public hearings. The American political system tends to be poor at solving conflicts, because of the strength of self-assertive values and the tendency to use adversary institutions. There is a tendency to let the parties fight it out among themselves and if the opposing parties do not reach an agreement, a neutral third party is empowered to decide for them; this gives the courts more power than in any Western European country. The adversarial proceedings for a specific policy sector or policy aspect are imposed by independent regulatory agencies, in charge of rule-making, rule implementation and enforcement. In contrast with Sweden, where policy making takes place informally in small groups behind closed doors, the devising, implementation and enforcement of rules in the USA is open to public participation.

At face value the Swedish system seems to be good at solving conflicts: the deferential values and small group negotiations in accommodationist institutions promote broadmindedness and collaboration in a way that the US system cannot. In the US the presence of self-assertive values hinders the solution of difficult social conflicts and the use of adversarial trials does little to encourage agreement; in the words of Kelman (1981:232): 'parties speak to a judge rather than to each other'. The Swedish regime, however, has shortcomings too (Kelman 1981). The presence and use of normative inducements and accommodation are not always sufficient to persuade actors and enforce decisions, sometimes compliance has to be legally enforced. The Swedish system provides considerably less data and information to central decision makers than the American one does. The Swedish regime favours the search for consensus and compromise; this generates policy alternatives which are politically feasible and acceptable to all the stakeholders involved. The American system of adversary proceedings provides more information and evidence to central decision makers. In front of an arbitrator, investigating the dispute, each party has an incentive to find as many facts as possible to back his/her case. Another disadvantage of the Swedish system is that the interests of parties who do not like the terms on which the agreement is reached, are inadequately represented; the American political-administrative system is much more accessible to (relative) outsiders.

A further elaboration of Kelman's argument, that each nation exhibits a distinctive regulatory style, stemming from the political and social context in which public policies develop, can be found in the cross-national analysis of environmental policy making in the United Kingdom and the United States, as carried out by David Vogel (1986). He argues that a nation's approach to industrial regulation does not take place in a vacuum, but needs to be understood within the specific socio-political and legal-institutional framework within which the interaction patterns among government, business, labour and the general public are structured, and in which environmental policies evolve. Vogel (1986) concludes that the United Kingdom and the United States have adopted divergent approaches to environmental regulation. The reason for these strategic differences has to do with the fact that the two countries have different systems for making and implementing environmental policies. In Britain the relationship between business and government and their respective roles in the regulatory process can be described as an informal and relatively co-operative relationship between civil servants and industrial managers. The British regulatory system has a built-in tendency towards consultation and compromise-seeking between government officials and representatives of industry, and industrial self-regulation fosters a voluntary and informal system of environmental controls within the industry. In contrast with the flexible and informal British style of environmental regulation, the regulatory style of the US can be characterised as formal and rule-oriented, emphasising administrative discretion, legal enforcement and legislative supervision. The American regulatory system, being legalistic, stricter and more adversarial than other capitalist democracies, is open in that it provides more opportunities for political participation and makes more information available to the general public. The making and implementation of environmental regulations in the US takes place in a publicly accessible system, providing interest groups with various platforms on which to further their interests: Congress, lawcourts, the White House or the President's executive office.

Despite its theoretical and empirical contributions, the comparative public policy approach has its shortcomings too. For instance, the 'national style' concept over-accentuates the differences between national systems and to a large extent neglects the growing interdependence and harmonisation between various public policies. Especially in a growing world economy and against the context of European integration, domestic politics in Western Europe become more and more intertwined with international matters and, consequently, more convergence between national styles is needed and/or wanted, for instance in the area of economic and monetary union. Several authors have drawn attention to the growing tension between the increasing globalisation of business (e.g. internationalisation of home-based industries, penetration of foreign investors, international co-operation in R&D and standardisation) and regional integration (e.g. the coming into being of trading blocs) and decreasing national sovereignty (Ruigrok & Van Tulder 1995; Berger & Dore 1996; Hirst & Thompson 1996).

Another shortcoming of the comparative public policy approach is that it lacks a profound and clear differentiation between national and sectoral policies. The concept of national policy styles refers to certain distinctive forms of decision making, which are typical of the national political system at large (e.g. the French way). The relative importance of nation-specific *vis-à-vis* sector-specific factors in determining public policy for a specific sector is hardly ever spelled out (Freeman 1985). The concept of sectoral policy styles refers to certain characteristics that are typical of a policy sector as such, following its own policy pattern across national boundaries (Atkinson & Coleman 1989). This would mean that industrial development follows the same pattern across national boundaries (e.g. the universal trajectory of telecommunications reform). Originally, the concept of policy styles was used in a general sense to refer to national styles of public policy making. The search for a predominant policy style has become more complicated by the 'sectorisation of public policy making' (Jordan 1981, 1990). Policy sectors are increasingly made and ruled by their own élite of major stakeholders and such a disaggregation of the political system into semi-autonomous policy communities may imply a convergence of sectoral policy styles across countries. Instead of one single policy-making process, there tend to be numerous relatively narrow and self-contained policy subsystems with a high degree of group incorporation, by which actual policy shaping and decision making takes place. This trend towards sectorisation restricts the national governments' room for manœuvre and it can undermine a specific national policy style in favour of various 'sectoral policy styles', which display a certain convergence or similarity within certain policy areas that go beyond national boundaries.

Today's policy process, for instance, is increasingly riddled by debates and negotiations in various semi-autonomous sectoral networks. The comparative public policy approach underestimates the role of bargaining and strategic interaction between interdependent public and private actors and between national and international organisations. Just as styles change in architecture and painting, so too policy styles (can) change. Policy styles are subject to minor and major changes in the political-administrative and economic systems of countries, such as for instance changes in cabinets and political party dominance, changes in the formation and recruitment of élites, the emergence of new dividing lines in society, changes in technologies and markets, or a shift in the normal policy-making style as a response to critical events. National as well as sectoral policy styles can evolve and transform themselves over time. For instance, the coming into office of the Thatcher government with its privatisation programme produced a remarkable change in British policy style, and regulatory reform in the US airline industry illustrates that certain aspects of a sectoral policy style can also change.

2.3 Comparative organisation studies

Besides furthering their interests in the policy arena, either through a direct lobby or indirectly through industry associations, private actors, and especially large companies, position themselves strategically in particular industries and compete with other firms in the market place. The literature on business systems and comparative economic organisation, to be discussed below, is basically complementary to the comparative policy study approach because its primary focus is on companies, their internal operations and strategic partnerships, and their embeddedness in market structures and societies. Also in strategic management and organisation studies, there is a lively debate about whether 'nations matter' in an internationalising world economy with respect to the policy choices of governments and the corporate strategies pursued by leading domestic firms. For instance, Kogut (1991:44) has argued that companies wanting to survive in the global market place rely on their own capabilities and the institutional strengths of their countries of origin: 'The study of international competition is, in large part, the study of comparative management and societal institutions among countries.' Another related question is whether organisational and institutional capabilities vary substantially between countries and – if so – whether they have a different impact on national economic performance. Kogut has explained the occurrence of national differences in trade and foreign direct investment patterns by referring to a country's technological and organisational abilities, namely availability of natural resources, the national skill and knowledge base, and the role of supportive collective institutions.

In comparative organisation and management studies, a small group of scholars, including Lane (1989), Sorge (1991), Whitley (1992a, 1992c) and Mueller (1994) have investigated whether there is a correspondence between the presence of particular organisational forms and business practices on the one hand, and the institutional structure of society on the other. They argue that the occurrence of variations in socio-economic institutions and structures cannot be reduced to a single economic rationale or market competition selecting the most efficient patterns of economic organisation, but that instead, markets, firms and economic performance are socially structured and embedded in different contexts. Their basic assumption is that there are a number of different ways of organising economic activities successfully in a market economy, and no single pattern is necessarily superior to all the others. In that respect, they elaborate on the work of contingency theorists, such as Burns and Stalker (1961:125), who have argued that 'the beginning of administrative wisdom is the awareness that there is no one optimum type of management system'. The neo-contingency theorist Sorge (1991:163) has argued that 'the goodness fit' between business organisations and their task requirements and other environmental conditions is critical in understanding and explaining economic performance: 'Societal differences in organising and generating human resources, and the pursuit of different business strategies, are reciprocally related. An economy and society becomes populated by specific institutionalised organisational and human resource forms and practices, because economic niches and business strategies are different, and vice versa.'

In short, country-specific cultural characteristics and organisational and institutional forms and strategies in those product markets in which a nation and its domestic firms show particular strength, are mutually contingent. To describe the persistent and systemic influence of social institutions on the corporate strategies of and business practices within firms with the final result that company structures and processes reflect typically national patterns, Best (1990:145) has used the term 'institutional specificity of culture'. This 'cultural specificity' is not only expressed in, but also reinforced by national institutional frameworks that structure management and labour relations and the industrial enterprise in distinctive ways. Countries' socio-economic institutions and the particular distribution of power between management and labour have become so engrained in particular business styles and organisational practices that they may withstand the strong pressures toward economic convergence and the emergence of a globally integrated capitalist system (at least to some extent and temporarily).

Industrial sociologists like Maurice *et al.* (1980), have extended the cross-national study of organisations into the 'societal effects' of institutionalised human resources policies, socio-economic stratification and industrial relations frameworks on the shaping of particular corporate strategies and business patterns. For them the performance of an industry across countries is due to a congruence between a sector- or company-specific profile and the societal profile. A combination and eventually a congruence between specific types of product market, work pattern and corporate strategy on the one hand and a particular societal organisation and human resources policies on the other, make such diverse socio-organisational arrangements co-exist. In short, there are a number of different ways of organising economic activities successfully in an economy, and no single pattern is clearly superior.

In her comparative study on systems of industrial democracy, the organisation of work and vocational training in Germany, France and the UK, Lane (1989:292) presents evidence on the 'cultural specificity' of industrial relations, labour market policy and industrial strategies: 'Although the three economies and the business organisation, which constitute them, are being confronted with similar problems and opportunities, management, labour and the state in each society continue to interact in nationally distinctive ways to cope with these challenges.' Charkham (1994) has shown how companies in general and the way they are directed and controlled in particular, reflect processes and institutional arrangements particular to their society. Corporate governance regimes vary across countries and have a profound impact on the structure of firms: for instance, CEO-dominated and controlled versus collegiate decision-making structures, and centrality of stock market ('market for corporate control') versus supportive shareholding networks (preferred stockholders, cross-shareholdings, and co-determination). Corporate governance systems may also have an impact on the strategies of companies: short-term orientation versus longer-termism, and profit maximisation/high dividends versus market share growth/innovation.

In internationally comparative business studies, both the characteristics of national industrial relations policies and corporate governance frameworks, and the particular organisation of business systems are seen as important factors explaining variations in the relative success of firms and societies. Whitley (1992a, b, 1994a) has sought to explain the predominance of distinct business recipes and business systems by referring to the particular institutionalisation of dominant economic practices and values within and between firms and their general embeddedness in society-specific configurations. These national business systems contain organisational and institutional responses to three key problems in capitalist societies: how economic activities are to be structured, how markets are to be co-ordinated, and how operations within authority hierarchies are to be organised and controlled. Whitley (1992c:36) has defined business systems as 'relatively stable and cohesive configurations of hierarchy–market relations that have developed and remain effective in particular and separate institutional contexts, typically within the boundaries of the nation state.' In analysing dominant patterns of economic activities in sectoral and/or national settings, three features of business systems stand out:

1 the nature of the firm (e.g. ownership and management patterns, degree of diversification, development of skills etc.);
2 the organisation of the market (e.g. trust and loyalty relations, information sharing, different forms of interfirm relationships etc.);
3 authoritative co-ordination and control systems (e.g. organisation of work: degree of specialisation, decentralisation, formalisation, management style etc.).

In integrating authority structures, firm types, and interfirm relations into distinct forms of economic organisation, Whitley (1992c, 1994a) also includes a (re)productive role for background and proximate social institutions. The former refers to the absence or presence of trust relations and collective commitment patterns and the latter refers to the national system of property rights, the functions of the state, and the availability of and conditions governing access to financial and labour resources.

In his multi-level and integrative study of the economic organisation of business in various European and Asian countries, Whitley (1992a, b) has investigated and contrasted Europe's well-known mix of solid institutional pluralism, strong emphasis on individuality and its adherence to legal-rational principles with Asia's legacy of cultural homogeneity, strong group and/or family loyalty, combined with a general weakness of the legal system. In the social construction and institutionalisation of business systems, the cases of Japan, South Korea, Taiwan, and Hong Kong, where fast industrialisation and rapid economic progress were accomplished, offer interesting examples (Johnson 1982; Amsden 1989; Wade 1990).

Whitley (1992a) has looked at the economic organisation of business in its social context in East Asia and compared the various firms, market dynamics and interfirm relations, and the roles socio-economic institutions have played in the aforementioned countries (see Table 2.1). In Japan, the national business system is dominated by the *kaisha,* a large industrially specialised corporation, often part of an inter-market group and a wider mutual obligation network (i.e. banks, subcontractors, and other business partners). The company is normally run by company-grown managers with considerable autonomy from the firm's shareholders and a strong commitment to human resources policy (life time employment of 'core' workers, work group autonomy, promotion of training and internal promotion). The strategic focus is incremental change with a clear priority to further specialisation in a single industry and a clear distinction between keeping the key operations and processes in-house and subcontracting complementary activities. In Korea, the *chaebol* is the dominant business institution: this is a highly vertically integrated and diversified conglomerate, in which the founding entrepreneur/family still has substantial powers. By their effective control over credits and domestic banks, Korean government officials are also influential in the company's operations. Top managers are normally recruited because of their ties with the owning family, and show a very top down and directive management approach with low levels of consultation. Their corporate strategy is aimed towards furthering industrial growth and expansion into new technologies and industries. When they move into new areas, there is a willingness, unlike Japanese companies, to hire external senior managers. The Chinese family business, compared to the *kaisha* and the *chaebol,* is small and highly specialised in terms of both business functions and activities. It relies very strongly on kinship ties and family control. The corporate management style could be characterised as paternalistic. Most Chinese family businesses (at least those in Taiwan and Hong Kong) are involved in thick networks of extensive subcontracting, not only locally but also globally (through overseas Chinese).

With the increasing globalisation of economic activities, it becomes increasingly clear that there exists no universally applicable business recipe for corporate success and superior national economic performance. As convincingly made clear by Whitley, Lane, Sorge and Best, different cultures and institutional contexts can lead to very different but no less effective ways of organising firms, markets and societies. First and foremost they argue that different business practices can only be

Table 2.1 Business systems in East Asia

	Characteristics	*Kaisha (Japan)*	*Chaebol (Korea)*	*Chinese family business*
Nature of economic actors	• size	• large	• large	• small/medium sized
	• capital intensity	• varied	• varied	• low
	• owner control	• low	• high	• high
	• strategic/growth focus	• incremental and sectoral	• discontinuous and diversification	• opportunistic and volume expansion
	• integration of different activities	• minority shareholdings	• authority hierarchy	• personal ties and ownership
Differences in market organisation	• interdependence of firms	• high	• low	• low
	• commitment to particular partners	• high	• low	• low (except family connections)
	• reciprocity of commitments	• high	• low	• mixed
	• horizontal co-ordination between sectors	• high	• low	• limited to personal ties
Organisation of work	• long-term commitments to core staff	• high	• limited	• considerable
	• external recruitment of senior managers	• no	• yes	• rarely
	• importance of unions	• high	• low	• low
	• work group task autonomy	• high	• low	• limited
	• centralisation of decision making	• medium	• high (with family top managers)	• high (with family top managers)
	• managerial style	• supportive	• authoritarian	• patriarchal

Source: Whitley 1992a

understood with reference to the societies of which they are part. The internationalisation of business and the expanding reach of multinational companies, however, puts the comparison of European and East-Asian business systems and the (possible) transferability of management practices and organisational forms between institutional contexts on the research agenda (Whitley 1994b). At present, there is definitely a clear cross-cultural exchange of ideas and practices between European, American and East-Asian capitalist systems (e.g. the diffusion of Japanese lean production, the spread of US-inspired reform of regulatory and competition policies), but it is still too early to tell whether one global form of capitalism will eventually replace many nationally diverse business systems (Berger & Dore 1996).

2.4 Comparative political economy

The comparative political economy approach points out that the size of national economies, their place in the world market and foreign economic policies all have a serious impact on the domestic economic and industrial policy. For example, small open economies like the Netherlands and Belgium are highly dependent on the world economy; this means that there are a number of external influences that are beyond their control. World market dependency seriously reduces the efficacy of national economic policies. The sheer size of the national economy and the dependence on the world market (i.e. the particular location of a country in the international division of labour) are perceived as important elements when comparing economic and industrial policies. The comparative political economy approach acknowledges that differences exist in the domestic structure of economies and their dependency on the world market (Katzenstein 1978, 1985; Zysman 1983, 1994). Domestic policies are made up by the dominant coalition between business and the state in the political-administrative and economic system and in key policy networks (Katzenstein 1978). The character of a country's ruling coalition and the key linkages between state agencies, intermediate associations and other societal actors condition a government's foreign policy objectives and domestic industrial restructuring.

An important element in comparative political economy is the role of the state in advanced industrial countries. The degree of state intervention in advanced capitalist societies and the role of interest groups in the policy process are decisive factors. Modern societies can be placed on a continuum between the two ideal types of economic organisation (Lindblom 1977): one that emphasises the importance of decentralised market forces and exchange between the stakeholders in the economy (e.g. the United States to a large extent) and the other that embraces centralised economic planning by the state as a legitimate power (e.g. the former Soviet Union to a large extent). The mixed economies of Western Europe could be located roughly midway between the extremes of this in that they combine, to varying degrees, market capitalism with selective state intervention (Shonfield 1965). Within the group of capitalist societies, Katzenstein (1978, 1985) has made a further distinction between three types of political-economic regimes furthering industrial adjustment:

1 *liberalism*, based on macro-economic policies and market solutions. Countries like the USA and the UK follow a liberal economic policy aimed at stimulating market forces and reducing government intervention wherever possible. Their economic policies are characterised by a strong export emphasis, liberalisation of imports and a positive attitude towards foreign investment;
2 *statism*, aimed at pursuing a structural transformation of the economy by way of state intervention. Countries like Japan and France pursue strategies favouring active government participation. The state bureaucracy has a large number of instruments with which they can exert a direct influence on specific sectors and firms. Their economic policies promote exports to increase domestic competitiveness or to meet import needs, applying selective protection, and tolerating or even restricting foreign investment;
3 *(democratic) corporatism*, consisting of an ideology of social partnership at the national level, centralised policy making by way of a concentrated system of interest groups and voluntary co-ordination of conflicting objectives through continuous political bargaining between interest groups, state bureaucracies and political parties. The economic structure of countries like Sweden, Switzerland and Austria is not very well suited to a political strategy based on liberal or statist premises: with their small and open economies, highly dependent on the world market, they follow a strategy aimed at flexible economic adjustment (export-led growth) and niche specialisation, backed by stable politics and domestic compensation at home (demand management). For these economies, economic flexibility and political stability are mutually contingent (Katzenstein 1985).

Another distinction between states can be made on the basis of the power a state has in relation to its own society. Governments differ in the degree and ways they intervene in the national economy, varying between state-less societies and state-oriented societies or between strong and weak states (Shonfield 1965; Dyson 1980; Zysman 1983, 1994; Atkinson & Coleman 1989). The Anglo-Saxon societies are characterised by a liberal or *laissez-faire* tradition with business operating relatively independently from the state; in these countries the government is unable to formulate policy goals to change the behaviour of specific groups or the structures in society autonomous of particular groups. The German, Swedish and Dutch states are examples of so-called neo-corporatist systems with moderate state intervention, where the state works together with its social partners (organised business and labour), and sometimes with the financial community to reach co-operative agreements which benefit the individual parties involved as well as the economy as a whole. In France and Japan industrial development is government-led; these countries are characterised by a strong and close relationship between government and industry, making it extremely difficult to determine where government ends and industry begins. The governments in these countries are able to formulate policy goals and change the structures in society in collaboration with business or, when necessary, independently from particular groups in society.

The comparative political economy approach studies the national economy and the development of domestic industrial policies as embedded in the broader context of international political and economic structures. Scholars working in this tradition have accentuated the relationships between domestic actors and factors (e.g. the role of the state, interest groups, and economic ideology) and the international system (e.g. cross-national policy competition, foreign direct investment, trade policy), constraining policy choices at the national level. Others have referred to the historical timing of a country's industrialisation, causing variations in the process of economic development among countries, the task a particular society faces and the economic position it will occupy in the international division of labour (Gerschenkron 1962; Shonfield 1965; Gourevitch 1986). The timing of a country's industrialisation will determine which industries must serve as a growth engine, and how the social, technical and financial resources have to be mobilised and organised for the society to be successful. Early industrialisers and market leaders are more oriented towards free trade, because they have a competitive advantage in trade over those who come later. First movers, benefiting from superior technological, economic and/or political conditions, tend to favour and introduce trade liberalisation unilaterally (Ruigrok & Van Tulder 1995). Compared to the early industrialisers, with their entrepreneurial spirits and well-developed capital markets, disciplined labour forces and relatively autonomous firms, late industrialisers, in terms of their institutional infrastructure, are poorly equipped. According to Gerschenkron (1962), these societies are economically backward with private firms, the economic system and the institutional framework inadequately equipped to accomplish economic modernisation (e.g. predominance of pre-capitalist feudal forms). There are certain advantages in being backward, however, as Korea has shown with its successful catch-up strategy in shipbuilding, such as having lower labour costs and being in a position to build the most modern plants (Amsden 1989). Gerschenkron has argued that the later a nation industrialises, the more organised it must be (in terms of the concentration and centralisation of capital and the mobilisation of resources), in order to compete effectively with early industrialisers. Industrial followers, overshadowed by stronger competitors, have to pursue a shelter strategy and protect their weak key industries (the 'infant industry' argument) and build up competitive capabilities through state guidance and appropriate investment banks (or credit-based institutions) to generate funds and make longer-term loans to industry.

Traditionally, governments have been involved in influencing domestic industries through the formation and implementation of technology and competition policies. Industrial policy aims at identifying new technological and economic opportunities, switching resources from slower growing 'sunset' industries, such as steel, coal and mining, to faster growing 'sunrise' sectors such as information technology and biotechnology. In his comparative research on technology policy, Ergas (1984, 1986, 1987) has posed such intriguing questions as, *Why do some countries innovate more than others*? and, *Do technology policies matter*? Ergas has made clear that technology policies do matter in the sense that their effects depend on their institutional environment. The performance of national technological systems will vary according to the features of each country's socio-political and economic setting. Technology policy is part of a larger political and economic strategy aiming at international leadership, national security, general competition policies, or the provision of innovation-related goods such as education, training, standardisation and co-operative R&D. Drawing upon the distinction between exit and voice-types of societal change made by Hirschman (1970), Ergas argues that in the United States, Japan and Europe different mechanisms are used to bring about socio-economic change and innovation.

The US political economy represents the exit-based model, while Europe and Japan can be seen as representatives of the voice-based model for generating and structuring innovations. The US 'entry-and-exit' model refers to an open and competitive market, characterised by decentralised decision making, new entry, experimentation, and the eventual 'exit' of firms from the market. The resulting income shifts (i.e. future profits) will reward the growth of winners, and will squeeze losers out of the market. The American strategy of deregulating the telecommunications and airlines industries in the early 1980s was aimed at establishing an environment in which rapid resource (re)allocation could take place and thus create opportunities for businesses to enter or to leave these key industries. Most of the Western European political economies (with the exception of Britain) and Japan are illustrations of the voice-based approach to structural change. Here, the innovation and economic adjustment process is subjected to political negotiations between the various stakeholders and to a socially acceptable adaptation to the major changes in demand and technology. The ideal type of voice-based change stresses that change is induced through consensus and through the conscious weighing of interests and options, buttressed by mechanisms for redistributing income between winners and losers. Compared with the exit-based approach the voice-based approach carries significantly lower social costs, but it has the built-in risk of sacrificing change to the defence of vested interests (Ergas 1984, 1986, 1987).

Ergas (1986, 1987) has pointed out that governments have both political and socio-economic reasons for furthering innovation policies. The political rationale refers to the exertion of technological power for public or national security purposes. Governments are involved in technology and industrial policy in their search for international strategic leadership and first-mover advantages by protecting their domestic industry by way of large public procurement programmes. The fact that governments are leading-edge customers of new products (especially in aerospace, micro-electronics and nuclear energy) and that they feel a need to compete with other nation states in global markets, is an important reason for strengthening national technological capabilities. This can be a primary motive for government action to shape its technology policies. The second reason has to do with the domestic industry's dependence on the provision of innovation-related 'public goods'. The development and diffusion of advanced technologies requires an adequate infrastructure, made up of an advanced education and training system that provides the necessary skills for innovation, a legal framework for defining and enforcing property rights and product standardisation, and interorganisational networks for co-operative research and technology transfer. The socio-economic rationale for industrial/technological policies concerns the manner in which these 'public' goods are provided and the roles government and industry play in this respect.

The socio-economic and institutional 'embeddedness' of innovation policy also differs from country to country. On the basis of the two rationales guiding technology policies, Ergas makes a distinction between *mission-oriented* and *diffusion-oriented* countries. In the first category priority is given to the policy objective of promoting and securing national key industries through ambitious research programmes of national strategic importance; in the second category the policy objectives of increasing productivity and diffusing technological capabilities widely throughout the industry structure are emphasised. Examples of mission-oriented countries are the United States, the United Kingdom and France; examples of diffusion-oriented countries are Germany, Sweden and Switzerland, with Japan showing a combination of the two, deploying some selective ambitious programmes while at the same emphasising institutional capacities to diffuse innovation-related public goods.

Several scholars have analysed two opposing trends in the high-tech industries that seemingly pose a dilemma for both national governments and (multi)national corporations: the choice between *techno-globalism* and *techno-nationalism* (Stevens 1990; Ostry & Nelson 1995; Berger & Dore 1996). The development towards the transnationalisation of industrial activities, the emergence of borderless and stateless firms, and the promotion of free trade has been described as techno-globalism (Ohmae 1990; Reich 1991). International technological co-operation may take several forms, varying from soft collaborative arrangements like international trade, foreign investment, international subcontracting, to cross-border mergers and acquisitions, and global interfirm agreements with a more strategic character. The reasons for establishing techno-global arrangements can range from expanding markets, enlarging market access and increasing returns to R&D, localising production, allowing flexibility, obtaining technology synergies, economies of scale, sharing high R&D costs and spreading risks.

The trend towards techno-globalism has provoked reactions in the academic and policy communities. Scholars such as Hu (1992) and Ruigrok & Van Tulder (1995) have questioned the assumptions of the increasing globalisation of companies and the arrival of the borderless business world. Despite having developed many activities all over the world, the large majority of multinational companies is still very much dependent on their home markets and domestic governments for economic and political support. With the probable exception of bi-national companies (ABB, Royal Dutch Shell, Unilever), internationally operating companies have kept most of their staff, key managerial functions (R&D, corporate decision making, tax domicile, sales), and business partners (networks of suppliers, financial investors) at their home base. Also in terms of the nationality of top management and corporate legal nationality, 'global corporations' are in fact 'national firms with international operations' (Hu 1992).

The notions of techno-globalism have also been attacked by policy makers from protectionist countries such as France and Japan and replaced by the alternative of techno-nationalism. In these mercantilist views, in which particular objectives and interests of a society are safeguarded (e.g. employment, control of macro-economic policy, national culture and language) and domestic control of valued industries is retained (Gilpin 1987:404), key technologies are a country's strategic assets, to be developed and nurtured at home. Despite the rhetoric of free trade and globalisation, industrial democracies subscribe only selectively to the principle of free trade: they pursue strategic trade policies for the home market where foreign markets are protected and domestic manufacturers are losing international competitiveness (Milner & Yoffie 1989; Ruigrok & Van Tulder 1995). Like big transnational firms rivalling each other, governments also compete with each other in acquiring world market shares and seek to promote the competitive position of their domestic industries in the global high technology race (Van Tulder & Junne 1988; Roobeek 1990; Stopford & Strange 1991). In promoting large domestic firms in core technologies, European governments have shifted from trade protection to measures aimed at stimulating domestic competitiveness, such as joint R&D projects, fiscal policy, public procurement programmes, and export subsidies. Of course the main proponents of techno-nationalism and trade protectionism are governments, but they are often supported by national champions, relying to a large extent upon their home market. Multinational corporations like IBM, Philips and Siemens, however, operating on a worldwide scale, take advantage of both techno-globalism and techno-nationalism, following a twofold strategy and benefiting from both global operations and a protected home ground.

2.5 Gourevitch's integrative approach: five national variables

In his comparative study on economic policies in responding to large changes in the international economy, Gourevitch (1986) has illustrated that the policy packages and supportive coalitions that emerge in the political process of preparing an appropriate adjustment strategy vary from country to country. In explaining industrial adjustment policies to international economic restructuring, Gourevitch has distinguished the following national variables or contingencies:

1. *production profile*: the policy preferences of societal actors as shaped by the different resource conditions, the size and degree of openness of their domestic economy;
2. *organisation of interest intermediation*: the distribution of power among social groups (dominant coalitions) and the framework of interest representation through which intermediate associations have established particular linkages between societal actors and the state;
3. *structure of the state*: the organisation of the political and legal-administrative system, the balance of power between the executive, legislative and judiciary and the relationships between departments within the public administration;

4 *economic ideology and culture:* values, norms and blueprints for the socio-economic ordering of society, providing economic motives and social beliefs for societal actors;
5 *the positioning in the international political-economic system*: the place of a particular country in the global economy and its international political, economic and military affiliations.

Production profile

The first factor shaping the domestic response to international and techno-economic restructuring is the production profile of a country. Broadly interpreted, this includes all the tangible and intangible resources, capabilities and endowments of one economy compared to other economies and the way these domestic assets contribute to economic performance and give the economy a competitive edge. The relationship between the attributes of countries and their international competitiveness has been mainly investigated in the field of economics. Traditionally, national competitiveness was mainly defined through macro-economic variables, such as factor prices, aggregate demand, levels of savings and investment, exchange rates and trade performance, employment and wage levels, interest rates, profitability and government deficits (cf. Shonfield 1965). The causal link between national production factors, aggregate economic conditions, and a government's policies to stimulate the macro-environment and policy outcomes referring to a country's performance *vis-à-vis* other economies is, however, rather weak and not necessarily straightforward. Industrial economists and economic historians (Best 1990; Chandler 1990; Porter 1990; Nelson 1993) have suggested that intermediary variables located at the industry and the institutional levels need to be included; dynamic forces such as entrepreneurship, technological innovation, domestic rivalry and geographical concentration in the home market, also contribute to a country's success or failure.

Porter (1990) claims that, paradoxically, the increasing internationalisation of business activities and the intensification of worldwide competition makes nations more, not necessarily less, important. In other words, he argues that national prosperity is created, not inherited. Although a country's natural endowments and macro-economic conditions might be beneficial, national competitiveness ultimately depends on the capacity to innovate and upgrade. Differences in culture, economic structures, institutions and history, all influence the feasible options of countries on their path to competitive success, but ultimately competitive advantage is created by a country's leading companies and sectors. National characteristics and attributes apply to many if not all industries, but insufficiently recognise the actual economic organisation of an industry and sector-specific circumstances that constrain the choices of firms and the policy outcomes; in the words of Porter (1990:619): 'firms compete in industries, not in nations'. Given the fact that firms operate in industries with different levels of competition and that these rivalling firms are also embedded in a domestic institutional environment that might facilitate (or inhibit) innovation and advancement, one should examine the way a nation shapes the ability of firms to succeed (or fail) in particular industries.

The competitive strategy of firms, their particular capabilities and skills, the nature of competition in industries and the linkages with related and supporting industries, are crucial in explaining a country's competitiveness (Best 1990; Chandler 1990; Porter 1990). In the knowledge-intensive industries, the influence of the national institutional environment has become vital. National governments, but also local/regional authorities, have certain instruments at their disposal to enhance advanced factor conditions (e.g. education and training policies), provide collective goods (e.g. technology transfer mechanisms), and establish supportive institutions (e.g. science parks), and thus promote national (or regional) competitiveness. In addition to this, national governments can increase domestic rivalry (e.g. strong competition policy), improve the quality of the factors of production (e.g. training and education), upgrade demand, stimulate intra- and intersectoral linkages (e.g. joint R&D projects), and enforce tight product, safety and environmental standards. Whether the national government should take an active role in trade promotion, for instance by promoting free trade and/or through the pursuit of a managed trade policy (e.g. infant industry protection), is a point for discussion among industrial and political economists. While Porter (1990) favours the promotion of domestic rivalry and a minimalist role for the state, Ruigrok and Van Tulder (1995) argue that strategic trade policy, aimed at strengthening the techno-industrial base and international competitiveness of a country, are still dominant.

Further evidence that national capabilities still matter in an internationalising world economy can be found in Nelson (1984, 1993) and Ruigrok and Van Tulder (1995), who focus on the contribution of national systems of innovation and industrial complexes to the performance and competitiveness of countries. Nelson (1984, 1993) has compared high-technology industries across countries (e.g. electronics, aerospace, aviation, nuclear power) and related national systems of innovation of economies, varying in size and affluence. He has focused on the way public and private funding of R&D, large military and public procurement programmes and commercial spillovers in innovative and strategic sectors contribute to the larger objectives of national security, technological leadership and international economic expansion. Nelson has sought to link the domestic institutional arrangements that govern the activities of large firms, industrial R&D laboratories, user–supplier linkages, education/training institutions, government agencies, and leading-edge customers, with the technological and economic performance of economies. Ruigrok and Van Tulder (1995) use the term 'industrial complex' to describe the close relationships between a core company (i.e. the spider in a sectoral or technological web) and its stakeholders: suppliers, trade unions, distributors and governments. These industrial complexes are highly internationalised in terms of the transfer of goods, services, money flows and investments, but at the same time these corporate-industrial networks are solidly embedded in the core firm's home base. Large core firms appear to be relatively successful in controlling their domestic stakeholders, including their home government (e.g. benefiting from favourable trade barriers and trade policies). At the same time, they are able to play one state against another and thus negotiate favourable conditions (subsidies, fiscal measures, infrastructure provision, etc.), that will facilitate their international expansion.

Porter (1990:1) starts his massive study on the competitive advantage of ten nations in a large set of industries with the intriguing question: Why do some nations succeed and others fail in international competition? The level of analysis chosen in his cross-national study on the international success (and failure) of particular industries is the home base in which the competitive strategies of the leading firms develop, processes of rivalry and geographical concentration evolve, and where advantages in technology, factor quality and corporate capabilities are created. Porter (1990) introduces four determinants that contribute to a nation's competitive advantage:

1. *factor conditions*: each economy has certain endowments or factors of production that function as inputs to any industry. A distinction can be made between basic factors, such as land, natural resources and labour, and advanced conditions, such as transport and communications infrastructure, knowledge resources, educated labour, and capital;
2. *demand conditions*: the nature of home-market demand for the industry's product or service is a second determinant of national competitiveness. It includes the composition and character of the home market and the quality and quantity of domestic demand;
3. *related and supporting industries:* the third determinant of national competitiveness is the availability of networks of suppliers, distributors, or otherwise related industries (e.g. specialised banks or service firms), and the way they are linked with the leading domestic companies (e.g. loosely or tightly coupled);
4. *firm strategy, structure, and domestic rivalry*: the final determinant of competitiveness refers to the goals, commitments, and skills of the leading companies, the products or services they provide, and the degree and nature of domestic competition. Domestic rivalry will push local competitors to increase efficiency, improve quality and service, and create new products and processes.

Together, the four determinants of national competitive advantage make up a *diamond*, defined by Porter (1990:72) as a mutually enforcing system in which the effect of one determinant is contingent on the state of the others. A mix between domestic rivalry and geographic concentration will upgrade the entire diamond and intensify the interaction between the four elements, that contribute to the competitive advantage of nations. The four determinants are complemented in Porter's model by two influencing factors, namely chance and government (see Figure 2.2). 'Chance' includes unexpected developments, such as scientific breakthroughs and inventions, wars, external political developments, and other shifts in the world economy (e.g. oil shock). 'Government' normally influences the four determinants as catalyst, pusher and challenger without taking full responsibility; in the words of Porter (1990:128): 'Government, it seems, can hasten or raise the odds of gaining competitive advantage (and vice versa) but lacks the power to create advantage itself.' Van Den Bosch and De Man (1994) have criticised Porter

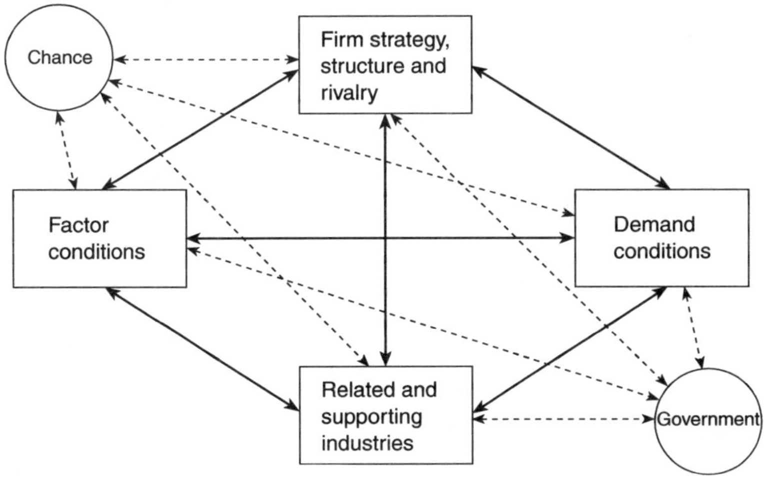

Figure 2.2 The complete national industrial system
Source: Porter 1990:127

for underestimating the direct impact of government on the business environment by referring to two specific roles governments may play to enhance a country's competitiveness. First, the role played by local and regional authorities in business development, innovation and industrial restructuring is not fully acknowledged in Porter's framework. Second, the role governments play over time in shaping firms and industries needs to be differentiated according to the phase of the life cycle an industry is in. The authors, therefore, suggest that government ought to be included as a fifth determinant of a nation's competitive advantage.

In the process of national competitive development, Porter (1990) distinguishes between factor-driven, investment-driven, innovation-driven and wealth-driven nations. Factor-driven countries emphasise the exploitation of the basic factors of production (e.g. natural resources and cheap labour), price competition and the use of simple product and process technologies (often acquired from abroad) (e.g. Singapore). Investment-driven nations appropriate new technologies and methods from abroad and improve them gradually (e.g. Korea). Given the large scale of investments needed and underdeveloped condition of the related and supporting industries, the strategy to catch up with the leading nations is only viable in some 'targeted' industries. Innovation-driven nations show an ideal mix between the four determinants with healthy competition, sophisticated and differentiated demand, a wide technological-industrial base, and overall favourable factor conditions (e.g. Japan, USA). Such a business climate fosters innovation and encourages the formation of new businesses and emerging industries. Wealth-driven nations, finally, are in relative decline: domestic firms lose their international competitiveness and the whole range of industries gradually erodes (e.g. the UK). With social programmes already laying a heavy burden on national competitiveness, the country's diamond is losing coherence with domestic rivalry, investment and innovation jettisoned and leading firms following risk-avoiding strategies, seeking government support and reducing competition by widespread consolidation.

The system of interest intermediation

Another factor influencing the formation and implementation of particular economic policies, accommodating domestic conflicts and coping with international and techno-economic pressures, is the system of interest intermediation. Political scientists and sociologists have promoted the comparative analysis of public policy by looking at the varying patterns of the exchange relationships between state agencies and the organised interests of business and labour as key variables in explaining cross-national variations of political and economic (in)stability and industrial performance. Firms, labour unions, consumers and farmers, all participate in policy making through their umbrella organisations acting at industry and state levels. Besides providing certain services to members, interest associations contribute to public policy making by functioning as an information and lobbying channel to articulate and transmit their private interests to political decision-making bodies, and carry out (some) administrative tasks on behalf of the government (e.g. setting codes of conduct in sectors). Three levels of interest intermediation have been distinguished:

1 the macro-level of national policy formation: the accommodations between the government and the peak-level organisations of labour and business (Schmitter & Lehmbruch 1979; Lehmbruch & Schmitter 1982; Olson 1982);
2 the meso-level of industries and regions: this intermediate level serves as the site for productive bargaining between state agencies and collective actors with a specific sectoral concern, and where interests are limited to one policy field or geographical area (Wassenberg 1982, 1990; Cawson 1985; Coleman & Jacek 1989);
3 the micro-level of individual corporate entities: the direct interaction between state agencies and the representatives of an individual firm (Grant *et al.* 1989; Cawson *et al.* 1990; x).

Two distinctive systems of policy advocacy and interest mediation can be distinguished: corporatist and pluralist frameworks. The distinction between these two ideal types relies upon the extent to which the activities of pressure groups are structured and officially recognised by the state.

The corporatist system is a comprehensive and integrated framework of institutionalised bargaining on economic policy issues (e.g. industrial relations and employment) between representatives of government, labour and management designed to create socio-political stability and economic growth. Examples of corporatist regimes can be found in the Scandinavian countries, Austria, Germany and the Benelux countries. Corporatism is characterised by strong organisational control of interest organisations on the basis of compulsory membership, clearly defined jurisdictions and the exercising of public authority by 'privileged' private groups (Schmitter 1979). In such a system the functional categories of labour and business have organised their interests through a representational monopoly within their own ranks either at the macro level of the state through peak organisations representing employers and workers, or at the industry or meso level through branch or trade associations and sectoral trade unions. On the basis of their representational monopoly, these interest associations are involved in political and economic exchanges with the state in the formation and implementation of public policies. The peak-level associations of labour and business contribute to the functioning of the state apparatus through their involvement in the policy process by rendering expertise and in the implementation of specific government programmes through commitments from their constituents. The role of the state in such a corporatist system is restricted to mediating and regulating the conflicts between organised labour and business. Public authority in policy formation and implementation, previously the exclusive responsibility of the government, has been transferred to 'privileged' private interest groups that now fulfil legitimate functions in the policy process (Offe 1981). Integrated interest associations, recognised by the government as its exclusive social trading partners, have led to a mixed governance mode of 'private interest governments, beyond market and state' (Streeck & Schmitter 1985).

Neo-corporatist regimes may reduce a country's capacity to adopt new technologies and hamper the adaptation of institutional structures to changing international and techno-economic conditions. Olson (1982) has argued that the longer a society enjoys political and economic stability, the more likely it is to develop powerful special-interest lobbies with a strong desire to preserve their established economic privileges over time. The special-interest organisations develop over time into powerful *distributional coalitions*, which dominate markets throughout the economy on the basis of their representational monopoly and political incumbency. After some time these accumulating networks of special interest organisations slow down collective decision making, reduce the mobility of resources, build up entry barriers and discourage institutional change just to protect their vested interest. The overall effect of these distributional coalitions is a decline in economic efficiency, slowing down economic growth and finally resulting in stagflation (Olson 1982).

The pluralist organisation of interest intermediation is characterised by a political order in which multiple and overlapping interest associations, representing individual preferences, compete for access to the political process (Richardson & Jordan 1979; Schmitter 1979; Van Schendelen & Jackson 1987, 1993b). Such a relatively open and dynamic system responds selectively to the demands of contending private interest groups in the political market place. Recently, researchers have begun to study European decision making and the organisation and integration of interest intermediation at EU level. Surprisingly, in view of the corporatist patterns found in many of its member states, there does not appear to be such a thing as a corporatist European Community (Streeck & Schmitter 1991; Greenwood *et al.* 1992; Streeck 1993). Although there were some initial steps towards Euro-corporatism with some tripartite structures in the 1960s and the early

1970s, the current institutional system of the European Community is qualified by Streeck and Schmitter (1991:159) as either an 'American-style pattern of disjointed pluralism' or as 'competitive federalism, organised over no less than three levels – regions, nation-states, and Brussels.' Like the United States of America, the European Union exhibits a dispersed decision-making structure, made up out of supra-European, intergovernmental, national and even regional layers and allowing for multiple channels of influence and persuasion.

The dispersion in policy and decision making is mirrored by the fragmented structure of organised interests, allowing various stakeholders to compete with each other in their attempts to influence EC decision making. Furthermore, the European-level interest organisations, hampered by overlapping jurisdictions and thus without a representational monopoly, are not (yet) integrated hierarchically in policy formation and implementation. In its consultations, the European Union seems to favour the representatives of national governments, organised business and individual large firms. Next to the relatively weak peak associations of organised European business and labour (UNICE and ETUC), we find a host of other collective actors like the European Round Table of large industrialists (the transsectoral ERT), various national and European branch associations, national champions, large American and Japanese firms, and dispersed national (sectoral) unions (Greenwood *et al.* 1992; Van Schendelen 1993b).

Lehmbruch (1982, 1984) has put the various systems of interest mediation on a scale from pluralism (or 'no corporatism') to strong corporatism:

- *pluralism*: characterised by the predominance of pressure group politics and the lobbying of government agencies and parliament by fragmented and competing interest groups and a low degree of effective participation in policy making (US, Canada, Australia);
- *weak corporatism*: distinguished by the institutionalised participation of organised labour in the formation and implementation of policies only within certain limited sectors of policy or by its participation only in specific stages of the policy process, with limited scope for collective bargaining (United Kingdom, Italy);
- *medium corporatism*: sectoral union and business participation with a wider scope for collective bargaining and some successful attempts at concerted income policies (Ireland, Belgium, West Germany, Denmark, Finland, Switzerland);
- *strong corporatism*: effective participation of labour unions and organised business in policy formation and implementation across those interdependent policy areas that are of central importance for the management of the economy (Austria, Sweden, Norway, the Netherlands);
- *concertation without labour*: unbalanced corporatist system, biased towards the integrated participation of business in policy making, at the expense of labour (Japan and France).

In the corporatist/pluralist research tradition there has been a strong emphasis on macro-economic policy making, with the focus on issues such as inflation, employment, growth, and tripartite bargaining between government and the peak associations of labour and business on formulating an appropriate economic adjustment strategy. Other mechanisms of interest intermediation, such as financial institutions governing money flows from households to industry and facilitating transfers between the state and industry, and among businesses, have been overlooked and underestimated (Zysman 1983; Fox 1986). Stock markets, state-controlled credit institutions and banks can also affect the corporate strategies of firms and the macro-economic and industrial choices of governments. Zysman (1983, 1994) has identified three different national financial systems and associated adjustment strategies: a capital market-based system with a company-led adjustment strategy, a state-guided credit system with an administered adjustment strategy, and a private credit-based system with an adjustment strategy based on negotiations.

The first institutional arrangement relies upon a dominant stock exchange, in which financial resources are allocated through competitively established prices and market forces (e.g. the market for corporate control). The adjustment strategy in those societies is company- or market-led, with the leading corporations in a country, their shareholders and other (potential) capital suppliers, taking the responsibility for industrial risk taking; discretion of labour and government in investment and production decisions is limited. The long-term development of the national economy is largely in the hands of corporate management (supposed to serve the owners' interest), which negotiates short-term loans and seeks to acquire funds from financial investors on the capital market. The USA and the UK are examples of a market-based financial system.

The second arrangement is a state-controlled credit-based financial system, in which discretionary control over the money flows and industrial credit allows the government to instigate business alliances and restructure industries. The process of industrial change relies upon strong ties between financial institutions and industry through large holding companies, and overall state guidance of economic development. The adjustment strategy is state-led: through administered prices and creative credit manipulation (together with state ownership of key firms), the central government actively intervenes in industrial affairs and is thus (re)shaping particular sectors. France is an example of a state-controlled credit-based system.

The third arrangement is a credit-based financial system, dominated by large industrial banks that often have a minority stake in the control of the nation's key corporations. The system is institutionalised in a social partnership economy that gives more or less equal political weight to the interests of corporate management, shareholders/investors, labour and government. In Germany, for instance, large universal banks, together with industry and the federal and state governments, are the pre-eminent actors in the transformation and allocation of financial resources. Banks have close links with industry: they hold substantial blocks of the shares of leading companies and are represented on companies' boards. The adjustment strategy in such a setting is negotiation-led: all the various social and economic constituencies are bargaining continuously over the terms of industrial change (e.g. wages, employment conditions, innovation, longer-term planning etc.).

In their comparative study on government–industry relationships in the British and German chemical industry, Grant *et al.* (1989) have found that this capital and research-intensive sector has traditionally been dominated by only a handful of large transnational firms: ICI in Britain and Hoechst, BASF and Bayer in Germany (the three IG Farben successors). These firms are highly internationalised in terms of their corporate structure as well as in the spread of their markets. Furthermore, the chemical industry is characterised by a tendency to resort to cartelisation (i.e. governing output, prices and markets), largely insulated from state intervention. The small number of key decision makers from all over the world meet regularly, giving the impression of a 'club-like' industry able to sort out its own affairs. Problems of industrial adjustment are solved by the industry itself, largely without state intervention; the overcapacity problem of petrochemicals in the 1970s, for example, was solved through bilateral and unilateral arrangements, reducing capacity by the chemical industry and its constituents (firms and industry associations).

In contrast with management of chemical firms and industry associations who operate both at the national and international level, trade unions, political parties and domestic governments still focus on the domestic activities of the industry. Trade unions still represent domestic interests and conduct their bargaining with transnational firms on a national basis. They lack the international orientation to develop an effective counterforce at higher bargaining levels. The insulation of the chemical industry from government and union involvement was seriously challenged with the entrance of various environmentalist pressure groups into the political arena. 'Green politics' became one of the main political confrontations for the chemical industry in the 1980s (Grant *et al.* 1989). In the highly international chemical industry differences between national patterns of government–industry relations tend to be rather small; global market forces and the increasing influence of international forums, such as GATT/WTO, OECD and the EU, have increased the need to harmonise European, American and Japanese regulations. And although the state is in a comparatively weak position *vis-à-vis* the chemical industry, national governments still play a role in implementing measures, dealing with environmental regulation, supporting R&D and so forth.

The policy networks of the British and the German chemical industries were made up of the same actors: senior managers of leading firms, the main industry associations, the section of the national or European government administration responsible for the chemical industry, and a minor role for organised labour. There are a number of striking similarities, but Grant *et al.* (1989) saw also some national variations between the two: the German network was more integrated, with tighter links between its parts than the British one, which was more informal. As a consequence, Germany was more able than Britain to manage industrial adjustment in the chemical sector, through an interorganisational concertation between the three giants (BASF, Hoechst and Bayer), their industry association VCI, the federal state, the regional governments (*Länder*), the financial sector and the labour unions. The British policy community lacked the German tradition of co-ordination through collaborative action and illustrated a tendency for autonomous action on the part of individual firms, unions, banks and government departments. As opposed to the German characteristics of partnership and co-ordination, the British system could be characterised by government factions establishing direct relations with the main companies, and keeping industry associations, banks and unions at arms' length (Grant *et al.* 1989). While German firms relied on their industry association VCI in dealing with the (federal) government, in Britain the public affairs departments of the large chemical firms were extensively engaged in private lobbying. Furthermore, the researchers found that the German banks have played a more active role than the British in the selection of high-ranking personnel and expertise in internal and external restructuring; the British chemical companies have a more distant and autonomous relationship with the financial sector. Another difference between the two countries has to do with the distinctive feature of the German federal system, i.e. the influential role of the *Länder*. At the regional or subnational level a German equivalent can be found to the direct and intensive relationships between individual firms and central authorities in Britain. Here, just as in the Italian knitwear industry, chemical firms based in the area and regional authorities work closely together in the upgrading of the knowledge base of the 'industrial district' in order to strengthen the position of its 'local' firms on international markets (Best 1990).

The role of the state

The role of the state in the national economy is another factor influencing a nation's particular response to techno-industrial developments and to changes in the world economy. In explaining the character of government–industry relations, Dyson (1980, 1983) has referred to the differences in legal-political traditions by distinguishing between state-oriented societies and state-less societies. The latter category is represented in the Anglo-American liberal tradition, placing an emphasis on values such as representation, voluntarism, self-regulation and scepticism about government competence; within this tradition there has been a preference for government at 'arm's length'. The former category is represented in the European continental tradition, emphasising values such as unity/integration, precedence of public over private interests, solidarity and objectivity. This state-oriented tradition has favoured the emergence of formalised institutional structures, where the state intervenes directly in the economy for strategic reasons, or where the state merely co-operates with the social partners in order to increase the welfare of its citizens. Evans (1995) has introduced the term 'embedded autonomy' for the mutually reinforcing relationship between bureaucratic capacity ('corporate coherence') and social connectedness ('dense state–society relations') as the precondition for rapid industrial progress.

A government may be involved in the economic process as a large producer of goods and services in a competitive market place, as an owner of enterprises in the key industries or as an important customer, buying expensive equipment for its defence, communications and utility systems. While officially states may govern society or stipulate the conditions for the functioning of the national economy, in practice they may be captured by powerful private interests and act as a vehicle or an agent of its domestic industries to attract (or prevent) foreign investment. Furthermore, governments are actively involved in the management of the economy as negotiators and facilitators, i.e. bargaining with the major stakeholders in socio-economic policy formation and implementation, and stimulating industrial development and the performance of leading firms via indirect means (e.g. subsidies, fiscal incentives, regulation).

In his study on the role of Japanese government bureaucracy in promoting and guiding economic development, Johnson (1982) has introduced the term 'developmental state', as opposed to the 'regulatory' state, as manifest in the US. The notion of a regulatory state relies on a combination of a market rationale orientation, a commitment to general welfare and allocative efficiency, and a modest 'rule-oriented' role for the government in the economy. Unlike the developmental state, where the bureaucracy directs economic development, the regulatory state does not concern itself with substantive matters (e.g. making decisions about which industries should receive top priority to be promoted or rationalised). The allocation of resources in a market environment should automatically lead to the most efficient and socially desirable outcome. State action should be restricted to setting the forms and procedures of economic competition and the protection of individuals. In carrying out its rule-making activities, the regulatory state is subject to close parliamentary supervision and the supremacy of law.

The top priority for a developmental state, such as Japan, is stimulating the nation's economic effectiveness, measured in growth, productivity and competitiveness objectives. Although developmental states are committed to private property (unlike state planning economies), their governments set substantive national goals for the economy and identify long-range industrial priorities for state involvement. Compared with the rule-governed regulatory state, the developmental state is purpose-governed: the direction of economic development and 'industrial catching up' with world leaders is determined by an élite economic agency (e.g. in Japan by the Ministry of International Trade and Industry). The state bureaucracy is often the major source of policy innovation in the political-administrative system and it makes most of the decisions (Johnson 1982). The government bureaucracy may use a combination of instruments to realise long-range economic objectives: tariffs, fiscal investment incentives, industrial reorganisation and targeting policies, public enterprises, controls on foreign investment, technology imports and so on.

Another instrument that played a substantial role in Japan's modernisation is *administrative guidance*, through which the government effectively fostered a common outlook for the nation's future development. Administrative guidance refers to bureaucratic co-ordination through continuous negotiations with industry in numerous consultative institutions and active persuasion to commit private stakeholders to techno-economic advancement by issuing requests, warnings, and suggestions to the key enterprise involved (Johnson 1982). Compared to the substantial discretionary authority allocated to the bureaucracy, the supervisory powers of the legislature are weak: as nicely put by Johnson (1982:10): 'appropriations precede authorisations'. So in the Japanese budgetary process, Parliament basically did nothing more than rubber-stamp the bureaucracy's budget. From the early 1980s, things have started to change in Japanese industrial policy making. As argued by Wade (1990), the strategic role of government in promoting economic development is seen to change over time, responding to the needs of domestic industry. After Japan's successful industrial catching-up strategy and the accomplishment of its economic goal to be a world leader, developmental functions were less needed and the government took up more regulatory roles.

In his comparative study on state structures and state–society relations and their particular contribution to industrial transformation and economic performance in six developing countries, Evans (1995) has distinguished between two ideal types: the predatory state and the developmental state. The first category of predatory states refers to a setting with a weak and incoherent bureaucracy and authoritarian private élites, and 'individualised' ties between the state and society, whereby national policies reflect the vested interests of the ruling cliques in society (e.g. corruption and clientelism). In other words, the pre-bureaucratic government and large parts of society are dominated and exploited by a small group of powerful constituents and, as a consequence, the state apparatus lacks autonomy and is incapable of formulating or implementing goals for industrial transformation. To Evans (1995:45), Zaire under the Mobutu regime is the 'archetype of the predatory state'. The second category of the developmental state refers to a setting with an autonomous and competent state bureaucracy, and an organised network of private actors, who provide useful intelligence and channels for decentralised implementation in society. The bureaucracy shows a strong corporate coherence and a merit-based status system, insulated from clientelistic pressures or traditional loyalties. The institutionalised linkage of state agencies with private élites with a clear interest in economic and political modernisation, fosters a long-term perspective on industrial development and a joint commitment to engage in large-scale and risky transformative investments. Japan and Korea can be seen as pure examples of developmental states (Amsden 1989; Wade 1990).

Developmental states can be further divided into four subcategories (Evans 1995). The first two, 'custodian' and 'demiurge', focus on different combinations between the conventional roles of regulator and producer. The remaining two, 'midwifery' and 'husbandry' represent variations in interaction patterns between public agencies and private entrepreneurial groups. The first subcategory of custodian assumes a regulatory role for the state in the national economy and in industrial transformation. Besides playing the role of a more generic producer of infrastructural goods and complementing private investments, the central government seeks to combine the functions of protection, regulation and promotion to advance industries and technologies. By preventing private capital from investing in undesirable or inappropriate activities (restricting foreign entry and imports) on the basis of the infant industry argument, the government stimulates the emergence and expansion of sectorally specific administrative agencies. The protective development of the information technology (IT) industry in Brazil and India are examples of a custodian role of the state. In the second subcategory of demiurge the state plays a creative and managerial role in sectoral transformation through the establishment of state-owned enterprises that compete in markets for private goods. The government takes on productive activities itself rather than leaving them to private firms because of a lack of private capital (local capital is unavailable and transnational capital unwilling) to develop a new sector, but also to replace or compete with private firms. The promotion of steel production in Brazil and Korea are examples where the state played the role of demiurge.

In the third subcategory of midwife, the state attempts to nourish new business groups or to urge existing entrepreneurial groups to venture into more challenging kinds of production or expand into a new sector. Through the use of sector-specific financial incentives, public procurement and strategic alliances with foreign partners who have access to new technologies, the government seeks to create a strong set of locally owned firms and build up a broad range of related industrial competencies. The successful catch-up strategy of Korean manufacturers to become suppliers to the leading American IT companies, facilitated by the supportive role of the Korean state, is an example of midwifery. The fourth subcategory of husbandry exhibits a combination of 'support and prodding' (Evans 1995:81), whereby the state seeks to move private entrepreneurial groups forward within their sector to produce more technologically complex products. It assists local companies in their internal growth and market expansion by taking on riskier complementary tasks, such as subsidising R&D. If successful in the end, these domestic companies might be strong enough to attract strong transnational allies and share technologies and marketing assets on an equal footing. The eventual success of Korea in becoming a leading nation in the IT domain and the recognition by IBM and AT&T of firms like Samsung and Lucky Goldstar as equal partners, is an illustration of the husbandry role of the Korean state.

Economic culture and ideology

In their response to major techno-industrial challenges, policy makers exhibit variations not just over time and between sectors, but also across nations. Industrial policies are embedded in and shaped against a distinctive cultural, ideological and structural context which varies from country to country. Dyson (1983) has defined these national variations according to the tendency of states to intervene, through the concept of industrial culture. Countries develop their own distinctive industrial cultures that reflect their tradition of public authority. Industrial cultures are expressed in attitudes and behaviour, but are also embodied in institutional settings, forming a mediating factor in the policy process. In explaining these variations, Dyson (1983) has proposed that a distinction has to be made between the objective context of the economic problems that confront various nation-states and the subjective context of contrasting political-legal traditions and industrial cultures. The objective context refers to the domestic industrial structures and to the question of whether the industrial structure is well fitted to accommodate the new technologies and the changing patterns of international markets. By focusing solely on economic and/or technological forces, however, attention is drawn away from the subjective context of socio-cultural routines and legal frameworks, in which industrial policy making is embedded. The subjective context of legal-political and socio-economic frameworks refers to the prevailing economic ideology, commonly shared norms, networks of interpersonal ties and mutual obligations, and traditions of public authority (Dyson 1983). National economies not only vary in terms of technological, economic and political capacities, they also show differences in their socio-cultural endowments. Although acknowledging the influence of cultural values and norms on the competitive advantage of firms and nations, Porter (1990) sees 'cultural factors' merely as working through the four determinants of his framework. Fukuyama (1995) has made it clear, that in addition to the techno-economic imperatives, differences in the levels of trust and sociability inherent in national societies and in the availability of 'social capital' (i.e. the ability to work together for common purposes in groups and organisation) will help to explain differences in economic performance and competitiveness.

There is evidence that culture, loosely defined as a general system of beliefs, norms, values and habits, shared and deep-rooted within a nation, has an impact on the particular structuring of its economic activities. Culture is, according to Hampden-Turner and Trompenaars (1995:4), 'the invisible hand that regulates economic activity'. The use of cultural variables in organisational and institutional research, however, is complicated. As Child (1981) has pointed out, the concept of culture has rarely been defined, the boundaries of a cultural unit are difficult to define and, as a consequence, the identification of variables as 'cultural' and measuring cultural attributes is problematic. Too little effort has been made to specify the subcomponents of culture that are relevant to particular organisations and their members. Although their precise impact may be difficult to determine, cultural factors certainly constrain the selection process of organisational forms by facilitating or inhibiting the formation of different types of governance mechanisms and helping to determine how effective and stable the established (and the new) domestic governance regime is (or will be in the future). Child (1981) has argued that the most essential definitions of culture all refer to normative and preferential conditions for action; the disposition to act in particular ways can be located at the personal level (the ideas, values, norms and meanings shared by members of a social entity) and the collective level (culture as an almost total way of life and socially transmitted behaviour patterns within a community). At the social-psychological level within organisations, Child (1981) has identified the following cultural value orientations: attitude towards human nature (good or evil), man to nature (mastery or subjugation), time orientation (future or past), orientation towards activity (being or doing), and relationships (individual or hierarchical preferences).

At the collective level of culture, Lodge (1987) has identified two ideal types of belief system among countries: the national ideologies of communitarianism and individualism. A national ideology, defined by Lodge (1987:2–3) as 'the collection of ideas that a community uses to make values explicit in some relevant context', is often implicit and subtle, and there may be differences between the prevailing ideology and the institutional practices in each country. The impact of ideology on people in society, however, is profound and it rarely undergoes sudden changes (except in situations of crisis, war and revolution). In a communitarian ideology, the (organic) community is seen as more than the sum of its individuals. Equality of result (or hierarchy), social rights and consensus about community needs and priorities (i.e. collective goods) are the dominant ways of directing and governing society. The ideology of individualism reflects an atomistic conceptualisation of society, in which equal opportunities for individuals, competition and the emphasis on contracts and property rights, are the prevailing elements. While the belief system of communitarianism assumes an active state in the co-ordination and planning of the economy (often in collaboration with recognised domestic organised interests), the individualist ideology thinks in terms of a limited state and active civil society of competing interest groups, all seeking to impose their 'private' priorities on society. In comparing the two ideologies, Lodge (1987) and Vogel (1987) found communitarianism to be more coherent and adaptable to its international economic environment, while individualism is more democratic and entrepreneurial, instigating change instead of responding to it.

An illustration of the individualism–communitarianism dichotomy can be found in Albert's *Capitalism against Capitalism*, in which he compares the liberal and dynamic Anglo-Saxon economies (the USA and the UK) to the 'social market economy' model of the Rhine/Alpine economies (Germany, Switzerland, the Benelux and Scandinavian countries and, to some extent, Japan). France is considered a mixture of elements from the two. Albert (1991) examines the strengths and weaknesses of these two capitalist models in terms of efficiency and equity, and looks at the feasibility of these alternatives in a modern dynamic environment, challenged by ambitious privatisation and deregulation programmes in the Western world, and the downfall of the Communist block. After the first age of 'capitalism against the state (1791–1891)' and the second age of 'capitalism disciplined by the state (1891–1991)', Albert argues that we have entered a new period which could be labelled 'capitalism instead of the state'. Two capitalist models, the individualistic, short-term-oriented Anglo-Saxon model and the socially responsible longer-term-oriented Rhine alternative, rival each other in terms of political support and commercial success.

The two ideologies have diverging views on the way risk and insurance should be organised in society and on the best way to anticipate and act on future developments. The Anglo-Saxon model promotes the organisation of insurance as a self-regulated market like any other where commodities are traded, and where companies provide a minimum of security and offer competitively priced products. The Rhine/Alpine system conceives of insurance as a narrowly defined quasi-public service, provided by regulated institutions in a moderately competitive environment, but with strong communitarian notions permeating the governance of that industry (e.g. risk sharing, redistribution and solidarity).

The Anglo-Saxon model is clearly embedded in an individualistic ideology, in which narrow individual or private interests take precedence over collective and/or public interests ('an open society'). It assumes an atomised market with no cohesion and/or government interference, which is conducive to the release of entrepreneurial spirits and fair competition. In the Anglo-Saxon version of capitalism the firm is treated as a commodity, traded by (potential) shareholders on the stock market. The far-reaching powers of shareholders *vis-à-vis* management and labour and the concern about the return on investments may have a negative influence on corporate performance: hostile take-overs, corporate raiding (asset stripping), and short-term financial gains can prevail over longer-term viability (e.g. cutting back R&D and training expenses, downsizing). The Rhine/Alpine ideology exhibits clear communitarian and egalitarian views, in which communities protect the individual and provide stability for the whole economy. The firm is treated as a (micro-)community of interests, emphasising a social partnership ideology between management, labour, capital (banks) and other relevant stakeholders (e.g. consumers, government). Another important element in the Rhine/Alpine version of capitalism is the longer-term concern of its constituencies for the upgrading of production techniques and substantial investments in training and R&D. At both macro and company level, close attention is paid to future financial commitments. Rhine societies have a high (consumer) savings ratio, producers can borrow on easy terms and at low interest rates from credit-based institutions (often through their *Hausbank)*, and governments are determined to invest in education, health, social security and innovation policies.

The positioning in the international political-economic system

The fifth factor constraining economies in formulating an appropriate policy response to industrial restructuring, is the positioning of a country in the international political economy. Nation states are embedded in a world economy and have political and economic affiliations. Governments take critical decisions concerning the strategic location of their key businesses in the international division of labour, and display particular preferences towards domestic market access, industrial and trade policy. They can also decide to participate in international organisations, dealing with governing (cross-Atlantic) security issues (e.g. NATO), settling cross-border money and capital flows (e.g. European Economic and Monetary Union (EMU)), regional market integration (e.g. European Union), or establishing global free trade (GATT/WTO). As a consequence of trade liberalisation in the GATT/WTO framework and the global integration of capital markets, together with deregulation in numerous countries, the world economy has become increasingly volatile and national economies have become more vulnerable to international shocks. By virtue of its considerable powers over financial capital flows, its large home market and technological leadership, the US has played a leading and active role in this trade liberalisation and deregulation process, and effectively challenged the mercantilist views and practices of the Europeans and the Japanese negotiators (Gilpin 1987; Strange 1988; Berger & Dore 1996). Another institution that has contributed to the internationalisation of economic activities is the multinational company. In industrial and trade policy making, national governments have to take the domestic and foreign interests of both home-based and foreign-based multinational companies into account. If the demands of transnational businesses are not met, they can always threaten to invest somewhere else and relocate production, R&D and central offices. In attracting foreign investments, governments are competing with each other in offering privileges (e.g. tax deductions, subsidies) and other favourable conditions to multinational firms.

Before 1985–87, when the Single European Act was implemented and the Internal Market programme was relaunched, the European Community (EC) was a loosely integrated system of central governments and domestic industries, based on the sovereignty of the nation state and the instruments of the 'national champions' in certain industries and large public procurement programmes (Vernon 1974). Domestic governments, business and trade unions still perceived the national level as the appropriate level to put forward their interests. As a consequence, the position of the supranational institutions was weak, compared to the strong position of member states and intergovernmental bodies, such as the Council of Ministers, in which national interests still prevailed. Organised business and labour perceived the organisation and promotion of industrial activity as a domestic affair with little influence from the supra-national European institutions (Greenwood *et al.* 1992). A large majority of the bigger European firms was, at that time, still national in scope with only minor business involvement abroad. The first steps towards *l'Europe sans frontières* were already taken with the enactment of the Treaty of Rome in 1957, which proposed an elimination of tariff trade barriers and quotas and which argued in favour of establishing common internal customs duties and trade policies. These measures were finally implemented in the 1960s and the overall European customs union was completed in 1969. After a dip in the 1970s, the process of European integration gained momentum with the introduction of the Internal Market project and the Single European Act in the mid-1980s. These two programmes aimed at creating a wholly integrated Common Market by 1993, to promote EC policies with regard to advanced technologies and to smooth EC decision making by subjecting the abolition of trade barriers to qualified majority voting instead of unanimous voting.

The process of economic integration in Western Europe, which finally gained momentum in the 1980s, has to be seen against the background of a techno-industrial base threatened by fierce competition from the USA and Japan. The trade deficit in high technologies rose substantially from the 1970s onwards, making Europe increasingly dependent upon the non-European supply of hardware and software for vital sectors of the economy. The European Commission was confronted with a widening high technology gap between Europe and its major international competitors (Van Tulder & Junne 1988). Especially in the fields of semiconductors, computing and consumer electronics, Europe was clearly lagging behind the USA and Japan. The existence of a considerable gap between the 'advanced' American and Japanese technology and Europe's lagging capabilities, clearly showed the need for a Community-wide approach. Most of the reasons for Europe's inability to compete with its American and Japanese rivals had to do with the dominant national outlook: the markets were highly fragmented, R&D was duplicated, and the dispersed European industry failed to achieve sufficient economies of scale to compete internationally. According to European business (ERT 1985), the Community's over-regulated business environment hampered international competition even further with high wages, social benefits, strict working conditions and high tax economies. The main goal of the business community became the stimulation of European integration, to be created by an internal market, the convergence of regulations, and a general business environment promoting flexibility and economies of scale, and a market in which cross-border industrial activities (e.g. the abolition of internal trade barriers, the promotion of cross-border mergers, pre-competitive R&D collaboration) would be facilitated.

In its plea for furthering European economic integration, business was supported by the Community's supranational institutions: the Commission, Parliament and the Court of Justice. The Commission's institutional support for the proposals launched by European business was inspired by its motive to increase its general influence on European policy making: the demands of European business would support the Commission in its effort to persuade the Council of Ministers to accept its White Paper on the Internal Market and to advance economic and monetary union (CEC 1985). Within the European Parliament, the idea for economic integration was also supported, as it was hoped that the implementation of the Internal Market Programme would help to reform the institutional structure of the European Community. The Internal Market Programme suited the Members of the European Parliament, who wanted to reform the Community by restricting the influence of member states (i.e. Council of Ministers) and giving the European Parliament more political influence. The European Court of Justice, watching over the enforcement of the Treaty of Rome (1957), had by that time established a solid legal base for a Common European Market. Originally the jurisdiction of the EEC rules over national legislation existed only in theory, but in the 1970s and 1980s the Court of Justice gradually increased its influence. The Single European Act officially recognised the exclusive jurisdiction of the Commission and the Court of Justice over EC competition policy and Internal Market affairs. In the 1991 Maastricht Treaty, economic and monetary union within the European Union was widened and deepened, henceforth curtailing the discretion of member states even more. For instance, in regulated industries and competition policy, close co-operative networks between national regulators and the European Commission have recently been put into place (Hancher 1996). In the area of monetary policy, peer group review procedures among the Finance Ministers have been set up, in which development of the budgetary situation, inflation levels and the stock of government debt in the member states are periodically monitored (Majone 1997).

As a consequence of the increasing internationalisation of trade, finance and foreign direct investment from the 1970s onwards, national economies have become increasingly interdependent. Traditional multi-domestic and insular competition, in which competition in each country functions autonomously of corporate behaviour and strategic rivalry in other countries, is gradually replaced by more open and global competition in which the competitive position of firms and industries in one country are significantly affected by their position in other countries or vice versa (Porter 1986). The level of economic and political interdependence between the economies of Japan, the USA and Western Europe has increased sharply over the last twenty-five years. Sectors like automobiles, financial services and computing have become more and more integrated into an emerging global economy with an ever-increasing free flow of information, goods, services, labour and capital around the world (Ruigrok & Van Tulder 1995; Hirst & Thompson 1996). As a consequence of rising R&D costs, converging consumer preferences, the search for economies of scale and comparative cost advantages, companies have embarked upon an internationalisation path towards expanding their businesses to neighbouring geographical areas or trading regions. In their striving for economic success and institutional legitimacy, multinational companies seek to combine global integration strategies aimed at co-ordinating geographically dispersed activities, with a national differentiation strategy, allowing them to respond to local customers, governments and circumstances (Prahalad & Doz 1987; Bartlett & Goshal 1991). In order to serve their customers worldwide and build up a global presence, companies need to work together with international partners. Especially in the high-technology industries (e.g. micro-electronics and bio-technology), with high R&D expenditures and shortening product life cycles, globalisation has manifested itself through a myriad of international strategic alliances, joint ventures and other collaborative agreements (Van Tulder & Junne 1988; Mytelka 1991; Hagedoorn & Schakenraad 1992).

Ohmae (1985, 1990) has stressed that if one wants to operate on a worldwide scale, one needs to operate with international partners. Within the global market a 'triad' of three important regional markets can be discerned: Japan (and South-East Asia), the USA and Europe. Every enterprise that wants to have a strong position in this triad needs to create a direct presence in all three regions. This foothold can be created by establishing strategic alliances, joint ventures or mergers with local/regional enterprises. By becoming an 'insider' in that regional market, protectionist trade barriers and a hostile business environment are avoided and new economic activities within that region can be developed. As pointed out by Ohmae (1985, 1990), transnational firms must be able to play this triad power game of penetrating and strengthening their position on the Japanese, American and European markets simultaneously. This process of *insiderisation* allows for an integration of world-scale activities and decentralised production in the key national markets around the world. This 'triad' logic of international strategic alliances, exemplified by European firms going transnational and becoming insiders in Japan and the USA, and Japanese and American corporations in turn moving into Europe, makes it increasingly difficult for national governments to control and manage their domestic industries. The sovereign role of national governments in globalising high-tech industries is challenged, because governments can no longer directly support national champions and domestic industries for employment and/or industrial objectives.

Ohmae's notion of national economies losing their power and influence both upwards and downwards (to regional trade blocks or subnational structures, respectively) has been strongly criticised as 'a myth' (Hirst & Thompson 1996). His defence of global free trade as the most effective allocation of resources to the production of goods and services, and his rejection of protectionism, however, overlooks other alternatives that may still be at the disposal of firms and their (home) governments. When confronted with the superiority of a hegemon in the techno-industrial domain and its interest in open trading routes, however, some nation states might decide to pursue strategic industrial and trade policies to foster technological autonomy and eventually develop a solid domestic industrial base. In addition to the possible strategies described above (furthering a global division of labour/production and following domestic sheltering strategies), governments and businesses still have the third option of 'glocalisation' (Ruigrok & Van Tulder 1995) or 'regionalisation' (Hirst & Thompson 1996). This is a geographically concentrated interfirm division of labour in the three major trading blocs in order to become relatively immune to trade barriers.

In analysing transnational business groups, Stopford and Strange (1991) draw a distinction between multinational companies and locally operating firms and between host and home governments. Stopford and Strange see today's interdependencies between firms and states and their strategic interactions as a *triangular diplomacy* by delineating three dimensions of bargaining: between states, between firms and between states and firms. The first dimension, interstate diplomacy, exhibits the traditional bargaining among governments striving for power and influence. In the domain of international competition this form of diplomacy leads to rivalry between governments looking to strengthen their national competitive advantage of domestic industries and firms by attracting foreign investments, policy competition and pressing for access to foreign markets for their home firms. The second form of bargaining, interfirm diplomacy, illustrates the competition between firms competing in the world market, the strategic partnering among large conglomerates and the collaboration between multinational enterprises and local firms. Bargaining between states and firms, or industrial diplomacy, focuses on government–industry bargaining. Firms seek permission from national governments to enter protected markets, while governments try to persuade domestic companies to locate their activities at home and attempt to attract investments from foreign companies.

2.6 Summary

Two alternative reactions of nation states and industrial sectors to international pressures have been discussed in comparative policy and institutional studies: international convergence (or techno-globalism), emphasising both international competition and transnational collaboration, versus persistent local diversity (or techno-nationalism), respecting national sovereignty and protecting domestic interests. The trend towards convergence manifests itself through the impact of structural forces, such as technological innovation, globalisation of markets, international deregulation and European integration, which challenge nationally and sectorally confined governance regimes and instigate the emergence of similar restructuring programmes across different countries. The persistence of local diversity refers to the cushioning of global impact by national structures and institutions, which continue to shape domestic policies in distinctive ways, reflecting the nation's particular policy traditions, priorities and practices.

We have extended our research model to explain industrial adjustment policies to international economic restructuring in telecommunications (as formulated in Chapter 1) on the basis of the literature on comparative government–industry relations. We have refined the national variables, originally formulated as different kinds of economic adjustment strategies (i.e. market-led, state-led and negotiation-led responses), by introducing additional institutional contingencies of the countries to be investigated. In this chapter we have discussed the following five national contingencies, as found in the work of Gourevitch (1986):

1 *production profile*: the policy preferences of societal actors as shaped by the different resource conditions, the size and the degree of openness and character of their domestic economy;
2 *system of interest intermediation*: the framework of interest representation through which intermediate associations have established linkages between societal actors and the state;
3 *structure of the state*: the organisation of the political and the legal-administrative system, and the balance of power between the executive, legislature and judiciary;
4 *economic approach* (ideology and culture): values, norms and blueprints for the socio-economic ordering of society, providing economic motives and social beliefs for societal actors;
5 *the positioning in the international political-economic system*: the place of a particular country in the global economy and its international political and economic affiliations.

3 Structural forces challenging the European national telecommunications regime

3.1 Introduction

For more than a century, a strong consensus prevailed in Western Europe on the provision of telecommunications services as the exclusive responsibility of nation states and the joint provision of international services by the national PTT administrations. Domestic governments stipulated the industry's relevant market conditions, socio-economic issues and trade policy. The state empowered the PTT with a statutory or *de facto* monopoly and supported the domestic equipment industry through R&D funding, public procurement contracts and export subsidies. The influence of other stakeholders in the telecommunications domain (such as computing firms, large business users and consumers) on policy formulation was restricted. Today's telecommunications has become a market-driven industry, in which the European national economies, regional trading blocs, public operators, service providers and multinational corporate users are more and more integrated into one global market. The established governance regime has been put under pressure by techno-economic and political-institutional forces. New technologies like cellular radio, cable networks and satellites have effectively challenged the traditional public monopoly. Multinational business users, demanding low-cost, high-quality and one-stop shopping telecommunications services, have successfully lobbied for more flexible and customised facilities from the national PTTs and a more integrated and harmonised European regulatory framework, in which the operational and administrative tasks of the PTTs become disentangled. The internationalisation of leading American telecommunications and computing businesses, together with the general retreat of the interventionist state across Western Europe, has made the national public monopoly regime even more obsolete.

In this research we will argue that the traditional organisation of the Western European telecommunications industry has been challenged by at least four developments that have curtailed the development of national sovereign policies:

1 technological (r)evolution: innovations in information and communication technology leading to a digitisation of all types of information;
2 globalisation and differentiation of communication markets: boosting growth

and differentiation of user demand, sectoral convergence, the rise of multi-media conglomerates;

3 international deregulation: deregulation in the USA, followed by the UK and Japan;

4 European integration: Community-wide policies, as developed by EC institutions, constraining the activities of member states' governments and businesses in telecommunications.

Before dealing with these structural imperatives for transforming the business environment of European telecommunications, we will discuss in section 3.2 the established institutional framework of actors and policies which remained highly stable for almost a century: the domestic monopoly/international cartel model. The aforementioned four structural forces, that have undercut the centralised hierarchical model of public telecommunications, will be discussed in the remainder of this chapter: technological (r)evolution (section 3.3), globalisation and differentiation of telecommunications markets (section 3.4), international deregulation (section 3.5), and European integration (section 3.6). Together, these centrifugal forces make for an evolution towards a more open, flexible and dynamic telecommunications polity, characterised by the removal of government-imposed entry barriers, and the introduction of competition and privatisation. Also new 'burning' policy issues, such as open access to the public infrastructure and network interconnection, price and profit regulation, and the safeguarding of universal service, come to the fore in the opening of the telecommunications market. National and European policy makers cannot rest on their laurels yet.

3.2 The *ancien régime* of telecommunications: public monopoly, PTT and rent-seeking coalitions

Traditionally, the provision of telecommunications services and equipment was operated by a state-controlled monopolist, subjected to strong government interference and parliamentary supervision. In Europe this telecommunications monopoly was organised in a variety of ways: functionally separated monopolies (i.e. between the operation of national and international services e.g. Italy), geographically subdivided defined monopolies (i.e. on a regional basis e.g. Denmark) and *de facto* monopolies (either statutory as in Belgium or exclusively operated by a semi-public corporation, as in Spain). In most cases, the telephone service was operated by the PTT, organised either as a separate department within the public administration or as a state-controlled public enterprise with more distant, but nevertheless close government links. Originally its exclusive task was to control the telecommunications network and to provide services on a break-even basis. The PTTs were, however, regularly used by the government to pursue other national objectives as well, such as fiscal policy, regional development, industrial/employment policy. In some European countries, the PTT monopoly included, besides the provision of public telephone services, exclusive control over related functional areas such as postal and giro services (Foremann-Peck & Müller 1988).

The telecommunications administrations collaborated in the provision of international services, thereby minimising the possibility of competition among national operators and reducing competitive entry by private outsiders. Through two intergovernmental bodies, the Conférence Européenne des Administrations des Postes et des Télécommunications (CEPT) and the International Telecommunications Union (ITU), the PTTs, agreeing upon joint service provision for cross-border traffic, established cartel-like arrangements on accounting rates and sharing revenues. The CEPT/ITU framework could be qualified as a *soft regime*, respecting the exclusive hegemony of its constituent members over domestic telecommunications systems (i.e. the sovereignty of national governments and their public operators) (Aronson & Cowhey 1988; Cowhey 1990). The multilateral ITU and CEPT bodies provided a forum in which the various national governments and the PTTs defined unanimous rules with respect to standards, interconnectivity, tariffs and the introduction of new services. Within the ITU, the International Consultative Committee for Telephone and Telegraph (CCITT) provided a platform for telecommunications officials (and often PTT engineers) from various countries to consult with each other on technical and regulatory matters dealing with enabling cross-border communications and to make recommendations about operating standards, interconnection, tariffs and accounting rates. Cowhey (1990: 176) argues that the ITU/CCITT acted as a virtual telephone cartel for the PTTs:

> The CCITT rules for international commerce in telecommunications services were (. . .) the anchor of a regime that facilitated bilateral monopolistic bargains, reinforced national monopolies, and limited the rights of private firms in the global market.

The PTTs, inspired and legitimised by economies of scale and the political objective of socio-economic integration, expanded the public network to achieve the goal of universal service provision. The PTT/public monopoly model was built on a system of burden sharing between customers, services and geographical areas: long distance telephone services were used to sponsor local services and postal services, and business customers to subsidise residential users. For almost a century, the organisation of the domestic telecommunications market was based on the natural monopoly of the PTT network; for instance, as convincingly argued by Holcombe (1911:441–2):

> In the telephone business competition is a failure. Considered as an automatic arrangement for maintaining an accurate adjustment of the supply of telephone facilities to the demand, it easily gets out of order. (. . .) For a while it is capable of bringing about low rates and stimulating a rapid development. Sooner or later, however, the self-interest of the competitors or the disillusionment of the public authorities will cause the termination of competition and the substitution of a régime of monopoly. This has been the result everywhere in Europe, where competition has once existed (. . .). Competition as a permanent status in the telephone business is neither desirable nor possible.

In addition to its public utility function, the PTT was assigned tasks, such as business-like, regulatory and supportive policy functions, that sometimes interfered with or ran counter to the provision of its public services. Instigated by the saturation of the telephone market and technological innovation at the end of the 1970s, the PTTs gradually developed a diversification strategy by which they sought to bring new facilities that were not directly related to fixed-line telecommunications under their exclusive monopoly. They adopted a more offensive and commercial attitude by entering promising markets like cable, satellites, switched data networks, fax, videotex, electronic mail and cellular radio. These new technologies were considered potential markets which could be very profitable in the future and prove complementary to the saturating public telephone service. Or, interpreted more tactically, such a pre-emptive strategy would also prevent competing service providers, like cable companies, publishers, computing firms and private network operators, from entering the telecommunications market. The PTT's regulatory and policy functions included the governance and supervision of the telecommunications industry through the setting of rules and standards, allotment of frequencies, equipment approval etc. Moreover, on behalf of the minister responsible, the PTT administration was actively involved in drafting legislation and implementing policy (see Figure 3.1).

For the supply of telecommunications equipment, European PTTs relied upon their government's public procurement programmes and the local computer/electronics industry (e.g. CIT-Alcatel, Philips, GEC and Plessey). In the USA and Canada, the telecommunications sector was vertically integrated within a single industrial conglomerate, in charge of both service provision and manufacturing (AT&T/Western Electric and Bell Canada/Northern Telecom). In Europe, the relationship between the monopolistic network operator and its monopsonistic

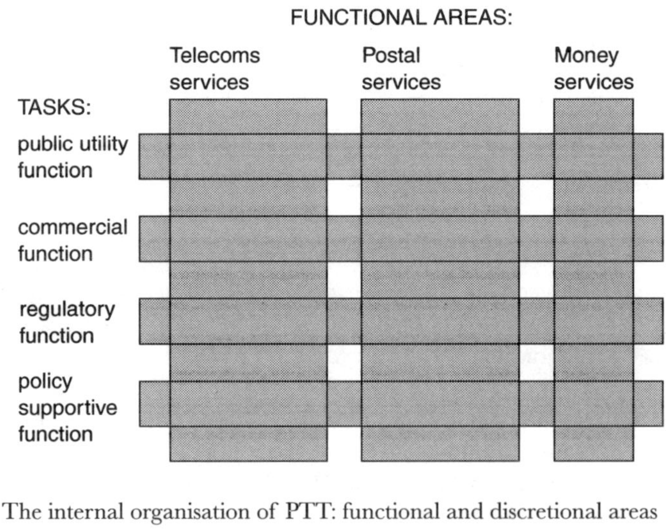

Figure 3.1 The internal organisation of PTT: functional and discretional areas

equipment suppliers could be described as quasi-vertical integration and exclusive trading (Dang-N'Guyen 1986). The strong ties between the public operator and its preferred supplier(s) were protected from new entry and foreign competition by long-term R&D collaboration, collusive standard setting and national certification rules. On a European level, the equipment market was highly fragmented. Substantial international competition among the local manufacturers outside their protected home grounds was virtually non-existent.

Noam (1987) has qualified the established telecommunications community as a *rent-seeking coalition*, that allows the extraction of selective advantages and the redistribution of the 'surplus' by a coalition of influential stakeholders in a well-organised and highly concentrated industry. In a regulated setting such as the telecommunications industry, characterised by exclusive market positions and/or restricted entry, government might run the risk of conferring the 'monopoly rents' and redistributional gains to privileged stakeholders, instead of serving the general welfare of society. In other words, public policies might reward specialised, well-organised and influential groups producing regulated goods and services at the expense of consumers (Buchanan 1980; Mueller 1989). The PTT and the domestic equipment industry were politically supported by central government, trade unions, the political left and rural/residential users (Noam 1987,1992; Cawson *et al.* 1990; Steinfield *et al.* 1994). The material and ideological reasons for these parties to support the *status quo* had to do with its tendency to protect key industries against imports and foreign market entry, and keep up employment schemes, job security and salary guarantees, and preserve universal service provision. To describe the tight relationships between the public network operator, the domestic private equipment industry, the trade unions, the political system, the residential (or rural) users and the international cartel-like arrangements among the PTT administrations, we will propose the concept of the *national public monopoly regime.*

The office equipment manufacturers, dataprocessing industry, large telecommunications users and transnational suppliers of switching equipment, questioning the public monopoly model and demanding flexible regulations, were to a considerable degree excluded from the selective benefits of the established PTT system (see Figure 3.2). Although it was involved in the introduction of office automation and the upgrading of dataprocessing facilities in the PTT administration, the computer industry gradually moved into telecommunications by providing datacommunications services and equipment to large international users, such as banks, news agencies, publishers, airline companies etc. Rather than being subjected to a monopolistic pricing regime by a public telephone company, offering mainly conventional telephony, these large users demanded cheap international and customised (data)communications services and facilities. They built and operated their own private networks on the basis of leased lines or relied upon new network operators, like IBM, GM/EDS and GEIS. The more internationally oriented manufacturers of switching equipment (e.g. Ericsson, Northern Telecom) also demanded a (controlled) liberalisation of national telecommunications markets. Although most of them participated in rent-seeking coalitions at home, these major equipment manufacturers were willing to enter the sheltered

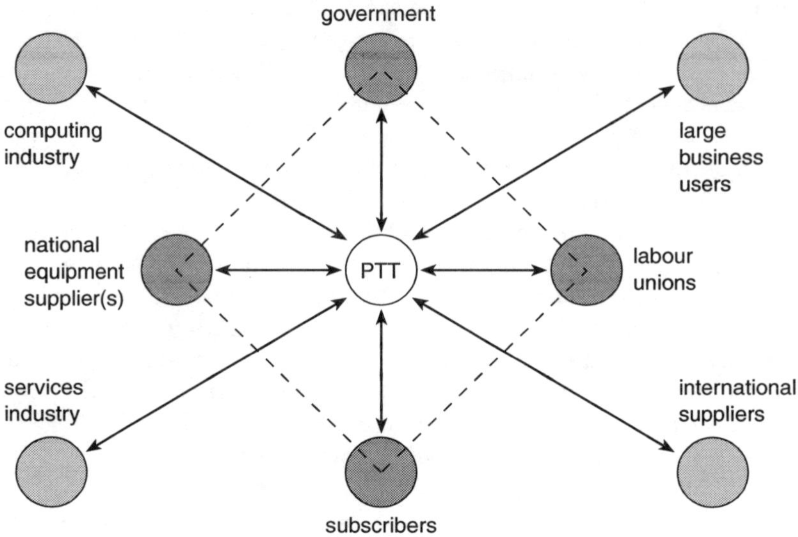

Figure 3.2 The rent-seeking coalition around the PTT monopoly

switching market abroad and compete with the privileged supplier(s) of the national PTT.

In short, traditional telecommunications could be characterised as relatively stable with measured network expansion and steady developments with regards to technology, markets and legislation, relying upon cost-sharing and redistribution. As a consequence of structural developments, the well-organised telecommunications polity started to destabilise in the late 1970s and an open and decentralised infrastructure of multiple networks emerged. The established national monopoly regime was seriously challenged by the powerful alliance of a globally oriented services industry and major business users, who were largely excluded from the rent-seeking system surrounding the PTT. This *information and services coalition*, having acquired expertise with private networking outside the PTT, questioned the cost-sharing and redistribution arrangements of the past and promoted the introduction of competition in the protected markets of equipment, services and network provision (Noam 1987, 1992).

As a consequence of the increasing dependence of telecommunications on micro-electronics, computing and software programming, firms like IBM, EDS and GEIS have made inroads into the industry and have become politically active in the policy process. The integration of office automation and telecommunications facilities has allowed the provision of value-added networks and new dataprocessing services, not necessarily provided by the PTTs. Large users like Reuters and GM, which operate on a global scale, have become important players with big stakes in telecommunications. Their dynamic and specialised demand for high-capacity and low-cost facilities on an international scale could not be met directly by the national PTTs. For these multinational companies, telecommunications has simply become too important, both in terms of costs and strategy, to be regulated on the basis of an international cartel between national monopolists. Therefore they have built large-scale private user networks, carrying company data traffic over circuits leased from the public operator. The services-information alliance was supported by those equipment manufacturers that were excluded from the collusive arrangements between PTT and its preferred supplier(s). Today's equipment industry, characterised by rocketing R&D costs, shortening product life cycles and the concomitant search for economies of scale and scope, requires entry to foreign markets in order to recover the huge investments made in the development of advanced switching systems. As a consequence, national entry barriers for aggressive international competitors have started to erode, world trade has expanded and there has been a process of strategic partnering and economic concentration among the national electronics champions.

In general, the restructuring of European telecommunications has implied a general shift from a stable market environment towards a more dynamic one. The distributional coalition of the traditional telecommunications regime, including PTT, residential users, domestic equipment supplier(s) and trade unions sharing the 'monopoly rents', has been challenged by a number of developments that could not be accommodated within the national public monopoly regime (Noam 1992). The public telecommunications monopoly was established to prevent 'bypassing and cream skimming' and put PTT in charge of the redistribution of charges and universal service provision. However, international equipment manufacturers, the information/computer industry (IBM and GEIS) and new carriers (e.g. MCI, C&W/Mercury, Vodafone) have responded to the differentiated demands of large businesses, professional users and consumer groups by providing alternative technologies and private networks that were more flexible, less expensive and cost-based. Technological, market and international developments have made the reshuffling of cross-subsidisation and the 'rent-seeking coalition' around the PTT administration necessary. The political claims for free entry, fair competition and cost-based tariffing, made by these new operators, service providers and large business users were supported by the European Commission with its plans for an integrated market. Although the aforementioned services and information industries by now are accepted as a legitimate stakeholder in the national policy domain, its claims so far have been crushed in the process of log-rolling between PTT and the other established players (Noam 1992). Therefore, a new kind of compromise has to be found that will accommodate both the 'established' parties, strongly in favour of step-by-step liberalisation, protecting the operator's core monopoly and securing its public service obligations, and (relative) newcomers seeking access to newly liberalised markets and responding to an increasing business demand.

In the newly emerging telecommunications system the co-ordination of economic activities has developed from an administrative hierarchy, represented by the public operator and its network monopoly, into a model of controlled competition. First the 'fringe' markets for value-added services provision and end-user equipment were liberalised, followed by the opening up of the 'core' markets for voice telephony and network provision. Furthermore, a new regulatory framework separating the operational and the administrative tasks of the former PTT administration was established. In order to oversee the market, ensure fair competition and further consumer interests, a system of regulation has been introduced to balance the various interests of suppliers, customers, employees and investors. This trend towards liberalisation and regulatory reform was accompanied by a transformation of the former PTT administration from a government department into a corporatised or privatised company, acting relatively autonomously from the government. The PTT administration was forced to rethink its corporate goals and transform its internal organisational structure in order to cope more effectively with the demands and pressures from its dynamic environment. The monolithic, hierarchical and centralised structure of the past has been replaced by one that is (more) market-responsive and profit-oriented. The newly established telecommunications companies – telcos – have decentralised their business activities to respond more effectively to the specific demands of their various customer groups (e.g. households, business customers). Nowadays, the former PTTs have established separate divisions, strategic business units, profit centres and joint ventures, all based on different product–market combinations and flexible operations. They have diversified into new markets such as dataprocessing, equipment manufacturing, cable and audio-visual production, enhanced services provision, network and facilities management etc. The telcos have also expanded internationally by seeking access to new geographical markets (often through acquisitions and strategic alliances), and serving multinational business customers.

The scope of the PTT's monopoly has come under close scrutiny from the globally oriented coalition of the large users, and new entrants in the equipment industry, supported by neo-conservative or centre-right governments. A policy debate was initiated to decide whether structural reform could meet both the demands of corporate users for efficient and adequate services, and that of the economy as a whole for highly advanced communications systems, while at the same time preserving the financial viability of the public operator and public service requirements. The large user groups, together with the computer and information industry, argued in favour of an organisational division between the operational activities of PTT (e.g. giro, mail and telecommunication) on the one hand and a separation between its operating and regulatory functions on the other (see Figures 3.3 and 3.4). The relative strength of the international services–information coalition in different countries *vis-à-vis* the more or less powerful position of the national public monopoly regime, manifested by the distributional coalition of PTT, the government, political parties, trade unions and the domestic equipment industry, determined to a large extent whether deregulation, liberalisation and privatisation measures were feasible and policy changes could be implemented.

In some countries the information and services coalition was supported by New Right governments, sceptical about the degree and scope of state intervention in the national economy and sensitive to budgetary restraint and pro-market

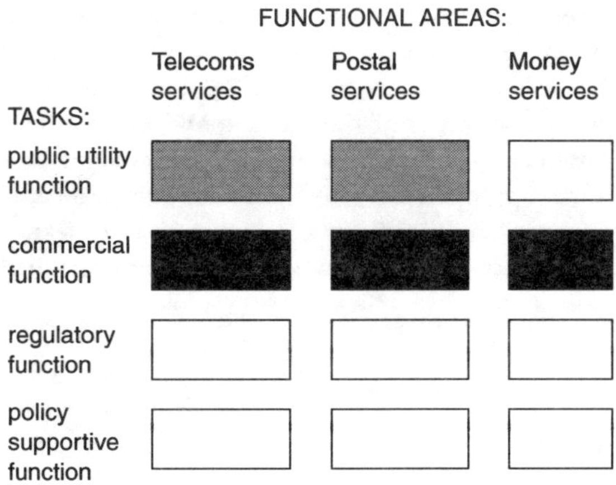

Figure 3.3 The disentanglement of the PTT functions and responsibilities

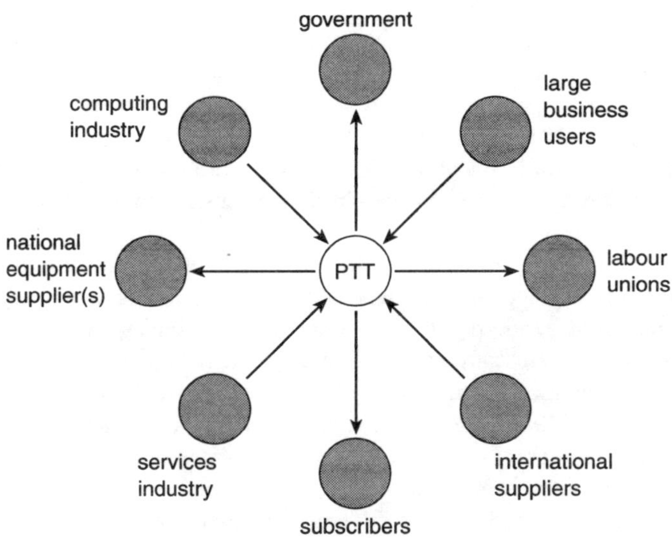

Figure 3.4 The changing industrial organisation of telecommunications

policies. The resurgence of the free market economy and the retreat of government was first of all promoted by the neo-liberal Thatcher and Reagan governments in the late 1970s, to be followed by the governments on the European Continent. The British and American governments, through the implementation of major deregulation and privatisation programmes, managed to improve the productivity and innovation of their domestic key industries, and hence their international competitiveness. In Western Europe, however, privatisation and deregulation were not nearly as drastic or on such an impressive scale as in the UK and US, where these programmes were inspired by the ideological notion of rolling back the frontiers of the state. In Western Europe privatisation was just one of the remedies for the problem of often inefficient and loss-making state enterprises and an appropriate way of financing the large-scale modernisation of the communications and distribution networks. European governments, however, decided to privatise only a few public enterprises and sell off a minority of shares of the public assets.

3.3 Technological (r)evolution

The technological (r)evolution in the telecommunications industry, as manifested by the availability of new transmission channels, digitisation and sectoral convergence, can be located at least on three levels (Mansell 1993; Davies 1994):

1. *transmission technologies*: the availability of high capacity fibre optics, in addition to the fixed twisted pair and coax infrastructure, and the emergence of alternative infrastructures, including satellite links, computer networks and cellular systems;
2. *switching technologies*: the replacement of electro-mechanical exchanges by digital switching, and the building of 'intelligence' into networks, make the processing, routing and storing of information more efficient and effective;
3. *terminal equipment*: a convergence of customer terminals into multi-media equipment, the diffusion of intelligent and conditional access technologies (e.g. smart-cards, decoders), located in the customer's home.

The first generation telecommunications infrastructure, consisting of narrowband networks and electro-mechanical switching technologies, was gradually upgraded to provide, in addition to the traditional telephone, telex, teletex and the new data and tele-information services. The large-scale implementation of digital switching and digital local loops in basic networks has created a second generation network in the mid-1990s, consisting of a digital network with a large bandwidth offering integrated services (voice, data, text), high-speed data services, Internet access, and video conferencing. The large-scale introduction of compression techniques, fibre optics and satellites in the next decade will make an integrated broadband services network possible. Such a third generation network will bring the additional facilities of videophony, interactive television and digital television.

In the mid-1980s the European PTTs started to develop a policy for the evolution of an infrastructure that would be more efficient and would rationalise the increasing costs of providing gateways between the specialised PTT networks. The PTTs launched the outlines for an integration between the separate analogue networks for voice, telegraph and datatransmission services within the so-called Integrated Services Digital Network (ISDN) for the 1990s, followed by an Integrated Broadband Communication Network (IBCN) at the end of the century.

These plans were part of a long-term strategy to upgrade and digitise the public infrastructure, to integrate the various dedicated networks for voice, telex and data, later followed by the integration of cable networks, into a single fully integrated broadband communications system under their control. Such an integrated network would provide all forms of telecommunications and information services: audio, video and high-speed data services. These plans still reflected the traditional PTT view of economies of scale and unified public service combined with a more adjusted diversification strategy in that it offered specialised and customised services for the business community on the basis of an integrated network. The plans were originally backed by national governments, which at that time still supported the leading and privileged role of the PTTs in the development of an integrated communication system for the future. In Europe and Japan there was at first a broad agreement on the extension of the PTT's monopoly on this integrated information/communication infrastructure by restricting entry and competition on the network level; competition on the higher levels of value-added services (teleservices) would be allowed. The ISDN/IBCN plans are very much technologically and long-term-oriented and underestimate the impact of contrary technological, economic, international and institutional developments which could seriously affect the plans for the future.

Alongside the trend towards integrating all the dedicated and separate networks for voice, data, telex and video transmission into one electronic pipeline, there is a strong undercurrent which suggests a future in which the telecommunications system is a patchwork of public and private, universal, specialised and differentiated networks (Noam 1987, 1992). Information technology also makes a flexible and customised use of the telecommunications network possible by 'unbundling' the functions of information distribution and the provision of teleservices (Mansell 1993; Davies 1994). Within the information distribution function a further distinction can be drawn between the physical infrastructure, service capabilities (leased lines) and basic (or bearer) services (telephone, telex and telegraph network). Within the service provision function a distinction can be made between value-added services (electronic mail, EDI) and information services (telebanking, on-line databanks). The flexibility in controlling the public network and developing network management capabilities for industry-specific applications, can be enlarged through the use of advanced software technologies and database management skills. Such an 'intelligent network' allows for the provision of 'individualised' services such as virtual private networks and calling-line identification, including itemised billing, call back and call forwarding services, credit card validation, and other customer-specific routing and charging services.

The transmission capacity in long-distance communications over the years increased dramatically thanks to the proliferation of optical fibre cables and satellites, eventually leading to surplus capacity and a steep reduction in tariffs. Transmission capacity was no longer scarce, and nor was it the exclusive domain of the PTTs, who set tariffs high above costs. The development of optic technologies and satellites was accompanied by the emergence of non-PTT service providers, claiming the same rights as the PTT to build their own private networks or lease capacity from the PTTs and resell it to their customers. The entry of new private service providers and the emerging competition between new distribution networks (satellites, cellular systems, interactive cable networks) as an alternative to the public-switched telephone infrastructure, has seriously challenged the public telecommunications system; the end of spectrum and capacity shortage (i.e. abundance of network capacity) and the growing insensitivity to distance of telecommunications costs have eroded the natural monopoly of the public telecommunications network (Pool 1983, 1990; Noam 1987, 1992). Originally the tariffs were based on a universal system whereby the accounting rate was the same regardless of which route the call actually followed. With the arrival of alternative transmission systems, however, the reason for exclusive intervention by the PTT is no longer valid. The adverse tariff system created sufficient incentives for large business users to bypass the regulated/monopoly system and facilitated the development of private/corporate networks and the emergence of new service providers beyond public control, such as MCI, Sprint, GEIS and IBM.

The development of alternative networks at reasonable and competitive costs has caused various countries to ask themselves whether a single telecommunications infrastructure remains the most efficient way to utilise resources. The traditional centralised hierarchical public telecommunications network, which became manifest in the traditional postal-industrial complex, is now being transformed into a transparent decentralised communications system of various networks, a system that is characterised by openness while at the same time securing competition between services, networks and operators. The key requirement for such an open network system is that openness of market entry, non-discriminatory access, transparent standards and interconnectivity are guaranteed. Such an infrastructure of the future, made up of a network of interconnected networks under decentralised control, has been delineated by Noam (1987:40):

> The future network concept is one of great institutional, technical, and legal complexity. The network environment will consist of an untidy patchwork of hundreds of subnetworks, serving different geographical regions, customer classes, and service types, with no neat classification or compartmentalisation possible. It includes a hodgepodge of participants, governmental and private, national and regional, general and specialised, narrow and wideband, terrestrial and satellite, tiny and vast, domestic and multinational.

Information and communication technologies have proved to be 'technologies of freedom' (Pool 1983) and 'technologies without boundaries' (Pool 1990), that erode the existing national telecommunications monopolies by making transmission almost distance insensitive and hence challenging the existing regulatory frameworks. On the other hand, these new systems also make flexible and customised use of the network possible by disentangling the functions of information storage, distribution and the provision of new services. The boundaries that once existed between press, broadcasting, mail, telecommunications and dataprocessing have now become blurred and one integrated multimedia industry is emerging with an ever-increasing number of available distribution media and the diffusion of new electronic services. Pool (1983) has labelled this phenomenon the 'convergence of modes'; Nora and Minc (1978) have put forward the term 'telematics' to describe the evolution combining and integrating telecommunications with information technology. The convergence between historically separated modes of communication, like broadcasting, publishing, point-to-point communications and dataprocessing, was made possible by the emergence of advanced micro-electronics and digital technologies: sound, text and images can be sampled, compressed, transmitted and processed as digital pulses and carried on the same electronic network. Three forms of convergence can be distinguished (OECD 1992a, b):

1 *technological convergence*: as a consequence of digitisation and related compression techniques, differences between the various subsectors in the IT domain disappear, facilitating a sharing of resources and leading to multimedia networks, multi-service networks, and multifunctional terminals;
2 *service convergence* (in format, function and transmission media): the development of new hybrid services, like datacasting, pay tv, videotex, audiotex etc.;
3 *corporate convergence:* the differentiation of revenue structures (licence fees, advertising, direct user payment, flat-rate tariffing), cross-media ownership (e.g. cable industry and telecommunications companies, computing and telecommunications, broadcasting and publishing) and cross-service provision.

In today's telecommunications networks a distinction can be made between the core public infrastructure, leased-facilities networks, alternative transmission technologies (e.g. cable networks, radio and satellite systems). Although its structure is still developing and far from definitive, the emerging multimedia market place seems to be characterised by a multitude of market players with varying interests and backgrounds (e.g. local carriers, national operators, international carriers, multiple service providers, hardware manufacturers etc.) involved in a complex game of antagonistic co-operation, including both competition in the provision of networks and services and collaborative efforts aimed at interconnecting different infrastructures and sharing networks and facilities.

The convergence of telecommunications and computers has created the rapidly growing market of telematics, which both the telecommunications and the computer industry wanted to enter and to conquer. This intermeshing of the dataprocessing and telecommunications sector generated substantial policy questions as to how and by whom services could be provided, whether under private, market-related control or public regulation. A recent example of convergence between telecommunicatons and computing is the rise and expansion of the Internet (Clark 1996; MacKie-Mason *et al.* 1996; OECD 1996). This is a packet-switched data communications infrastructure and structurally different from the public circuit-switched telephone network. The Internet is a high-capacity broadband network connecting multiple computer networks to one another around the world and, in combination with end hosts adding functionalities and applications, it provides new communication and information services, such as remote log-in services, videoconferencing, e-mail and file transfer. Its architecture, based on sharing communication lines and routing traffic dynamically over multiple paths and between most pairs of nodes, allows for balancing the load and improving network reliability. In the late 1980s and early 1990s the load of the Internet on telecommunications networks was negligible, since it was mainly used by public research institutions, universities and libraries. Its tariffication scheme was simple and straightforward, relying on non-discriminatory prices for 'bit transport'. The rapid diffusion of Internet across industries and households from the early 1990s has put the established technical and institutional framework under pressure. The increasing number and diversity of users has caused an explosive growth in traffic and – despite the recent upgrading of the backbone infrastructure from megabit to gigabit – has resulted in an overloaded and congested network.

With today's dramatic growth of network traffic on the Internet and the diverse nature of most traffic, new charging schemes and allocation mechanisms are being developed to minimise network congestion and allow users to specify different service needs under the assumption that they will be differentially priced. Network operators, Internet Service Providers (ISP) and academics are currently considering alternative charging schemes: either priority-based, usage-based, content-related (or premium-rate) pricing schemes or combinations of these. Initially, the incumbent operators simply collected fees for supplying the specialised ISPs the high-speed links they wanted for service provision, but recently PTOs themselves have started to offer Internet services to consumers and businesses. As the routing of voice telephony over data communications networks and high-speed Internet access via cable modems are potential threats to the investment plans and pricing schemes of PTOs, they have set up Internet access ventures or have acquired autonomous ISPs and cable operators. With their deep financial pockets, their critical control over the public infrastructure and access points, and their branding and 'one-stop shopping' advantages, the incumbent operators must be able, as in the past, to fend off the bypass threat through the Internet successfully.

3.4 Globalisation and differentiation of communication markets

Besides technological trends, changing market developments in the communications industry have made necessary a reappraisal of the existing communication policies. Traditionally the telecommunications market has been a dominant monopoly in which infrastructure, services provision and peripheral equipment were the exclusive domain of the PTT. The market for network equipment was indirectly dominated by the PTT through a quasi-vertically integrated relationship with its preferred suppliers. This integrated market, dominated for such a long time by the PTT, has fallen apart into several differentiated segments, in which private control and competition have increased substantially. In general, the markets for peripheral equipment, value-added networks and information services and switching equipment were the first to be opened up to more domestic and foreign competition. Control over the physical infrastructure and the basic telecommunications services initially remained the exclusive domain of the PTTs. The virtually simultaneous emergence of several different distribution technologies, such as cable networks, cellular systems, fibre-optic networks and satellites, has made it extremely difficult for the PTTs to exert total control. In order to control the various transmission alternatives, the PTTs sought to assert control in new areas like datacommunications networks, cable networks and broadcasting networks. The PTTs also faced a market for voice telephony that started to become saturated; by the 1980s the goal of universal service had largely been achieved in Western European countries (Garnham & Mansell 1991). Although the market for public telephony was still the most important one, the business market for leased lines and managed data services started to grow rapidly. The PTTs turned from basic services for high-cost-low-revenue customers towards advanced and customised services for multinational companies; they started to develop new services such as dataprocessing services, value-added networks, videoconferencing and global end-to-end services. As a consequence the existing structure and the organisational status of the PTTs was put under strain. PTT employees started to see themselves as belonging to a high-technology sector and wanted to emulate the computing industry in their salary expectations. The telecommunications division within the PTT administration felt increasingly constrained by their civil service status in their investments and social policy (recruitment and salary levels) on the one hand and the traditional organisational integration of the postal and telecommunications services. A breakaway from the mail division would bring an end to the profit-making telecommunications services subsidising the often loss-making postal services. Furthermore, the telecommunications managers of the PTTs began to recognise that a change in status and eventually privatisation would give them more flexibility and independence in developing their own corporate policy concerning personnel policy, investments and pricing decisions.

The current changes in international telecommunications are forcing the public telecommunications administrations to transform themselves from government bureaucracies without profit orientation into truly commercial corporations. Nowadays, the PTTs adapt their organisational structure and tactics and develop new strategies to cope with the fresh technological, commercial, international and regulatory challenges; they promote productivity, diversification, innovation and an emphasis on customer services. The PTTs have started to restructure their internal organisation towards more innovation and entrepreneurship by creating business units and profit centres, thereby diversifying into new businesses (e.g. information services, value-added networks, Internet access and multimedia applications). They also develop into multinationals by strengthening their international products divisions, acquiring new subsidiaries abroad, and establishing international alliances with other national public operators.

The market for network equipment has been highly protected; every PTT had its own preferred supplier(s) with their own switching systems built according to the prevailing national standards. Entry barriers were extremely high. The preferred equipment manufacturers were largely dependent on the investment plans of the PTTs for the expansion of the public telephone system. The stable picture of every country having its own switching systems conforming to national idiosyncratic standards no longer exists; increasing transnationalisation of supply and demand, high R&D investments in digital exchanges and pressure for international deregulation force domestic equipment vendors to look for new markets. National markets, although still vital for many of the equipment manufacturers, are no longer big enough to cover the R&D costs of developing new digital switches and produce minimum economies of scale. It has become crucial to win orders outside the home market. Since almost all national equipment manufacturers face the same challenge competition is intense and a shake-out has been the result. The present minimum market size in switching equipment to recover R&D and marketing costs is only attained by a few of the present players, notably AT&T (now Lucent), Northern Telecom, NEC, Siemens, Alcatel and Ericsson (Roobeek 1988; Roobeek & Broeders 1993).

Current digital switches have huge research and development outlays (at more than $2 billion), new technologies can only be developed by companies with a marketing, production and service presence in many national markets. The liberalisation of protected markets has led to major regroupings in the telecommunications equipment industry. There has been a shake-out of various big companies who either have hived off their telecommunications divisions completely (e.g. ITT, GTE and Philips) or necessarily or deliberately restricted themselves to specific niches in the market (e.g. Alcatel: public switching equipment and Motorola: cellular radio). The remaining national champions have regrouped into international consortia to achieve economies of scale and benefit from different geographical locations, thus becoming less dependent on their domestic markets. As argued by Roobeek (1988) and Roobeek and Broeders (1993), paradoxically the movement towards liberalisation results in a further concentration in the telecommunications equipment industry (e.g. GTE, Italtel and Siemens, Alcatel and ITT, Siemens/Plessey/GEC, AT&T and Philips, and Matra/Ericsson/CGCT). The equipment market is being restructured from various protected domestic markets into a (more) globally integrated and concentrated market with major mergers and joint ventures between AT&T and Philips, Alcatel/ITT/CGE, Siemens/GTE, GEC/Plessey/Siemens, Matra/Ericsson/CGCT. Apart from the emerging markets in Eastern Europe, the evolving world market for switching and terminal equipment has become tight, competitive and uncertain. The well-organised character of the domestic equipment market of the PTT's preferred supplier(s) is threatened by the inroads made by foreign equipment suppliers such as Japanese manufacturers who have become especially strong in business telephone systems, fax machines and computer gateways. The high R&D costs of developing new digital switches and the reduction of the product life cycle from almost thirty years to ten years, and the growing importance of software and automation expertise, are reasons to link up with foreign partners and computing companies. Only then is it possible to achieve the necessary scale and expertise and rationalise companies by shedding labour (for instance, employment in the

European telecommunications equipment sector decreased from 360,000 in 1980 to 250,000 in 1992 (Roobeek & Broeders 1993)).

The PTTs loosened the close relationship with their domestic equipment manufacturer(s) by looking for second or third suppliers of switching equipment. By trading with more suppliers, the PTTs became more aware of the often overpriced equipment offered by their traditional supplier(s) and could invite more competitive bids. But although the ties between the PTTs and their equipment manufacturer(s) will become weaker in the near future, at the moment they are still dominant. In the words of Roobeek (1988:316), the character of the close relationship between the national PTTs and their domestic equipment manufacturer(s) will change from 'a good marriage into a modern living apart together (LAT)-relationship'. The ties between domestic manufacturers and the national PTTs will nevertheless remain intact for reasons of historical mutual dependency, longer-term contracts and national idiosyncratic standards. They are further enhanced by the national governments, which still protect domestic markets to some extent for strategic reasons (i.e. industrial policy and national security) through public procurement programmes and large R&D subsidies.

In the past, telecommunications was ancillary to dataprocessing, banking, retailing, manufacturing, publishing and other activities. These days, however, telecommunications has become an important business opportunity for gaining significant and sustainable competitive advantages through increasing corporate and sectoral efficiency and encouraging the development of new goods and services (Keen & Cummins 1994). Originally, businesses were mainly interested in reliable telephone and telex services needed for their operational activities, but they have gradually discovered that advanced and flexible telecommunications systems not only cut back costs considerably, but also facilitate the co-ordination between dispersed corporate units such as R&D, marketing, production, distribution and sales (Davies 1994). Besides furthering operational efficiency, telecommunications systems were also important for increasing business effectiveness; with the help of telecommunications links and computer facilities companies built up specialised expertise and started to develop new value-added services, like dataprocessing services, financial services and EDI services. Some multinationals have succeeded in changing their internal telecommunications network from a cost centre to a profit centre by reselling excess capacity (e.g. Merril Lynch), leveraging in-house services into new markets (e.g. GEIS). Corporate networks of multinational firms allow for the penetration of geographic areas beyond the traditional reach of the firm and beyond national sovereignty (Irwin & Merenda 1989). Telecommunications has enhanced the internal as well as the external co-ordination of the firm by on the one hand facilitating the further integration of various production functions and on the other by linking the firm with its business environment of dealers, suppliers, contractors and customers. The management of the firm's internal and external facilities is sometimes contracted out to specialised system integrators like IBM and GM/EDS.

For large transnational users such as banks, airline companies and news agencies, telecommunications has become one of the biggest corporate expenses. For the international co-ordination of their activities these large users have developed a specific need for globe-encompassing information services at low cost, leased lines, customised services and tariff cuts in long-distance telecommunications. However, the requests of these large transnational users could not be met directly within the inclusive PTT system, which was built on national priorities, stable preferences, burden sharing and cross-subsidisation. Instead of a hodgepodge of national answers to their communication problems and needs, as Schiller (1982:104) has made clear, 'transnational users wanted transnational solutions'. In the 1970s and 1980s, the large corporate users organised themselves in national and international interest associations of large customers to further their collective interests *vis-à-vis* the PTT: in France, CIGREF; in the UK, TMA, in the Netherlands, BTG; in Europe, ECTUA, and internationally, INTUG. Although their views on pricing and related cost issues differed substantially, the representatives of large corporate users often worked together with traditional consumer organisations, representing the interests of residential customers in the policy process, advocating the promotion of the widest choice of services and facilities (going beyond conventional telephony), to be delivered cheaply and at high quality levels. The large users and specialised service providers were more radical in their political demands: they wanted the removal of legal or tariff barriers hampering transborder traffic and the introduction of cost-based service charges (i.e. an abolition of cross-subsidised pricing with its adverse tariffs for large businesses) (e.g. ECTUA 1987; McKendrick 1987). Consequently the large business users have become an important force in telematics policy making on a national as well as on a global scale.

The large corporate users developed alternative private systems outside the existing public infrastructure, rationalising their telecommunications costs by building and operating their own (inter)national telecommunications links. Together with new private service providers, such as GEIS, IBM and GM/EDS, the group of transnational users started to seek exclusive control over the nature, diversity and costs of their telecommunications services by developing alternative private telecommunications systems. These 'bypass' networks make use of an internal corporate or branch-like communication system made up out of high-capacity fibre optics, microwave technologies and satellites, supplemented by leased transmission capacity from the PTTs, to meet the increased and customised demand for high-volume dataprocessing services. Examples of global alternative networks for internal use are: Societé Internationale de Télécommunications Aéronautiques (SITA) and Society for Worldwide Interbank Financial Telecommunications (SWIFT), the co-operative networks for airline reservation and international banking services, respectively. Also the traditional telecommunications companies have started to go beyond their domestic jurisdiction and are seeking to build up a presence in each other's national markets and form international strategic alliances, such as Unisource, Global One and Concert. In order to secure the economies of scale and provide their multinational customers with complete end-to-end services around the world with one-stop shopping facilities, a combined global/local outlook is needed (e.g. Prahalad & Doz 1987). The newly created 'global carriers' have to respond effectively to local demand (often from their constituents and associated partners), while simultaneously integrating their activities and networks on a global level.

In short, new technologies have eroded and redefined sectoral boundaries; satellites and microwave technologies have opened up possibilities for competition in long-distance services markets. The PTT's monopoly has come under close scrutiny by what Noam (1987, 1992) has called the internationally oriented 'services–information coalition'. This new influential alliance of large service users and computer manufacturers, supported by those equipment firms previously excluded from domestic PTT procurement programmes, has successfully challenged the established order. As a consequence, the PTTs, policy makers and the protected manufacturing industry have been put under pressure to respond by upgrading the telecommunications network to offer highly advanced communications systems and adequate basic and – more importantly – valued-added services for specific business user groups. With the rapid growth in corporate demand for international telecommunications, the powerful lobby of internationally oriented computer manufacturers and the entry of rival carriers, Noam (1987:44) is right in considering international telecommunications the 'soft underbelly of the domestic service monopoly'. The revenues from international telecommunications have always been a major contributor to the traditional system, but with rival service provision, excess capacity, and a drastic reduction of tariffs, the PTT's system of cross-subsidisation has started to erode. The increase of competition in international telecommunications seriously threatens the monopoly/cartel regime. Modern telecommunications is moving in the direction of cost-based pricing, quality and price differentiation, enabling consumers more choice but less equity (Von Weizsäcker 1986). The traditional cost structure of the public network, where business customers were used to subsidise residential and rural services and international traffic to pay for domestic communications, is being reversed with international and business services becoming cheaper and costs for residential and rural users rising.

3.5 International deregulation

In addition to the techno-economic imperatives to reconsider traditional telecommunications policies, Western European states were also confronted with far-reaching changes in the international telecommunications regime. The process of worldwide liberalisation, that started originally in the USA, followed by the UK and Japan, and increasing global competition, dealt another blow to the domestic national monopoly framework. An important step towards deregulating international telecommunications was made in the early 1980s when the national champions of US telecommunications and computing, AT&T and IBM, were both allowed to penetrate one another's markets. The carrying out of deregulatory measures implied, however, more than just instigating domestic competition between the two giants. The bigger aim was to reinforce US techno-economic world leadership and dominate the fast-growing markets of datacommunications and telematics. To achieve these objectives, the American government decided to break up the AT&T/Bell system into separate Regional Bell Operating Companies (RBOCs) and a slimmed down AT&T, and liberalise the supply of telecommunications equipment and value-added network services. In a separate action, the anti-trust suit against IBM was dropped, and the computer company was officially given permission to enter the markets of telecommunications services and network equipment (Fisher *et al.* 1983).

Before the break-up of AT&T, the American telecommunications industry was organised as a private regulated monopoly, in which the telephone services were provided by AT&T and its associates, also known as the Bell System. This conglomerate consisted of twenty-two Bell Operating Companies in charge of the local/regional monopolies, the subsidiary AT&T Long Lines (the monopolistic provider of long distance and international telephone services), the famous Bell Laboratories and the vertically integrated equipment manufacturer Western Electric. This private conglomerate was subject to formal regulation, divided between the Public Utility Commissions (PUCs) (acting at the state-level), the independent Federal Communications Commission (FCC), anti-trust authorities as located in the Department of Justice and the courts. The FCC, installed after the 1934 Communications Act and subject to Congressional scrutiny, was the central administrative body for controlling broadcasting and telecommunication: its regulatory activities consisted of licensing services, setting price controls and limitations on profits. AT&T was prohibited from providing telegram/telex and enhanced telecommunications services. These services were provided by Western Union International, ITT World Communications and RCA Global Communications. AT&T was also banned from selling switching and terminal equipment abroad; the sale of international telecommunications equipment was left to ITT and GTE.

In the USA the markets for telecommunications and computing had been regulated separately since 1956, when the Department of Justice with its Consent Decree envisaged AT&T as a regulated monopoly in telecommunications protected from competitive entry, but at the same time AT&T's entry into computer services was prohibited. Subsequent legislation attempted to draw increasingly finer lines between telecommunications and computing, leading to a gradual process of deregulation. Until the 1970s, there was essentially no real competition for AT&T and the Bell companies in the local and long-distance markets. The deregulation of American telecommunications started cautiously in 1959, with the Above 890 Decision, when the FCC allowed the construction and operation of private microwave transmission systems, and the entry of new alternative operators, such as MCI. The approval by the FCC in 1969 of MCI's application for specialised private micro-wave lines opened the door to competition in long-distance communications. According to AT&T, MCI was a 'cream skimmer': MCI 'bypassed' the national public-switched network and by contributing less to the flow of cross-subsidies it could set its prices for long-distance services lower than a regulated AT&T/Bell system (Brock 1994). The next step towards deregulation was the 1968 Carterphone Decision, in which the FCC stipulated that the Bell System should allow equipment that was not produced by AT&T itself to be connected to its network. The 1971 Specialised Common Carrier Decision was the first step towards competition within the Bell system. This measure included the allotment of licences to specialised carriers (e.g. MCI), granting them the right to offer switched telecommunications services with direct connection to the existing network of local telephone exchanges.

With the 1971 First Computer Inquiry, dataprocessing and information services were removed from traditional common carrier regulation and subjected to competition. AT&T, as a regulated common carrier, was allowed to provide dataprocessing services as long as they were offered through a separate corporate subsidiary. The next step towards deregulating American telecommunications was taken in 1976 when the Equipment Registration Programme was implemented, in which procedures were developed to liberalise the terminal equipment market. In the 1980 Second Computer Inquiry the FCC abandoned the 1971 decision to treat the regulation of telecommunications and dataprocessing services differently, in favour of an attempt to draw a distinction between basic transmission services ('common carriage'), not altering the content transmitted, and enhanced (or value-added) services, including information and dataprocessing services. Furthermore, the market for terminal equipment was deregulated when AT&T was permitted to enter the market through an independent subsidiary. Although AT&T was still ferociously defending its integrated Bell System, it gradually became clear that the technological changes and market developments, eroding AT&T's monopolies and blurring the boundaries between telecommunications and dataprocessing, needed regulatory adjustment.

With the 1982 Modified Final Judgement many of the distinctions between telecommunications and computers were abandoned in favour of deregulation. It stipulated that in return for divesting itself of its Regional Bell Operating Companies (RBOCs) and hence losing its monopoly status as far as local and regional telecommunications were concerned, AT&T was allowed to keep its long-distance services, Bell Laboratories and Western Electric (manufacturing). As a *quid pro quo*, AT&T was allowed to enter international telecommunications, as a provider of both services and equipment, and to offer value-added services. The twenty-two BOCs were reorganised into seven regional carriers which enjoyed a regulated monopoly for the operation of local ('geographically restricted') networks: Ameritech, Bell Atlantic, Bell South, Nynex, Pacific Telesis, Southwestern Bell, Bell South and US West. These RBOCs were required to provide equal access to their networks to other carriers for the provision of non-local services, while they themselves were prevented from providing inter-exchange services, long-distance services, information services, cable services and manufacturing equipment. However, with the decision based on the 1986 Third Computer Inquiry, the RBOCs were allowed to offer terminal equipment and enhanced services (as part of the condition to provide an open network architecture to allow other providers of enhanced services to interconnect). In the course of the deregulatory process, new carriers like MCI and Sprint continued to make inroads into the markets for long-distance and international voice and data services, taking away traffic from the established international carriers AT&T, RCA and Western Union. As a compensation for the loss of its monopoly on international telephone services, AT&T was granted the right to enter the lucrative markets for value-added services in the US (with the exception of electronic publishing) and to sell terminal and switching equipment abroad. In 1991, the market for value-added services was further opened to competition, when the RBOCs were also allowed to provide value-added services.

In the Modified Final Judgement, the established rules of keeping the telephone companies out of computing and vice versa were no longer considered viable. In addition to its remit as a long-distance and international carrier, the divestiture of the Bell System implied that AT&T was permitted to enter the lucrative but highly competitive dataprocessing industry. However, computer firms like GEIS and DEC were now allowed to enter the markets for telecommunications equipment and tele-services as well. After the anti-trust suit against IBM was dropped in 1983, the company was allowed to move into the telecommunications services and equipment market. Between 1982–4, AT&T and IBM, free to enter one another's markets, were in a position to compete for the booming markets of value-added network services and information services. The two rivals developed diversification strategies, focusing their marketing efforts on the large American firms which demanded competitive and high-quality leased lines, corporate data networks and a global coverage of their communications services.

The decision to deregulate telecommunications was part of a straightforward industrial and trade policy aimed at increasing the efficiency and competitiveness of the US telematics sector (Schiller 1982; Hills 1986; Pool 1990). The US government reorganised its domestic telecommunications industry with the general aim of promoting competition between its domestic telecommunications and computing industry and the transnational giants IBM and AT&T at both national and global level. Domestic liberalisation was aimed at stimulating the productivity of the national economy at large and strengthening the competitive position of American industry and thus challenging the dispersed and protected telecommunications markets elsewhere and transforming them into an integrated global market place. With America's sheer market size and technological leadership (e.g. in telecommunications R&D, equipment manufacturing, computing, software), a first move to deregulate and encourage competition in the communications industry would strengthen the American position *vis-à-vis* its European and Japanese contenders. Bruce (1981) was right in predicting that regulatory change in US telecommunications would have significant domestic and international implications, affecting all countries. Large American telecommunications firms such as IBM, AT&T, the RBOCs, MCI, GEIS, GM/EDS and Sprint have diversified into various new geographical and product markets, a development that has had an adverse effect on Japanese and European protectionist policies and practices.

The preservation of Bell Laboratories and Western Electric as integral parts of AT&T as expressed in the Modified Final Judgement and the abandoning of the anti-trust suit against IBM, indicates that domestic deregulation was just a lever in maintaining America's world leadership in telematics. A vertically integrated AT&T could have an international competitive advantage both as an equipment and as a network supplier with Bell Laboratories' R&D expertise, Western Electric's manufacturing capabilities and a computing division in the making. After its divestiture AT&T reacted by spinning a large international web of joint ventures, participations and acquisitions to complement its shortcomings in computing and information services, and to assure access to protected markets (e.g. Philips, NCR, Olivetti, Italtel). IBM adopted a contrary strategy by diversifying into telecommunications. From its strong foothold in computing and value-added services, IBM moved into telecommunications service provision and equipment manufacturing. It acquired shares in the long-distance and international carrier MCI, with the equipment firm Rolm (an American PABX manufacturer), and set up its own Satellite Business System (SBS) to provide dataprocessing services.

After a decade of substantial losses and difficulties with regards to integrating the various activities, however, most of the investments and joint ventures of AT&T and IBM in new technologies and across-sector boundaries were discontinued. AT&T predominantly focused on telecommunications equipment supply and international services provision (in close collaboration with a group of Asian and European PTOs) and started to sell off or rationalise the majority of its stakes in the computing industry and non-domestic equipment manufacturing abroad. In 1996, AT&T demerged its computing activities (NCR) and equipment branch (now Lucent). While AT&T's first divestiture was justified by the anti-trust division of the Department of Justice as a case of abuse of market dominance, the second break-up of AT&T was a clear illustration of a company focusing on its core competencies in long-distance and international service offerings. In the late 1980s, IBM focused on its core computing business and sold its participations in telecommunications service provision (MCI) and equipment manufacturing (Rolm).

Finally, in 1996, a long time after the liberalisation of the markets for long-distance services and equipment, the US telecommunications market was opened to full service competition with the adoption of the 1996 Telecommunications Act. The new legislation overhauled the 1934 Communications Act by removing the regulatory separation between the telecommunications, cable and broadcast industries and between long-distance and local telecommunications services. Telecommunications companies responded by diversifying into new markets or consolidating their activities: for instance, WorldCom acquired MFS, UUNet, and the network facilities of Compuserve and AOL; the RBOCs, SBC and Pacific Telesis merged in 1997 and the other two, Bell Atlantic and Nynex, are still negotiating the terms of their integrated business with the regulatory authorities. In the same period, by seeking to link up with the SBC, AT&T tried to enter the local phone market. Due to strong opposition from the FCC, the proposal for a re-integrated Bell system was abandoned.

The bigger goal of deregulation was that it would not only make US industry more competitive at home, but also strengthen its position on the world market. It was believed that international competition in an open market place was the best way to combat European and Japanese protectionism. Although deregulation in US telecommunications gave foreign firms the right to invest in the American economy, the position of the US in the various segments of the telecommunications market at that time was so dominant that foreign investment did not pose a real threat. It was only in the telecommunications equipment market that the

American firms lost some ground to Canadian, European and Japanese firms. The opening up of its domestic markets to non-American firms gave the American government and firms the right of reciprocity to push for deregulating telecommunications elsewhere and in turn enter the European and Japanese markets. A large US business user group formulated it as follows: 'International telecommunications provides the pipeline which enables US industry to extend its enterprise to the vast world markets' (quoted in Schiller 1982b:99). The focus of attention from a US point of view was the entry barriers erected by other countries, notably the member states of the European Union and Japan, with their foreclosed markets and restrictive government procurement policies and preferred treatment practices.

In the US deregulation was used as an explicit industrial/trade policy, on the one hand promoting competition and efficiency gains in the domestic market, and on the other hand pressing for international liberalisation and to support its internationally leading industries. The unilateral policy of extending deregulation of domestic telecommunications into the international market was, as Hills (1986) has pointed out, an attempt to impose its domestic policy on the world and hence enforce American law on a global scale. The strategy of accompanying domestic deregulation with intensified international competition turned out to be effective. The deregulation strategy in the mid-1980s was successfully exported to the United Kingdom and Japan, who also started to liberalise their domestic telecommunications markets and privatise their national PTTs. The Americans effectively managed to put liberalisation and regulatory reform on the agenda in Europe which, eventually, resulted in the cutting back of protectionist measures in the sheltered Continental markets in the 1980s and early 1990s. The leverage of domestic deregulation to improve its position in international markets meant that the lead the Americans already had in satellite transmission, value-added services, electronic superhighway initiatives and Internet technologies was reinforced. Confronted with the (international) deregulation policies of the world leader in the telecommunications market, followed by two other leading countries, many Western European countries realised that a revision of their traditional policies and institutional structures was inevitable.

Today the trend in international telecommunications is towards increased competition between competing public and private service providers, and between rival global or regional carriers. For decades, the rules set by the ITU and CEPT for the joint provision of international services were appropriate for a stable environment, where technology was changing gradually, demand for services was predictable and largely domestically confined, and roughly similar regulatory conditions for service provisions existed. In the early 1980s the ITU and CEPT were confronted with various technical possibilities to (re-)route communications flows around the globe to circumvent regulatory barriers and restrictive practices in particular countries, and increasing competition in cross-border service provision. The global and European cartels in administering international services, ITU and CEPT, were challenged by alternative institutions that promoted divergent perspectives on commercial policy and standards setting. According to Rutkowski

(1991), even the regulatory world is becoming institutionally competitive. The General Agreement on Tariffs and Trade (GATT) and its successor the World Trade Organisation (WTO), and the International Standards Organisation (ISO), have joined the ITU at the centre of regulating and standardising international telecommunications. The new international institutions represent a different and broader perspective on international telecommunications, seeking to guarantee free trade and open access to standardisation. In 1993 the ITU responded to the institutional threat by widening its consultations with stakeholders in the telecommunications area to include the participation of non-public alternative operators of telecommunication networks in its policy process.

In the past the regimes of international telecommunications and international trade were different, but from the late 1970s onwards, American policy makers acknowledged that the international telecommunications services were indispensable in a global trading system. Irritated by protective measures in Europe and Japan (e.g. public procurement, idiosyncratic standards setting, cross-subsidies), the Americans strongly advocated the subjection of the provision of international services to GATT jurisdiction (Woodrow 1991). Since 1986, when agreement was reached to include trade in services in the GATT framework, the liberalisation of international telecommunications services has emerged as a politically sensitive issue in negotiations in the GATT and the WTO. In 1997–8, the seventy leading countries in the world agreed on a global pact to liberalise trade in basic telecommunications services and to establish a domestic regulatory environment which allows for the proper functioning of competition and free markets (i.e. the adoption of rules to ensure market access and fair competition, to use scarce resources, to interconnect, and to establish independent regulatory authorities). The role of the ITU in the standards domain has been challenged by the emergence of regional standards-setting bodies, like ETSI (Europe), T1 (USA) and TTC (Japan), and the increased importance of the ISO at world level. For a long time, standards setting in telecommunications was the responsibility of national administrations, supported and legitimised internationally by the ITU. Over the years, the convergence of telecommunications, computing and broadcasting, and the increased internationalisation of technologies and markets, facilitated an alternative and more open three-tiered regime: the ISO/International Electrotechnical Commission, the three aforementioned regional standards organisations, supported by a network of standards bodies at the national level.

3.6 European integration

Before the publication of the 1987 Green Paper, preparing the ground for liberalisation in the European Community's member states, the governance of the European telecommunications sector was regarded as a matter of national sovereignty and organised through a public monopoly. Although the articles of the EC Treaty could have been applied earlier to the telecommunications area (as a 'service', one of the free movement categories, and subject to the competition provisions), it was only in the early 1980s that the EC institutions became actively involved, when linking it with the overall objectives of the Community (i.e. industrial, competition and commercial policy). The legal basis for the moratorium in the application of European law to telecommunications between 1957 and 1987 which protected the sector from competition and the EC public procurement regime was found in article 90 of the EC Treaty. This key provision referred to public undertakings entrusted with the operation of key economic services to which member states granted special rights. As argued by Scott (1994), the EC Treaty is contradictory as applied to the utilities and telecommunications sectors: on the one hand it supports free movement principles and the internal market objective, while on the other hand advocating the goals of co-ordination and integration to foster R&D and network development (as then justified by the natural monopoly argument). Sauter has summarised the relationship between the telecommunications administrations and policies and European law in the pre-Green Paper period: 'The PTTs were under Community law excluded *de jure* from the liberalisation of public procurement, and *de facto* from the application of competition policy' (Sauter 1995:96).

In the early 1980s the worldwide telecommunications industry was clearly in turmoil: the dominant monopoly regime had already been questioned in the USA, Japan and the UK (Hills 1986). The European Commission was aware of the American and Japanese challenge and the vulnerability of Europe's telecommunications industry: in a highly fragmented market with a plethora of incompatible standards domestic markets were protected by preferential treatment of local producers, and the industry depended on Europe's weak information technology base (semiconductors and computing) (CEC 1984). The Community responded by developing a joint telecommunications policy that would increase its competitiveness by supporting joint R&D programmes and standardising new technologies in close co-operation with the national authorities, the CEPT framework of PTT administrations and the European standardisation bodies CEN-CENELEC. In 1983 the Commission installed a Taskforce on Information Technologies and Telecommunications, made up of EC officials and industry experts. The aim of the Taskforce was to monitor the EC technology programmes and develop a common strategy for the future in co-operation with the Ministers of Industry, PTT administrations and representatives of the telecommunications industry.

The Taskforce was installed as an autonomous Directorate General of the Commission for Telecommunications, Information Industries and Innovation (DG XIII) in 1986. The activities of the Taskforce were supported by the Senior Officials Group on Telecommunications, SOG-T, which had been established in 1983 to represent the EC member states and their national operators. This sectoral Committee of Official Representatives played an advisory role with respect to EC telecommunications policy, mediating between supra-national and national authorities. In its 1984 Action Programme, the Commission set six targets for an integrated European telecommunications policy that would set medium- and long-range goals at EC level, define and implement a research and development programme, expand the market for terminal equipment by mutual recognition of standards, co-operate closely to create a future telecommunications infrastructure, use modern information technologies in underdeveloped regions within the EC, and open up hitherto protected public procurement contracts (CEC 1984).

The European business community found itself inadequately equipped to cope with two external threats (e.g. ERT 1986; McKendrick 1987). First there was the high-technology threat from the US and Japan and their rising global dominance in vital sectors to the exclusion of Europe, second there was the low-end technology threat from the newly industrialising countries with their low cost base and high degree of flexibility, damaging Europe's industrial interests even further. The reasons for Europe's inability to compete with its global rivals were manifold, but they all had to do with the predominantly national outlook of Europe's industry: the markets within the European Community were highly fragmented, R&D was duplicated, the dispersed European industry was failing to achieve the economies of scale needed to compete at the international level. Despite its superiority in R&D, European industry was slow in bringing its innovations to the market (markets that were still being seen as domestic instead of European or global), and the socio-economic environment in the European Community with its high wages, social benefits, strict working conditions and high tax economies hampered international competition even further. The main goal for the business community became the stimulation of European integration, to be created by an internal market, the convergence of regulations (deregulation at national and re-regulation at European level), and a general business environment in which flexibility and economies of scale would be promoted and cross-border industrial activities (e.g. the abolition of internal trade barriers, the promotion of cross-border mergers) facilitated.

In the mid-1980s, the Commission expressed its intention to complete the Internal Market before 1993. The White Paper on the Single Market declared that diverging national rules and standards were hindering the creation of a common European market (CEC 1985). These trade barriers within the Community handicapped Europe's competitiveness *vis-à-vis* Japan and the USA. The Commission introduced a broad package aimed at stimulating competition and harmonising standards, rules and regulations throughout the Community. Also, telecommunications policy was linked to the new goals set by the Commission in the 1986 Single European Act: to complete the Internal Market, achieve a European Research and Technology Community and strengthen European cohesion. With its Green Paper on Telecommunications, an industry-specific application of the White Paper, it responded to the major complaints of European (and American) business users about big differences in services in terms of technical interfaces, tariffs and regulations within the Community, frustrating network expansion and the demand for new facilities and services (CEC 1987). Within the overall investigation into the costs of trade barriers in Europe by Cecchini *et al.* (1988), the costs of a non-integrated European market for telecommunications services and equipment, caused by a patchwork of varying standards and protected suppliers that limited the size and dynamics of the telecommunications market were estimated to be about 5bn ECU.

The Commission's Green Paper was aimed at liberalising the Community's monopolistic telecommunications market by the following measures:

- liberalising terminal and network equipment supply and provision;
- opening up the services market (with the temporary exception of public voice and basic network operation);
- separating the operational and regulatory functions of the public operator (PTT) in order to create sound and transparent market structures;
- ensuring open access conditions to networks and interconnection;
- stimulating European standardisation, with the creation of the European Telecommunications Standards Institute (ETSI);
- promoting and creating pan-European compatible networks and services (ISDN);
- fully applying competition regulation to the sector (i.e. reducing protectionist procurement policies and harmonising tariffs).

The Green Paper did not suggest any provisions with respect to the following market conditions: the regulatory structure to supervise the industry, the ownership structure of the telecommunications administrations (i.e. private or public ownership) and the introduction of competition in the domains of network provision and the supply of reserved services. The only explicit provision with respect to the restructuring of PTT administrations was that a clear separation between its regulatory and operational functions was required. This implied that the PTT administrations were no longer responsible for ensuring universal services, frequency management, charging schemes, standardisation, supervision of the industry and so forth. The member states were allowed to make a regulatory distinction between the basic infrastructure and reserved services to ensure public service goals and competitive services. Network provision and voice telephony fell under the jurisdiction of member states who could grant exclusive rights to the national operator; the provision of reserved services was subject to the Commission's liberalisation path set for the mid-term.

The Green Paper was intended to encourage the debate on telecommunications deregulation in the Community and to attract comments from a broad spectrum of stakeholders (Council, Parliament, national governments, equipment industry, services industry and the PTTs). Its aim was to break up the nationalistic and fragmented organisation of the European telecommunications market and to foster an integrated market through a controlled deregulation of the domestic markets. However, the core segments of the telecommunications market, network provision and voice telephony, were temporarily excluded from the liberalisation plans to ensure the long-term financial viability of the PTT administrations and public service provision (CEC 1987). After wide and intensive consultations and policy discussions, the Commission's Green Paper was endorsed by the Council of Ministers one year later in 1988. In a Communication, the Commission set out a programme of action, in which priorities were set and strict deadlines for implementation were proposed: mutual recognition of type approval (before 1988), liberalisation of the terminal equipment market (before 1990) and services (with the exception of voice) (before 1992), and implementation of accompanying measures (i.e. separation of operational and regulatory functions, open network provision and the creation of a European Telecommunications Standards Institute) (CEC 1988). The Commission also identified 'politically sensitive' areas where a comprehensive consensus still had to be worked out: the liberalisation of satellite communications, the promotion of pan-European services, the definition of common tariff principles and trade aspects of the Community's telecommunications industry.

The Community's market-driven strategy was a clear compromise between preserving the public operators' monopoly on infrastructure and basic services provision, while meeting the demands of large business users by introducing competition into the markets for enhanced services and terminal equipment. The liberalisation and deregulation measures proposed by the Commission had a direct relevance to the policy process in the member states, because they constrained the formation and implementation of policy adjustments in national telecommunications. The Commission started to develop policies which would use its jurisdiction to facilitate the modification of national telecommunications monopolies and to liberalise the terminal equipment and services markets and open up public procurement to competition.

In this process, key roles were played by the Commission and the European Court of Justice. In 1988 and 1990, the Commission ignored the persistence of article 90 of the Treaty and for the first time the Commission used its powers to implement two directives to push open the markets for terminal equipment and services (Sauter 1995). Due to this 'contested' policy innovation, the member states were required to withdraw all special and exclusive rights on the provision of services and terminal equipment; for a transitional phase they were allowed to retain their exclusive rights for voice telephony if public network access was transparent and non-discriminatory. The Court of Justice became a prominent actor in the area when in the 1985 British Telecom case it confirmed that the European competition rules were fully applicable to the telecommunications sector. The adoption of the 1988 Terminal Equipment Directive (88/301/EEC) and the 1990 Services Directive (90/308/EEC) meant that the member states had to take all measures necessary to ensure that companies would be allowed to supply terminal equipment and telecommunications services (with the exception of voice telephony).

In short, the Commission effectively extended the application of the competition rules and the fundamental freedoms to those telecommunications undertakings granted exclusive rights. This also implied that any proposal for an economic alliance between public telecommunications operators likely to affect trade in the Community had to be referred to the Commission and became subject to its Merger Control Regulations. The Commission has tried to bring greater transparency in awarding public contracts and opening up collusive procurement policies in the utilities sectors. Initially telecommunications and other public utilities were exempt from the Community's rules on competitive tendering. In 1990, after the implementation of Directive 90/521/EEC ended this exclusion, the

Commission managed to introduce more competition in the highly protected public procurement market. The adoption of a Commission's Directive on public contracts by the Council implied that the member states had to establish more transparent bidding procedures before 1993. The Community also promoted co-ordinating the construction of pan-European and new services: the introduction of digital cellular and paging systems, the digital cordless telephone, ISDN/IBCN, high-definition television and Trans-European Information Networks and so on. Another step towards establishing effective competition in the Community-wide telecommunications market was the regulation of the relationship between operators and users to ensure open access to public networks and abolish discriminatory and unfair restrictions on networks access and usage. The politically sensitive issue of the division of responsibilities between the Commission and member states in the domain of Open Network Provision (ONP), was temporarily resolved in an agreement, stating that regulation was primarily the responsibility of the national regulatory authorities, leaving the Commission with a minimal administrative and basically supportive role in Community policy making. The Commission has also become active in telecommunications pricing by attacking presumed cartel and monopoly price fixing and a speedy introduction of cost-oriented charges and a reduction of intra-Community tariffs. The Commission encouraged member states to implement measures that were aimed at the introduction of price cap formulae and the permission of bulk discounts to large-scale users.

At the time of the official accomplishment of the Single Market in 1992, the openness of European telecommunications to competition was hardly satisfactory. With the exception of the UK, where an effective duopoly had been established, competition in the Community was still restricted to the 'fringes' of the industry. The key voice services market was still excluded from any liberalisation measures and the implementation of the EC directives on mutual recognition, services, equipment and ONP were only partially implemented. Nevertheless some major achievements had been made by the Commission, the Parliament and the Court of Justice, the driving forces behind the integration of markets and orchestration of new EC-wide structures, turning Brussels into the 'new capital of European telecommunications' (Bessières 1989). However, the process of economic integration came to a standstill after 1992, when member states, through the Council of Ministers, dampened the ambitions of the Commission in its preparations for the establishment of a fully integrated European telecommunications market. The fear of furthering economic and political union and the subsequent centralisation of Community decision making, were the main factors behind a Community-wide adoption of the subsidiarity principle, implying a *de facto* increase in legislative powers for the member states at the expense of the Commission. While no mention of subsidiarity was made in the 1992 draft version on the review and future prospects of EC telecommunications, the Commission's final document included one paragraph on the decentralisation of policy making from EC to national levels. The Commission had to accept a delay in the introduction of competition in voice telephony, water down its ambition for an open and transparent public framework and compromise on the level of regulation by the creation of a high-level committee of national regulatory officials and Commission representatives.

The European Community has become increasingly divided over the desired telecommunications strategy and the time frame involved in implementing it. On the one hand, there are the 'liberalisers' (e.g. the United Kingdom and the Scandinavian countries), in favour of speeding up the process of market integration. On the other hand, there are 'protectionist' countries (e.g. France, Spain, Italy, Greece) who argue that state intervention and a core public monopoly alongside liberalised 'fringe' markets are temporarily justified. A straightforward liberalisation strategy might result in a fragmented telecommunications infrastructure, which would imply that expensive interfaces are required to interconnect these subnetworks. Another drawback is that the principles of redistribution and universal service, which assume some degree of government control over the telecommunications system, will be lost. The 'controlled deregulation' path followed on the European continent may have the disadvantage that, by giving priority to the universal service provision, it may be unable to meet differentiated business demands. Compared to the saturating market for residential users, the market for customised and enhanced services is very promising and profitable. The protected public network runs the risk of being bypassed by alternative private network operators or the entry of resellers targeting the profitable segments of the public monopoly. Another disadvantage is that competition may prove a better environment than monopoly for stimulating innovation. The 'protectionist' countries have suggested a step-by-step liberalisation process and a gradual adjustment to the new market conditions. One reason for implementing incremental change is the concern national governments have for preserving employment, which is at a relatively high level in the telecommunications industry. The position of Germany in all this has been somewhat unclear. Traditionally, the German government was strongly in favour of economic integration and the effective abolition of trade barriers. In the telecommunications domain, however, the position of the Deutsche Bundespost (DBP) was leaning towards the protectionist stance. Especially after German unification, the German government and the DBP gave clear priority to modernising the infrastructure in the new *Länder* over the restructuring of its domestic telecommunications and the reorganisation of its national public operator.

These intra-EU differences have always existed, but after the acceptance of the Maastricht Treaty, it seems more likely than ever that the European telecommunications sector will develop at two different speeds. According to the obligations set out in the Services Directive, a review was scheduled for 1992. Before publishing the Review, however, the Commission, as before, extensively consulted the industry's stakeholders. In the 1993 Review the Commission identified several problem areas in the establishment of a more integrated European market (CEC 1992, 1993): the need for tariff rebalancing (intra-community tariffs were still high in relation to costs, especially cross-border calls), the high charges and poor service provision of high-speed leased lines, the growing diversion of traffic within member states and between the Community and other countries, and fierce competition from low tariff operators. Furthermore, a more precise interpretation of closed user group networks, necessary for the liberalisation of private business communications, was required. The Commission proposed four policy alternatives regarding future telecommunications liberalisation (CEC 1992):

1 freezing the liberalisation process and maintaining the *status quo*;
2 introducing extensive administrative regulation of both tariffs and investments;
3 liberalising all voice telephony: both national, EC-wide and international calls;
4 an intermediate option of opening voice telephony between member states to competition.

After the consultative process, the Council decided that the exclusive rights to public voice telephony would be removed before 1998. The Commission found itself in a difficult situation, balancing between liberalisation and harmonisation. On the one hand a full implementation of the competition rules and the freedom to provide services was required, while on the other hand the interoperability between technically diverging national networks needed to be ensured. The Commission also had to safeguard the financial viability of PTT organisations. Originally, the Commission proposed an immediate liberalisation of the network infrastructure and services markets by 1996. However, this proposal was perceived by the protectionist countries, France and its allies, as too radical. Then the option of a two-tiered liberalisation was given full attention: first the deregulation of infrastructure provision (i.e. cable networks and alternative 'public' networks) before 1996, then, two years later, in 1998, the remainder of the market, i.e. fixed-voice telephony, would be opened up to competition (CEC 1994, 1995). Finally, the member states agreed upon full market liberalisation by 1998. Spain, Portugal, Ireland, Greece and Luxembourg were granted a derogation to delay the implementation of full voice telephony competition, because of lagging economic and infrastructural development.

The Maastricht Treaty on Political Union (1992–3) promoted open and competitive markets while at the same time ensuring interoperability and access to public networks and requiring the construction of Trans-European (information) Networks. A High Level Group on the Information Society, in which the captains of European industry were represented (Siemens, Philips, Olivetti etc.), was established in 1993 to draw up a plan for the creation of a Euro-electronic highway. This group, headed by Commissioner Bangemann (DG III), investigated the areas of technological applications and their market possibilities, and the regulatory and political aspects of multimedia development (High Level Group on the Information Society 1994). However, telecommunications and cable TV operators, the driving forces behind multimedia development, were not represented. According to the influential Bangemann report, there is a wide variety in the availability, functionality and prices of digital infrastructure and services across the EU: for instance, cable penetration varies between more than 90 per cent in Belgium and the Netherlands to 1–2 per cent in Greece, and the prices of services differ enormously between the UK and France on the one hand and Italy and Greece on the other (High Level Group on the Information Society 1994).

European business users are not only dissatisfied with this divergence within the European Community, but also with the huge differences between Europe and the USA. An expensive and sub-optimal EU communications network puts EC firms at a disadvantage. For instance, the European Roundtable of Industrialists (ERT) (1994:5–6) complained that:

> Europe today is a patchwork of incompatible communications networks marked by high costs, low quality of services, and very limited interoperability between systems. European communications costs are up to ten times higher than in the US and present a major obstacle to the introduction of new applications. National monopolies still persist in most European countries, restricting innovation and competition, whereas deregulation has brought about a dramatic development of the markets in the US and Britain. The European Union has launched a process of liberalisation, but it is too slow, and in many countries is being applied inadequately or not at all.

The European Commission and the related Bangemann group were susceptible to the complaints raised by the European business community. The High Level Group on the Information Society (1994) asked for a full liberalisation of service and infrastructure provision in Europe, with clear deadlines and timetables, and the establishment of a single European regulatory authority in charge of the implementation of and compliance to telecommunications policy. To some extent, a new European regulatory framework is already evolving as a consequence of the adoption of the ONP Licensing and Interconnection, and the Services and Terminal Equipment Directives. It basically includes a two-level policy network with representatives of both the National Regulatory Authorities (NRA) and the Commission involved in regulatory policy making at various stages of the process (Sun & Pelkmans 1994; Sauter 1995; Hancher 1996). General policy formation may take place at the EU level, the detailed implementation of telecommunications policy may be left to the member states and their NRAs. Monitoring, peer review and benchmarking are carried out jointly by a network of NRAs, while the Commission is responsible for overall policy enforcement and compliance, acting as arbiter of last resort. Between 1995–7, the European Union issued the following key measures for the opening of the Internal Telecommunications Market. For instance, the Full Competition Directive (96/19/EC) aims at safeguarding fair competition; the Cable Directive (95/51/EC) forces dominant operators with extensive cable operations to establish their CATV interests as separate entities; the Licensing Directive (97/13/EC) seeks to establish common principles for licensing and the equal treatment of applicants for accessing frequencies and numbers; and the Interconnection Directive (97/33/EC) develops a framework for negotiations of interconnect agreements between operators. At present, the Commission and the NRAs are occupied with the final task of translating the aforementioned directives into national legislation.

3.7 Summary and concluding remarks

Most Western European countries used to have a common institutional framework in which the provision of telecommunications services and facilities was the exclusive responsibility of the state-controlled PTT administration. The PTT's traditional business view was essentially that of an engineer with a strong emphasis on orderliness, system continuity and centrally planned end-to-end services which met economic infrastructural needs and served a social function of redistribution and universal service. The public telecommunications monopoly was broadly defined. The PTT offered standardised products and services to households and was largely insensitive to meeting the demands of business users. The telecommunications environment was for a long time stable and well organised: technology was simple, demand patterns were clear and legislation was perceived as adequately organised at the national level. However, the established government-sanctioned monopoly, which included exclusive control over the infrastructure, service provision and equipment supply and a supportive *iron triangle* of PTT, its monopsonistic equipment suppliers and the domestic government, was effectively challenged by technological, economic and international developments.

In this chapter we have discussed four structural forces that have eroded the natural monopoly structure and reduced the possibilities for national governments to control their telecommunications sector in the traditional way: technological (r)evolution, the globalisation of communication markets, international deregulation, and European integration. The governance regime of the European telecommunications industry has moved from a quasi-hierarchically integrated polity, based on a state-controlled monopoly and the obligational networks of the PTT and the domestic equipment industry, to a new industrial structure, that seems to be less exclusive and protectionist and more favourable to new entry, competition and innovation. Nowadays, the telecommunications industry is characterised by the development of a whole range of new products, services and networks, an evolution towards open access to the key infrastructure and interconnected public and alternative private networks, fast-growing demand for specialised business services and increasing competition at both the national and global level.

The majority of EU member states have prepared their telecommunications sector for international competition by transferring their PTT administration to the private sector. The formerly protected telephone administrations are changing from production-oriented bureaucracies into flexible multi-divisional companies, which are (more) service-oriented and responsive to their customers and market demand. The administrations are developing an 'enterprise culture' and have started an active search for innovative services, profitable market segments and economies of scale and scope. The bigger telecommunications companies have embarked upon a strategy of horizontal and vertical integration aimed at securing their home markets and penetrating new geographical, product or customer markets. European telecommunications operators have diversified into manufacturing, information services, cable and broadcasting, and have established new cross-national alliances, penetrating each other's markets in Europe and elsewhere. There is also a trend towards a global approach aimed at providing worldwide services to large business customers. At the same time, the former PTTs are confronted with emerging competition from computing firms, information services providers and new network operators (cellular radio, cable companies, satellite service providers), all of which have recently entered the liberalised telecommunications markets.

As a consequence of technological innovations and changing market conditions, new players like the computing industry, services industry, new equipment suppliers and business users have entered the traditional telecommunications domain. The emergence of international deregulation, strongly pushed by the American government, together with the process of European integration, were other factors behind the growing need for a redefinition of the traditional institutional framework of European telecommunications. There was a trend towards 'privatising' the national communications network; as an alternative for or supplement to the existing public network, characterised by strong government intervention and universal access, new private networks for closed (business) user groups were created. Large telecommunications users and the (American) computer industry started to lobby for a more market-based organisation of the telecommunications industry. The big users demanded an extension of specialised managed data services and asked for the provision of network capacity at cost-based prices. The computer industry argued for a liberalisation of the protected national markets of telecommunications equipment. The coalition of big telecommunications users and the computer industry first became politically successful in the United States in the late 1970s, followed by the United Kingdom and Japan at the beginning of the 1980s. The governments in these countries responded to these pressures by liberalising their domestic markets. The main reason for following such a deregulatory move was to stimulate the domestic information and telecommunications market in the short run and to have a global competitive advantage in the telecommunications sector in the long run.

A new balance has to be found between the promotion of new and alternative networks and the competitive supply of advanced services for the small 'big business' communities and a state-controlled or regulated public infrastructure and the provision of basic communication services, universally available and accessible for everyone. In implementing institutional change, governments have to match the political and economic demands of three constituencies:

1 the domestic preferences of established stakeholders: e.g. 'sheltered' corporate restructuring and concerns regarding the loss of market shares, universal service provision, and jobs;
2 the international ambitions and trade interests of its home players: e.g. the revitalised PTT negotiating access to foreign markets and seeking strategic partners, promoting the competitive position of equipment manufacturers

and the information-intensive services industries *vis-à-vis* foreign contenders;

3 the claims of new (foreign) entrants (supported by the representatives of large users), that competition in the home market will stimulate efficiency, innovation, customer responsiveness and the availability of cheap and high-quality services.

4 The liberalisation, privatisation and regulatory reform of telecommunications in the UK

In care of the market?

4.1 Introduction

The industrial adjustment strategy adhered to by UK governments over the years has been characterised by a market-led or company-led pursuit of efficient markets and new technologies, and achieved by promoting free enterprise and withdrawing the state from the national economy (Katzenstein 1978, 1985; Zysman 1983, 1994). This strategy has sustained the openness of the British economy and forced domestic companies to compete in world markets. The typically British market-oriented or liberal policy response was taken to its extreme in the 1980s and 1990s, when the 'Thatcherite' Conservatives implemented their 'big bang' programmes, aimed at a rapid removal of remaining barriers to free enterprise. Before 1979, the character of economic policy in the UK was less profound, alternating between market-led and interventionist adjustment strategies, as promoted by 'modest' Conservative and Labour governments. The study of the restructuring of telecommunications in the UK is a highly interesting one, because the country has played 'the role of policy laboratory for the world' (Garnham 1990:7). The UK programme, characterised by radical privatisation, far-reaching liberalisation and concomitant regulatory reform, became exemplary in Europe both as an inspiration and a deterrent. Since 1981, competition has been introduced in nearly all segments of the UK telecommunications market, including equipment, enhanced services, mobile communications, voice telephony and basic networks. The public operator British Telecom was separated from the Post Office and eventually privatised in the mid-1980s, and the independent Oftel body was established, taking responsibility for monitoring and regulating the liberalised telecommunications market. Britain's radical strategy was clearly ahead of the more gradual plans of the European Commission to open the Community's telecommunications market, as laid down in the 1987 Green Paper.

The dominant coalition, which played a decisive role in changing the institutional structure of UK telecommunications, was made up of the Conservative government and large business users, to some extent supported by the privatised administrations of British Telecom and Cable & Wireless. These actors argued that the (gradual) deregulation of domestic telecommunications would promote a favourable business environment, enhancing efficiency, choice, market responsiveness and technological innovation, to the benefit of both large and residential customers. Such a favourable business environment would have a positive effect on international trade in services and strengthen the position of the UK in the world by attracting international traffic and international businesses critically dependent upon adequate, flexible and efficient telecommunications facilities. As a consequence, all markets were opened up to competition and British Telecom shares were sold on the stock market between 1984–93. The preparation of the privatisation and the actual flotation triggered a multitude of business activities in the City of London, not only for the financial services industry, but also for lawyers, public relations/advertising and consultancy firms. The privatisation strategy received further support from BT's management, employees and the wider public, who were also actively involved in the purchase of (underpriced) state assets.

Before discussing the traditional structure and changes in the British telecommunications industry, we will examine those political and economic contingencies at the national level that might have had an impact on the policy process. The sector's wider political and economic setting, and the structural adjustment policies followed by the UK government in the 1980s and 1990s, are introduced in sections 4.2 and 4.3 respectively. Next, the focus shifts to the industry level, as we discuss the trend towards demonopolising UK telecommunications that eventually led to the restriction of the public monopoly, the introduction of competition in the domestic market and the privatisation of the Post Office/British Telecom (section 4.4). The processes of liberalisation, privatisation and regulatory reform are treated in more detail in sections 4.5, 4.6 and 4.7. Finally, we close with a brief summary and some concluding remarks on the major shifts in UK telecommunications policy (section 4.8).

4.2 The wider political-economic setting of UK telecommunications

Before discussing the established structure and the institutional changes in British telecommunications polity, we will examine the wider political and economic setting. To describe this institutional environment we will briefly touch upon a few national variables that are characteristic of the relationship between government and the business community in the UK: the production profile, the organisation of government, the system of interest intermediation, approaches to economic policy, and the British economy's international dependency.

Production profile

The British economy has traditionally shown an international outlook: a focus on overseas (colonial and European) markets, a globally oriented services sector (banking, insurance, trading and media), large overseas investments, and large domestic foreign investments. The City of London is a leading international centre, housing a large share of the world's commercial, banking and financial operations and the most important financial institutions and providers of business services (e.g. accountancy, marketing, consultancy etc.). Compared to the dominant role the UK occupies in the services industries (and in chemicals and pharmaceuticals), manufacturing is characterised by a poor overall economic performance, contraction and relative decline. Over the last twenty years, the UK has increasingly fallen behind its major foreign competitors. Ever since it led the world in industrialisation, Britain has seen its supremacy eroded by fierce international competition, eventually giving way to massive de-industrialisation and radical restructuring (e.g. British Leyland and British Steel). Britain's industrial and technological base proved rather one-sided, lacking in innovative flexibility and suffering from short-run profit maximisation and asset trading (Walker 1993). Compared to its major competitors, who are all more or less specialised in high-quality value-added products, British manufacturing found itself at the lower end of the market with many low-unit value products and a lack of sophisticated goods.

In the 1970s the UK found itself in a crisis characterised by major economic inefficiencies, high levels of inflation and unemployment, rising public expenditure and a large budget deficit (Gamble 1994). Both Labour and Conservative governments had been unable to modernise 'smokestack' industries and promote the interests of promising sectors in an orchestrated way. The industrial policies followed by the various governments before 1979 gave a clear priority to ailing 'sunset' industries over 'sunrise' industries and to the international interests of the City over domestic manufacturing. The government addressed the manufacturing sector selectively, giving most attention to the defence-based and defence-related industries (i.e. aviation, aerospace and computing). Given Britain's strong commitment to domestic security and with defence accounting for more than 50 per cent of government R&D expenditures, the leading defence electronics firms, British Aerospace, GEC and Plessey, have greatly benefited from the government's investment plans (Walker 1993). These protective measures, however, also shielded these firms from international competition. One reason for the erosion of the manufacturing base in the UK could be found in the strategy followed by the post-war governments of high exchange rates and an overvalued currency. This policy gave clear priority to the worldwide interests of the financial sector to secure London as an international centre. The consequence for domestic manufacturing was that it was saddled with high export prices while competitive imports and foreign investments in the already suffering manufacturing industries were encouraged (notably by Japanese companies).

Organisation of the state

The British political system represents a majoritarian model of democracy, characterised by a concentration of executive power (one party and bare majority cabinets), fusion of power and cabinet dominance, a one-dimensional two-party system and, as a rule, a unitary and centralised one-party government (Lijphart 1984). This system, based on the principle of 'winner takes all', tends to produce a process of policy making that offers little scope for consensus seeking and coalition cabinets. The most powerful body in government is the cabinet, which is usually composed of members of the party with an absolute majority in the House of Commons. 'Bare-majority' cabinets coming into office are able to implement the policies they advocate rather easily, without consulting the large minority in Parliament. As a consequence, most cabinets have used the state apparatus to impose their party manifestos on British society. The British political system is characterised by a party duopoly whereby the winning party has a virtual monopoly on the powers of government (Willis & Grant 1987).

Until 1979, political decision making tended to be incremental, with a slight shift in emphasis every time political power changed hands (Jordan & Richardson 1982). Often, policies implemented by the previous government were 'corrected' by the new one. After 1979, the British policy style changed from emphasising negotiation and accommodation, as promoted by the Labour party and the moderate sections within the Conservative party, to a deliberately confrontational style of politics, advocated by the radical Tories under the leadership of Margaret Thatcher. According to the Tory government that came into office in 1979 the cause of the economic recession was clear-cut and simple: state intervention and bureaucracy had gone too far and trade unions had far too much power in British society. A large public sector with a considerable number of nationalised enterprises (10.5 per cent of GDP and 8.1 per cent of the total workforce) and an expanding welfare state, had put a strong fiscal burden on the Treasury. From 1979 onwards, the radical Tories effectively reduced the degree of state intervention in the national economy by restoring the market mechanism (liberalisation), cutting back government subsidies to industry and the Keynesian welfare state, and implementing an impressive privatisation programme.

The UK government has traditionally followed a liberal strategy aimed at stimulating market forces and free trade, to a large extent delegating the formation and implementation of economic policy to the responsibility of powerful private actors. Compared to the state societies on the European continent, Dyson (1980:viii) referred to Britain as a state-less society, lacking 'a historical and legal tradition of the state as an institution that "acts" in the name of public authority.' Given the rather weak position of the state *vis-à-vis* society, the UK government is entrusted with a powerful executive: a strong cabinet supported by an insulated bureaucracy relatively invulnerable to direct pressure from Parliament and organised interest groups. In general, the British government has treated the management of the economy not only as a business to be run by supportive macro-economic and trade policies, but also as the domain of businessmen (Grant 1991).

The UK has been categorised as a 'company state', exhibiting an arm's length relationship between government and industry (Willis & Grant 1987). Britain has a tradition of state-approved self-governance by the industry concerned, relying upon privately enforced negotiations and voluntary compliance to regulate the industry, and supported by gentlemen's agreements between industry and government. Compared to the adversarial, bureaucratic and open character of US regulation, UK-style regulation is more flexible, informal, co-operative and closed (Vogel 1986; Baggott 1989). From the early 1980s onwards, however, the tradition of self-regulation in the UK underwent some changes as a result of the implementation of both deregulation and regulatory reform measures. The Conservative government liberalised various sectors and reduced controls and inhibitions in the fields of labour markets, industrial relations, corporate tax rulings and so on. Regarding public utilities, the government replaced far-reaching state intervention by a more detailed and formalistic US-like regulation through the creation of several quasi-administrative agencies to regulate and supervise the denationalised industries. These regulatory agencies are listed as non-ministerial government departments, and are staffed by civil servants. Originally, they depended on annual contributions from their sponsoring department and the Treasury, supplemented with revenues generated by the issuing of licences. Today, however, most of the regulatory agencies are fully self-supporting (e.g. through selling franchises).

System of interest intermediation

The system of interest intermediation in Britain can be classed as pluralistic, whereby the government actively seeks to mobilise the support of powerful private groups. In practice this has often led to direct negotiations between government departments and large firms to bring corporate behaviour in line with public policy objectives (Grant 1984). The use of intermediary channels of the interest associations of organised business and labour has usually had a lower priority in the formation of Britain's industrial policy. The Thatcher government no longer sought quasi-corporatist solutions with the traditional social partners, notably the TUC and the CBI, and deliberately bypassed the established industrial relations system by encouraging company-level negotiations and giving more influence to the less consensualist Institute of Directors. Britain's workforce is relatively well organised, but in a highly dispersed and decentralised way. At the end of the 1970s, about 53 per cent of the active workforce was unionised; under Thatcher union membership fell below 40 per cent (Crouch 1990).

The workforce is organised into more than one hundred trade unions, loosely integrated in the Trades Union Congress (TUC) with viewpoints varying from radical socialism to more moderate views on industrial relations. The TUC is a confederation of trade unions (many of them still craft-based) with only limited bureaucratic means and sanctions to force its constituents to comply. The trade unions have a special and direct link with the Labour party, contributing ideologically and financially to the party's activities. The Conservative governments of the 1980s regarded the high level of trade union militancy and union power in sectors like coal, railways and other public sector industries (with unionisation rates of more than 90 per cent) (Parris *et al.* 1985) as counterproductive to the economy and effectively cut back union influence in socio-economic policies. Business is organised along the same lines as the workforce, in that it is both highly organised and decentralised. The Confederation of British Industry (CBI) is a relatively weak confederation of loosely integrated semi-independent units and heavily divided between financial and industrial capital. Unlike the close relationship between the Labour party and the TUC, the relationship between the Conservative party and the CBI is more distant. The Conservative party has closer links with the financial sector than it has with the manufacturing industries.

Unlike France and Germany, Britain does not have a tradition with regards to the transfer of money flows to critical sectors through deficit spending by an active state and the allocation of large credits by national investment/industrial banks to sponsor domestic industries (Zysman 1983). The merchant banking system in Britain has prevented the evolution of strong alliances between state-controlled or private investment banks and manufacturing. The collaboration between state, the financial community and industry found in France and Germany, working through personal contacts, interlocking directorates, large shareholdings and close financial supervision, is absent in Britain. The British government has pursued a different strategy by transferring financial resources for key industries through the national budget and by controlling public expenditure. Instead of utilising the banking community to accomplish economic policy objectives, the government gave clear priority to the interests of the large clearing banks, unit trust companies, pension funds and insurance companies, with the London stock exchange as the major international financial hub (Hall 1987). The 'City nexus', which includes the commercial banks, the Treasury and the Bank of England, has traditionally fostered free trade, stable exchange rates and balance of payments, without taking explicit responsibility for the financing of domestic industrial developments or helping to prevent the contraction and the loss of competitiveness of domestic manufacturing (Capie & Collins 1992). The imperatives of the London Stock Market favour a rather one-sided orientation towards the profitability, share prices and annual dividends of companies over a clear commitment to corporate planning and long-term industrial restructuring. The involvement of the British banking community in industrial management is partial and temporary: institutional investors are primarily interested in short-term portfolio management rather than the future viability of businesses (Grant 1991). Consequently, longer-term investments in R&D, equipment and human capital have suffered. Because of this short-termism, corporate management have had to rely heavily on retained profits and short-term loans provided through the capital market (stocks and bonds) (Hutton 1996).

Economic culture and ideology

The economic approach of the UK combines a core 'liberal' doctrine, i.e. leaving major decisions to markets and firms, with a relatively centralised political system. There is a remarkable difference over time, however, between the post-war consensus among alternating Labour and Conservative governments and organised business and labour, and the strongly business-oriented and confrontational Tory governments from 1979 onwards. The Thatcherite creed was strongly inspired by a 'get the government off our backs' philosophy, a desire to break trade union powers and a firm belief in the overall functioning of markets as efficient means to achieve growth and innovation. Despite the free enterprise principles of Thatcherism, i.e. the decentralisation of socio-economic activities and non-intervention, the government showed decisive leadership in particular areas. To provide the goods that the market could not provide (defence, security, law enforcement and price stability) and to support and police the 'free economy', a 'strong state' was needed (Gamble 1994). The policies of the Thatcher and Major governments were directed towards privatising state equity, deregulating labour markets and financial services, furthering competition and reducing public spending and corporate taxation. The British approach towards competition policy has been characterised by a pragmatic spirit and an *ad hoc* application of the public powers of the Monopolies and Mergers Commission (MMC) and the Office of Fair Trading (OFT). Obviously, competition policy was given more priority under the Conservative governments than under Labour. The four Thatcher governments (1979–91) set forth clear and new objectives, such as the encouragement of new entry and fair play conditions in formerly state-owned utilities, the introduction of price caps and asymmetric regulation and the tightening of merger controls. Another economic goal of the Conservative governments of the 1980s was the promotion of free trade. The UK has been more open than, for example, Germany or France in its acceptance of inward investment of subsidiaries of American and Japanese firms.

Traditionally, British industrial culture has been strongly divided between business and labour, and business between manufacturing and commerce. The division between manufacturing and the services sector (finance, trade and commerce) is deep: it resembles not only the divergence between the industrial North and the financial and trade centres of the South, but it also exhibits major differences in terms of attitudes towards economic modernisation (Wiener 1981). British manufacturing has exhibited a pattern of business behaviour, favouring stability and giving little attention to technological innovation, economic growth and international competition. Its counterpart, the services industry, strongly promoted market innovation, high growth strategies and international free trade. While the former has displayed a lack of international competitiveness (i.e. outmoded work practices, overmanning, industrial conflicts and foreign direct investment), the latter has expanded successfully: in the 1980s and 1990s the City of London strengthened its position as Europe's centre in finance, trade and commerce. Organised labour in Britain shared the manufacturers' conservative attitude towards industrial modernisation and has maintained a defensive and even hostile attitude towards technological and industrial change. Another major rift in British economic ideology lies in the definition of the state's role in society. Until the coming to power of the Blair cabinet in 1997, a strong difference existed between the interventionist policies of Labour governments, emphasising nationalisation and rationalisation of firms and industries, and the economic policies pursued by Conservative governments, emphasising a mixture of free trade, monetarism and a low level of state intervention in society.

Positioning in the international politico-economic system

Britain owes its position in the international division of labour to the fact it was the first nation to industrialise its production and internationalise its trade in the eighteenth and nineteenth centuries. After World War II, Britannia no longer ruled the world and post-war governments had to lower what ambitions they had of dominating international trade and investment patterns. The Commonwealth had started to disintegrate, the US and the USSR were the new policemen of the world and the US, Japan and Germany became the world's leading economic powers. Britain had to find a new role for itself in the international political economy, striving to combine the objective of securing the nation's autonomy with the aim of fostering collaboration with the US and Western Europe. After joining the European Community in 1973, Britain's relationship with the European Community has reflected the dilemma of promoting an internal market based on free trade principles, without sacrificing the nation's sovereignty in social-economic matters, foreign affairs and security/defence policies. Fed up with the excesses of the Common Agricultural Policy, administrative red tape and high bureaucracy compliance costs, there was the persistent Euro-hostility of former Prime Minister Thatcher and other Euro-rebels, who had strong doubts about the extension of the Internal Market programme to include economic, monetary and even political union. Leaving the European Community, however, was never a real option, since this would imply an effective loss of UK authority in the world. Blair's Labour government, which succeeded the Major government after winning the 1997 elections, has displayed a more positive and constructive approach to the European Community.

Opposing any proposals to transfer sovereignty to supranational authorities, the UK government argues in favour of extending the free trade area to include Central and Eastern European countries: 'widening instead of deepening the Community' (Bulmer *et al.* 1992). The British government's economic policies, however, hardly arouse the wrath of the European Commission (with the exception of its social-economic and labour policies) and its industrial and trade strategy is more in line with the Treaty of Rome and the Internal Market Programme than the policies embarked upon by most of its European partners (e.g. France, Italy and Spain). Actually, Britain has one of the best records of implementing EC directives and regulations (Bulmer *et al.* 1992; Greenwood & Jordan 1993). Furthermore, the UK economy is relatively open and welcomes exports and foreign investments from both European and non-European multinational firms.

In addition to European collaboration, the UK maintained strong trade links with its 'natural' ally the US, as well as a strong presence in the Pacific Basin (Hong Kong, Australia). Britain's historical background as a former imperial power and its Commonwealth has played and still plays a decisive role in the formation and implementation of its macro-economic and trade policies. Although less important than before, the UK remains the pivot in a network of international economic relations, manifesting itself through vast monetary and military commitments in its overseas colonies, the pound sterling as an important world currency, a greatly internationalised services sector and the City of London as a global banking centre. Furthermore, the Conservative governments of the 1980s and 1990s strongly advocated the notion of establishing an open international trading environment ('a global playing field'), that would strengthen the international competitiveness of

British firms. To further the integration of Britain into the world economy, the UK governments gave full support to the implementation of the Single Market Programme and to a quick settlement of the free trade negotiations in the GATT/WTO framework.

4.3 The politics of economic adjustment in the UK

After World War II economic policies in Britain were dominated by two issues: the widely supported objective of achieving full employment through Keynesian macro-economic policies and a preoccupation with Britain's relative economic decline. It was only with the coming into office of the Thatcher government in 1979, that public policy clearly changed from tripartite settlements to an explicit *laissez-faire*, aimed at creating a more favourable business environment and giving priority to economic over social policy objectives. The Labour government (1945–51) developed and implemented an ambitious interventionist programme of nationalising the commanding heights of the economy: the Bank of England, railways, airways, utilities, the National Health Service, coal and steel and Cable & Wireless (the overseas telecommunications operator). Under Labour rule the welfare state was expanded to safeguard adequate levels of housing, health care, education and social security for all members of society. The Conservative government (1951–64) reversed Labour's decisions on the nationalisation of the iron and steel sectors, but adopted Labour's full employment programme and expanded the welfare state through Keynesian demand management. In 1962, the National Economic Development Council (NEDC) was established as a tripartite forum, chaired by the Treasury, for consultation between government, business and labour on longer-term economic and industrial planning with the purpose of achieving more balanced and rapid growth. The NEDC was supported by twenty-one Economic Development Committees (EDCs), consultative bodies set up at industry level exchanging information between government, business and labour. The NEDC structure, however, was badly integrated within the established policy structures and lacked clear responsibilities: it was more a consultative forum than an agent actively involved in the formation and implementation of economic policies. The NEDC turned out to be an institutional failure, lacking support from the Treasury. The latter was more interested in short-term macro-economic issues (e.g. inflation, balance of payments, public expenditure), than in active technology and industrial policy making, and in longer term indicative industrial planning.

The Labour government (1964–70) developed a more active industrial policy, (re)nationalising British Steel and restructuring British industry by promoting mergers among domestic companies ('picking winners') and developing technology policies ('the white heat of the new technological revolution'). In 1966, the Industrial Reconstruction Corporation (IRC) was established to rationalise certain industries: as a result of the IRC's intervention British Leyland and ICL Computers were created as national champions. The importance of industrial policies, relatively independent from macro-economic policy and Treasury influence, was only recognised in the mid-1960s, when at cabinet level a new

Department of Economic Affairs and a Ministry of Technology were established. The first was responsible for long-term national economic planning and the second was in charge of R&D policy and the promotion of high-tech industries. The Labour government attempted to develop balanced industrial policies that were both directed towards encouraging new enterprises, advanced technologies and promising industries, and towards reversing economic decline by rationalisation of size and production, organisational restructuring and revitalisation of management and equipment. This effort, however, proved ineffective as policies tended to favour sunset over sunrise industries.

At first, the Conservative Heath government (1970–4) made a strong effort to follow a free market and monetarist approach and intensify competitiveness by putting forward measures like tax cuts, trade union restraint and the reform of the social services. It also attempted to reduce state intervention in industrial activities by abolishing the Ministry of Technology and the IRC. This radical economic programme triggered a period of social and economic unrest, which was reinforced by the worldwide consequences of the Bretton-Woods aftermath. When the Heath government realised the unfeasibility of this first attempt at Thatcherism, the radical pro-market programme was abandoned and a U-turn made away from market principles towards state intervention and compulsory prices and wages control. When British manufacturing was hit hard by international competition and economic recession, industrial policy primarily concerned itself with keeping up employment in declining sectors and the nationalised industries through heavy subsidisation and reorganisation schemes, rather than with stimulating technological innovation. The state holding National Enterprise Board (NEB) for instance, established in 1973 to promote innovative and prosperous manufacturers by providing investment capital and taking substantial shareholdings, ended up in charge of reorganising the declining industries and preserving ailing firms from bankruptcy.

The Labour government (1974–9) tried a mixture of Keynesian macroeconomic management, indicative planning and tripartite arrangements. It continued with the Keynesian policies of voluntary controls on the growth of wages and prices. The withdrawal of support from the TUC for a centralised incomes policy, together with a revaluation of the pound, led to rapidly rising labour costs and a substantial deterioration of Britain's international competitive position. The Labour government extended the NEDC framework of industrial reorganisation through the creation of Sectoral Working Parties (SWPs) in thirty-nine industries, in addition to the already established twenty-one EDCs. These interventionist measures proved inadequate to cope with the economic recession that emerged in the mid-1970s. In 1976 the country was forced to accept loans and advice from the IMF. The rising levels of unemployment and inflation, a government powerless to combat industrial unrest and a disruption of the provision of essential services, eventually led to the downfall of the Labour government.

In 1979 Labour was succeeded by the Conservative government of Margaret Thatcher, who no longer considered the prevailing Keynesian demand management policies an appropriate instrument to combat recession. As Walker and Sharp (1991:271) have argued, Thatcherism may be regarded as an opportunist response to the perceived failure of Labour-dominated eras (Sharp & Walker 1991; Walker & Sharp 1991). The radical Tories, having gained control of the party from traditional moderate Conservatives like Edward Heath, were in favour of a radical *laissez-faire* approach, with policies such as deregulation, tax reform, monetarism and trade union reform. The Thatcher government effectively argued that the high degree of state intervention in the economy, and the growing influence of trade unions in economic policy making had contributed to rising public expenditure and climbing inflation levels. Especially after the 1978–9 'winter of discontent', the prestige and performance of the nationalised industries in Britain was low, and the hostility towards the public sector unions high. Due to political interference and the absence of a clear institutional framework for the nationalised industries in terms of democratic accountability, effective consumer and worker representation, the separation between the administration of public policy and the day-to-day management of state-owned companies proved ineffective, resulting in enduring problems between management and the government over external borrowing, personnel and pricing policies and representation of workers and consumers (Prosser 1986; Veljanovski 1987; Graham & Prosser 1991; Foster 1992). When the nationalised industries asked the Treasury to cover operating losses or finance modernisation programmes in the 1970s, the arm's length relationship between nationalised industries and the political system was eventually transformed into one where the nationalised industries were subject to strict financial controls. The strict rules on Public Sector Borrowing Requirement and the External Financing Limit, and the further tightening of financial controls in the mid-1970s were reasons for management of public sector enterprises to consider new ways of governing nationalised industries or more radically to escape from the public sector altogether.

The Conservatives of the late 1970s perceived the role played by government in the economy as too large and ambitious, causing major inefficiencies, excessive public sector deficits and a discouragement of private initiatives. To revive the economy and establish a favourable business environment, the following remedies were suggested: the implementation of deregulation and privatisation programmes, the abolition of the guided income and price policy, the curbing of trade union powers and an end to the tripartite consultations in the NEDC and IRC framework on economic/industrial policy making. These Thatcherite policies were supported by a fiscal policy that articulated tight monetary controls aimed at lowering the budget deficit and a proposed reduction in both the rates and progressiveness of taxation. The Thatcher government attempted to reform industrial policies by abandoning the established strategies of 'picking the winners' and 'subsidising lame ducks', identified by Burton (1983) as 'picking losers' and the 'industrial subsidy morass' respectively, by shifting from defensive to positive policies (aimed at electronics and new technologies) and by promoting supportive measures for business-like competition policy, deregulation and tax cuts.

The Thatcherite Conservatives had a strong belief in the smooth functioning of markets and the competitive process as the best way to maximise allocative efficiency in society. In the words of John Moore MP (1986:93): 'Our main objective is to promote competition and improve efficiency. Less government is good government. This is nowhere truer than in the state industrial sector.' The Thatcher government regarded private ownership as far more efficient than public ownership, and consequently insisted that the role of the state in the economy should be minimalist. The privatisation of nationalised industries would increase the economy's performance: a substitution of political control by market discipline would make state enterprises more efficient and generate substantial consumer benefits (Littlechild 1981; Beesley & Littlechild 1983). The ideology behind the deregulation and privatisation programme strongly referred to the Hayekian notion that selling state assets would enhance economic freedom throughout society and contribute to the realisation of a shareholder democracy with extensive equity ownership on the part of citizens and employees (Hayek 1944; Veljanovski 1987). The sale of state assets and local council homes to individual shareholders, employees of the privatised companies and former housing tenants would contribute to the neo-conservative notion of 'popular capitalism': the nation and the firm as a group of shareholders.

Another reason to pursue a programme of privatisation was that it would raise huge revenues for the government through the sale of shares and annual tax revenues and dividend yields, which could be used to reduce the government's budget deficit and finance its tax-reducing aspirations (Thompson 1990). A hiving-off of ailing state enterprises such as British Leyland would reduce public sector borrowing and public expenditure even further. The Thatcher government pleased the City and private enterprise by deregulating the capital market and reforming the corporate taxation system. The process of flotation has been criticised as furnishing the City (merchant banks, institutional investors, securities houses, advertising and PR bureaus etc.) and individual investors with windfall profits. As convincingly argued by Vickers and Yarrow (1988), the equity of British Telecom, British Gas and British Airways was undervalued and sold off in chunks that were too large. The sale of undervalued state assets would benefit the City institutions (financial investors, underwriters, brokers, auditors, advertising agencies etc.) by providing them with large contracts and huge profits (TUC 1985).

The privatisation would also serve the straightforward political goal of re-election. The sale of state assets and low-priced council homes enabled the government to give the voting public electoral advantages such as income tax cuts and instant profits. For the Conservatives, privatisation was part of a broader strategy to break the power of the Labour party and the trade unions, still quite strong at that time. The level of unionisation was by comparison very high in nationalised industries and public utilities. The Conservatives considered the widespread distribution of shares and share ownership among the employees of privatised firms, as opposed to state ownership, a major political obstacle to a possible return of a Labour government and its plans for renationalising the recently privatised industries. This strategy proved successful, and share ownership increased from three million people in 1979 to around ten million in 1996 (PI 1997). The Thatcher government succeeded in disciplining and curbing union demands through a confrontational economic adjustment policy, leading to rising levels of unemployment and cutting back unionisation rates substantially. The Thatcher government imposed far-reaching restrictions on closed-shop arrangements, union labour-only contracts and the introduction of secret ballots.

The first Thatcher government (1979–82) did not have a master plan with respect to privatisation. The privatisation alternative did not even feature in the 1979 election programme of the Tories; it only emerged, rather accidentally, midway through their first term in office. When its monetarist polices to combat the recession proved unsuccessful in 1981 (e.g. soaring unemployment and inflation rates) and the large-scale sale of council houses to residents had only turned into an enormous success due to the fact that they were sold far below their market value, the Thatcher government began to consider large-scale privatisation as a viable alternative. Notwithstanding the demand for a U-turn in economic policy by some leading economists and business leaders, the Thatcher government continued with its radical pro-market strategy. The sale of state assets and the political ideas behind privatisation fitted perfectly within the rather opportunistic practices of curtailing public expenditure, breaking the power of the Labour party and the trade unions and the radical Tory thinking about free market individualism (Steel & Heald 1982; Chapman 1990).

Under Thatcher rule more than one million council houses were sold to sitting tenants and several major state-owned enterprises were privatised: British Petroleum (1979–87), British Aerospace (1981–5), British Airways (1987), Cable & Wireless (1981–5), British Telecom (1984–93), British Gas (1986–8), British Steel (1988), the Central Electricity Generating Board (1989), and the Water Authorities in England and Wales (1990). In the 1980s, almost one million jobs were transferred from the public to the private sector and a large majority of the workforce of privatised industries became shareholders in the newly established companies (Hyman 1989). Although the replacement of Margaret Thatcher by John Major as Prime Minister in 1990 was a consequence of a fierce power struggle within Conservative ranks, it did not entail a major shift in public policy. Major acted as a unifier within the party and led the Conservatives to their fourth election victory in 1992. The privatisation programme was continued under his government: the remainder of BT's shares were sold in the summer of 1993, followed a year later by a traded sale of British Coal, a further privatisation of the energy industry (involving National Power, PowerGen, the National Grid and British Energy), and eventually, just before the 1997 elections, British Rail (after a major break-up of the company into more than thirty separate companies). Other candidates for privatisation included the London Underground, Royal Mail and the national Air Traffic Control System. Between 1979–96, the Conservative government orchestrated a total of fifty-nine major public sales, raising £65bn, and eighty-eight private sales bringing in an additional £6.7bn (PI 1997).

One effect of market liberalisation was the sharp increase in foreign investment (especially in consumer electronics and car manufacturing), which led to a boost in imports and net job losses. The plan of the Thatcher government to restructure the economy was based on two partly conflicting ideas: encouraging the services sector and reversing the decline of Britain's technological leadership. To emphasise the importance of innovation, the Thatcher government appointed a special Minister for Information Technology to superintend schemes that would support R&D in micro-electronics, fibre optics, information technology and space in order to bring the UK back to the forefront. With the 1980 Industry Act, the NEB was transformed from a state holding to a facilitator of private sector investments in the 'high-tech' industries. As a consequence NEB was instructed to dispose both of most of its stakes in profitable industries (Ferranti) and of its holdings in the 'lame ducks' Rolls Royce and British Leyland. The slimmed down NEB merged with the National Research and Development Council (NRDC), a self-financing public body fostering scientific contacts and contracts between industry and the universities, to form the British Technology Group (BTG). The new conglomerate was entrusted with the overall task of promoting and supporting commercially viable innovations.

In general, privatisation frees the corporation from bureaucratic scrutiny with regard to its operational and strategic activities, replacing it with a new regime in which the corporation's management is accountable to its customers and shareholders (of whom the government may still remain a very important one). However, one could have one's doubts about the large privatisation programme enacted by the Thatcher and Major governments. First, there is the argument that in some cases 'selling the family silver', as former Prime Minister Macmillan (quoted in Chapman 1990:10) saw it, was not necessary. For example the privatised companies now performing very well under private ownership, were already economically successful under public ownership. The Conservative governments have followed a 'picking the winner' strategy in their selling-off of (potentially) profitable state enterprises. British Rail and British Coal, both poor performers, were not included in the original privatisation programme. Second, a shift from a statutory public monopoly to privately owned utilities did not automatically guarantee more competition. To bring about genuine competition, new rules and adequate regulatory controls had to be established to compensate for the loss of direct government control. As the persistently dominant position of British Airways, the electricity generating and distribution companies, and British Telecom in their respective markets illustrates, the short-term profits of getting these institutions out of the public sector, turning the large-scale and complex flotation into a success and hence generating proceeds for the Treasury, seemed to be more important than the long-term aim of increasing competition and efficiency (Richardson 1994). With the exception of the National Bus Company, British Rail and the Central Electricity Board, the utilities were transferred to the private sector without altering their corporate structure (e.g. no divestitures) and new entry and competition were restricted to some fringe market segments. Because the industries to be floated were not broken up into competitive parts (as had happened in the USA), the government only had a few instruments to ensure a minimum level of control over these strategic industries e.g. licensing requirements and 'golden shares' in the company (Graham & Prosser 1991). A government's golden share usually contains certain exclusive provisions in the company's articles of association and a right of veto to preclude any changes. For instance, it could impose restrictions on foreign ownership, limitations on the voting shares, government-appointed directors, special shares preventing hostile take-overs, British nationality of the CEO, disposal of assets, restrictions on the issue of new voting shares (Hyman 1989).

To improve customer orientation and consumer choice and to increase efficiency, the public administration was reorganised in the 1980s and early 1990s (Hogwood 1994). Policy formation and implementation were separated by the creation of small core policy departments and executive service delivery agencies with substantial managerial autonomy. The responsibility for determining and enforcing the delivery of public services to agreed targets was delegated within a regulatory framework (e.g. the implementation of the Citizen's Charter). Market testing was introduced for all executive agencies to identify whether any potential savings could be obtained. In order to create and police a more competitive environment, a new regulatory framework was established. Independent regulatory agencies were set up, such as Oftel (Office of Telecommunications), Ofgas (Office of Gas Supply), Ofwat (Office of Water) and Offer (Office of Electricity Regulation), whose responsibility it was to secure fair competition among market players and to protect consumer interests by controlling prices and ensuring a certain level of quality was maintained. These single industry agencies have the power to lay down and enforce specific terms in the utility's licence, but they have no powers with respect to the licensing itself or general competition legislation. In case of detected unfair competitive practices and/or abuse of monopoly, the regulatory agency has to refer these practices to the Office of Fair Trading (OFT), which has the power to refer to the Monopolies and Mergers Commission (MMC). Initially, these regulatory agencies were poorly equipped in terms of expertise and staff to control the large privatised agencies, but after some teething problems they have developed a certain responsiveness to complaints about poor quality of service and unfair competition.

Some of the newly created agencies have established themselves as strong autonomous powers in the supervision and regulation of the privatised utilities and the promotion of industry-wide competition (Graham & Prosser 1991; Veljanovski 1991a, b; NCC 1993). Ofwat and Ofgas, for instance, were faced with public criticisms of unreasonably high profits, poor performance and large increases in the basic salaries and fringe benefits (e.g. share options) of chief executives and directors (Baldwin 1995; Flemming 1996). At the same time, employment levels in gas were reduced from approximately 100,000 to 50,000, and public health was put at risk when water companies refused to invest in improving water quality and ignored environmental regulations. Oftel and Ofgas were challenged by continuous technological developments and persistent market dominance by the incumbents blocking the introduction of workable competition. After coming into power in 1997, the Blair government has announced a thorough review of the functioning of the aforementioned independent regulators and the balance between the interests of shareholders and customers. The objective of this exercise is to investigate whether there should be any major structural changes in the regulatory system, and how the co-ordination with the MMC and OFT can be improved. Renationalisation of the public utilities is not on the agenda, but the Blair government proposed the introduction of the 'windfall profit tax' and a new regulatory framework based on a combination of price regulation and 'profit ceilings' where excessive returns are shared between shareholders and citizens.

4.4 Demonopolisation of telecommunications in the UK

The established telecommunications regime

Until the early 1980s, the Post Office had exclusive control over the provision of postal, giro and telecommunications services. Traditionally the Post Office was a civil service department with a workforce of about 400,000. Before 1969 it was headed by a Minister, the Postmaster General, who was responsible to Parliament. The Post Office was politically accountable in every aspect of the company's practices, operations and planning. In practice this meant that its investment schemes, industrial relations and tariffing were subject to detailed political supervision by Cabinet and Parliament. The Department of Trade and Industry (DTI) was the major regulator and sponsor of the communications sector in Britain and the Department of Prices and Consumer Protection (later incorporated in the Department of Industry) was in charge of monitoring and controlling the prices set by the Post Office. The Treasury determined the Post Office's budget in competition with the financial requirements of the other government departments. The investment schemes of the Post Office were constrained by the Public Sector Borrowing Requirement and the External Financing Limit, linked to the overall level of proposed public expenditure. A statutory body existed alongside each national industry with the function of safeguarding the interests of consumers, namely the Post Office User's National Council (POUNC) (Prosser 1986).

Another major player in UK telecommunications is Cable & Wireless (C&W), officially established as a merger between Eastern Telegraph and Marconi in 1934. Originally, C&W was one of the largest international telecommunications enterprises in the world, running telegraph services between the various parts of the British empire and the Commonwealth. After the phasing out of telegraphy in the 1960s, C&W moved into intercontinental cable and satellite links and became a leading operator in Britain's overseas territories with a strong base in Hong Kong and the Caribbean. C&W was nationalised by the Labour government in 1947 and privatised by the Conservatives in 1981. C&W can be considered as the opposite of the Post Office/British Telecom: it was involved in overseas activities and – although in public ownership between 1947–81 – never subjected to public scrutiny and political intervention. After its privatisation in 1981, C&W became actively involved in the home market through its subsidiary, Mercury.

The unionisation rate of the Post Office/British Telecom's workforce, like the other nationalised industries and public utilities in Britain, has always been high. In 1984 98 per cent of British Telecom's staff was unionised: the Post Office Engineering Union (POEU) (engineers) represented about 131,000 members, the

Society of Telecom Executives (STE) (managerial and professional staff) about 33,000, the Union of Communications Workers (UCW) (rank and file grades) 39,000, the Civil and Public Servants Association (CPSA) 32,000 and the Communications Manager's Associations (CMA) (managerial staff in operating and ancillary services) was relatively small, representing 5,500 BT personnel (De Zoete & Bevan 1984). In 1984/85, the POEU incorporated the clerical workers and became the NCU. In the period before privatisation and liberalisation, the industrial relations system of the Post Office/British Telecom was characterised by a high degree of co-operation with the unions keeping a nearly 'clean record' as far as industrial action was concerned (Smith & Terry 1993). There were elaborate consultative and consensus-seeking procedures, dealing with collective bargaining, the introduction of new technology and the marketing of new products and services. The Post Office/British Telecom unions had long-established and close links with the TUC, in which they participated, and with the Labour party, for whom they sponsored three MPs.

BT purchased about 90–95 per cent of its switching and peripheral equipment from the ring of British manufacturers (POEU 1964). The monopsonistic equipment market was made up of three major manufacturers, GEC Telecom, Plessey and STC, which had established cartel-like arrangements, dividing BT's bulk contracts among themselves and joint standardisation with the Post Office/British Telecom. In equipment manufacturing the UK's record was mixed, with an excellent position in R&D and a rather mediocre performance in production, sales and exports. The interests of telecommunications users were represented by four organisations: large businesses by the Telecommunications Managers Association (TMA), smaller businesses by the Telecommunications Users Association (TUA) and residential consumers by the Consumers' Associations (CA) and the National Consumer Council (NCC). The TMA and TUA were especially influential in public policy making, successfully persuading the government to embark upon a more commercial and competitive approach in the telecommunications domain and urging BT to improve its quality of service by providing its high-capacity and customised business services at more competitive prices (Cawson *et al.* 1990:357).

In the 1960s, the Post Office's management clearly wanted to be put at arm's length from the government administration and, consequently, to become a separate nationalised industry. The idea to convert the Post Office into two or three nationalised industries and divide it into separate public corporations for the postal, giro and telecommunications services, was regarded as too radical. The government, strongly supported by the large UPW union, opted in favour of maintaining one unified departmental organisation. In 1967, the government finally decided that the Post Office must be managed along (more) commercial lines if it was to run efficiently. It suggested the creation of a single statutory corporation headed by a single executive board in which the telecommunications and postal businesses could operate relatively autonomously. At the Post Office's central and regional headquarters, the two functions were placed under separate managing directors, each entrusted with their own personnel, finance and technological support.

The enactment of the Post Office Act in 1969 transformed the Post Office from a government department into a public corporation with statutory powers and duties, subject to the financial controls of a nationalised industry. The Post Office was released from direct political accountability by the division of the function of the Postmaster General between the sponsoring Minister of Post and Telecommunications (in 1974 replaced by the Secretary of State for Industry), and the chairman of the Post Office Board. The responsibility for personnel was transferred from the Civil Service Commission to the new public corporation. The Post Office was established as a self-financing public service with more discretion in its corporate investment and, consequently, external borrowing on the stock market was allowed (although Treasury approval was still required). It was believed that the new Post Office's corporation would establish more market-based relationships with its three preferred suppliers, henceforward increasing the efficiency, dynamism and performance of the British telecommunications industry (Pitt 1980; Hills 1984). A new body was set up, called the Post Office Users' National Council (replacing the Post Office Users' Council of 1966), to represent the interests of customers. The 1969 Act stipulated that the POUNC members were appointed by government and that the council had to be consulted whenever important decisions had to be taken by the Post Office Board.

Although the new framework gave the Post Office more autonomy in its corporate functioning, the persisting problems of capital shortage, over-centralisation, slow decision making and poor service provision were inadequately addressed. While allowing Post Office management a measure of commercial freedom, the government would ultimately determine which strategy the Post Office should follow. In 1970, the Tory government put forward some sketchy plans to liberalise monopolistic markets and privatise the Post Office. Minister Chataway announced the opening up of some parts of the posts and telephone monopoly to private competition and the denationalisation of some parts of the Post Office organisation. The then Chairman of the Post Office, Lord Hall, totally disagreed with the Minister's point of view, favouring an opposite strategy of enlarging the Post Office's control on telecommunications by diversifying into manufacturing. The Post Office seriously considered the alternative of acquiring a (public) stake in the privately owned equipment industry (notably by taking over STC). In its plea for entry into equipment manufacturing the Post Office was supported by the largest telecommunications union, the Post Office Engineering Union (POEU), which suggested the Post Office be allowed to set up its own production unit and enter the exchange market in competition with private suppliers. At that time the two contradictory proposals of liberalisation and vertical integration were regarded by the policy community as too controversial. The political disagreement led to a stalemate in policy making with the Minister's plans postponed and the eventual dismissal of the Postmaster General.

As a consequence of the 1972–4 price restraint policy of the Heath Government, the Post Office had slipped from a relatively comfortable position to a large financial deficit. It was constrained by excessive government intervention in its plan to modernise the telecommunications network and increase the productivity of its services. The decision of the Chancellor in December 1973 to impose an immediate 20 per cent cut in the 1974–5 investment programme without consulting the Post Office, adversely affected the company's financial performance and staff morale. The statutory price restraint policy forced the Post Office to keep its prices down, even to loss-making levels. To restore the Post Office to profitability, a return to more commercial pricing was implemented in 1975. Corby (1979:242) characterised the government's behaviour towards the Post Office as a 'penny wise pound foolish' approach: price rises and cuts in service provision were allowed or blocked according to the political needs of the moment rather than customer demand. Besides political interference and shortage of capital, other causes for the poor performance of the Post Office and its telecommunications subsidiary were the highly bureaucratic and centralised practices of its management and fundamental weaknesses in marketing. The Post Office's organisational structure was inadequately differentiated to perform its tasks with respect to the provision of efficient and effective postal and telecommunications services (Pitt 1980).

In 1976, the Labour government appointed a Committee of Wise Men to review the internal structure of the Post Office and the institutional framework established by the 1969 Act and to make recommendations on these aspects. This Review Committee, also known as the Carter Committee, consisted of eight members, drawn from the Civil Service, trade and industry and the labour unions. The Carter Committee (1977) made it clear that the Post Office suffered from several organisational weaknesses and managerial shortcomings. It regarded the relationship between the Post Office and the government as too close, chaotic and lacking in consensus on strategic objectives and long-term planning. The Post Office's investments, wages and tariffs were subordinated to the macro-economic requirements set by the government. The Post Office was further hampered in the pursuit of its social and commercial objectives by detailed political and bureaucratic meddling. A related weakness of the Post Office's structure was the lack of delegation, limited commercial freedom of local management and a low market responsiveness, that led to over-centralisation, slow decision making and risk-averse management.

The Carter Committee recommended that the Post Office be divided into two separate corporations, and – where possible – power be delegated to local operating units. The Committee suggested that these new companies be entrusted with the 'public' objective of breaking even and minimising costs rather than the 'private' objective of maximising profitability. According to the Committee, the differences between telecommunications and postal services had become too large and too fundamental for them to be marketed effectively by one integrated PTT administration: the former was capital-intensive with expanding volumes of supply and demand and the latter was labour-intensive with static and saturated economic characteristics. The postal and giro services on the one hand and the telecommunications and dataprocessing service on the other, needed to be separated into two autonomous state-owned statutory corporations. The Carter Committee furthermore proposed a Council on Post Office and Telecommunications Affairs, that would fulfil an advisory function to the Post Office Board. The Committee believed that such a 'neo-corporatist' Council, in which government, trade unions, consumer organisations, industry and independent experts were represented, could fulfil a consultative role with respect to the examination of the corporate plans, strategies and capital investment programmes of the two businesses as well as any matters of national concern. The Council would also be responsible for the representation of consumers. At that time the consumer watchdog POUNC had been ineffective, showing major shortcomings in the representation of interests with no formal political powers attached to its recommendations. The Committee argued that the integration of POUNC into the Advisory Council would strengthen the position of consumers. Additionally, the Committee proposed certain relaxation of the restrictions on the supply of PABXs and other peripheral equipment.

The Post Office Board (PO 1977) observed a close commonality between the proposals of the Carter Committee and views within its own ranks. The Post Office agreed with the suggestion of dividing the telecommunications and postal services into separate authorities. However, it was against the installation of an advisory Council on Post and Telecommunications Affairs because this would touch upon its corporate autonomy and would further complicate its already complex and overcentralised decision-making process. Instead, the Post Office suggested widening Board membership to include employee and consumer participation. The Post Office Board was susceptible to the Committee's plan to improve its management style and corporate structure by delegating decision making to decentralised business functions and promoting experiments with industrial democracy at the corporation's local and regional levels. To prevent the government from severe capital cutbacks and tariff changes and to allow for appropriate infrastructural planning, the Post Office demanded that financial targets be set for the mid-term. The Post Office strongly opposed the Committee's sketchy proposal to liberalise some parts of the terminal equipment market. In its view the interests of all users and the safety and compatibility of its network would be best protected by all apparatus being supplied, installed and maintained exclusively by the Post Office (excluding answering machines, modems and dataterminals).

The POEU (1977), representing mainly engineering and clerical servants in the telecommunications branch, supported the Carter proposal of a functional separation between the telecommunications and the postal business. With respect to industrial relations the POEU argued for greater union participation and representation in the Post Office's corporate planning. Such an involvement would stimulate the evolution of an official tripartite planning system, involving government, management and the unions. The other unions (UPW, now UCW, and POMA, now CMA) were in favour of keeping the postal and telecommunications functions integrated within the Post Office (Carter Committee 1977b; NCU 1992a). Their arguments against a separation of the Post Office referred to the extra costs and complications associated with dividing R&D, property, personnel, industrial relations systems, transport equipment and supportive business services, among autonomous telecommunications and postal entities. All the unions were against any liberalisation of equipment supply, because this would raise problems of systems compatibility and lead to an attempt by private enterprise to cream off the profitable areas of the Post Office's business. The POEU, supported by its union partners and to some extent by the Post Office itself, promoted the idea of the Post Office acquiring a substantial manufacturing capability of its own (e.g. by acquiring at least one of its manufacturers). The Post Office's main suppliers, organised in the TEMA association, were against the idea of the Post Office diversifying into manufacturing, because the industry's production was already sufficient. The equipment manufacturers, basically proposed a prolongation of the traditional close ties with the Post Office and the national government (Carter Committee Appendix 1977).

The then Labour government deferred judgement in its response to the Carter recommendations, partly as a consequence of the strong division within the labour unions on the separation issue. After more than a year of extensive consultations with the major stakeholders, the Labour government responded to the Carter Report with the publication of the 1978 White Paper. At first the government reaffirmed that the postal and telecommunications services would remain under the statutory monopoly and that a future Post Office would remain in public ownership (DoI 1978). The government accepted the Committee's proposals aimed at increasing the powers of POUNC and introducing more effective monitoring and performance standards. It rejected, however, the alternative of a final division within the Post Office's structure between the postal and telecommunications businesses and the installation of a new Consultative Council. The Labour government suggested postponing a decision on separation because of the Post Office's two-year experiment with industrial democracy that had already started and the ongoing process of decentralising management (DoI 1978:5).

For the Labour government the pilot scheme within the Post Office corporation was a test case to find out whether the existing system of industrial relations in state-owned corporations could be reformed and made more co-operative by the implementation of various bipartite and tripartite measures. The experiment included an extension of membership of the Post Office Board to include trade union representatives, a senior civil servant from the Department of Industry and experts familiar with consumer affairs, followed by an expansion of the participation arrangements for workers at local and regional levels of the corporation. The government was against the creation of a new Council on Post and Telecommunications Affairs: this would introduce an unnecessary level of decision making and 'detract from the role of the Board and blur responsibility' (DoI 1978:7). The government did not react to the suggestion made by the Carter Committee to liberalise the market for peripherals.

From public monopoly to managed competition (1979–91)

In their 1979 election manifesto the Conservatives identified the powerful role of the state as the major cause of Britain's economic malaise. After its 1979 election victory the Thatcher government at first developed a policy of running nationalised industries more efficiently and reducing the state's holding in nationalised industries by selling off profitable state-owned companies (e.g. Cable & Wireless).

At that time (1979–81) the alternative of privatising and deregulating public utilities was not yet acknowledged. With respect to telecommunications, the aim of the Thatcher government was to hive off the telecommunications function from the Post Office and to introduce competition to make British Telecom (BT), the Post Office subsidiary and the domestic equipment industry more efficient. The liberalisation measures were instigated by BT's outmoded infrastructure and poor record in innovation, service delivery and overall performance (Wood 1994) and the domestic electronics industry's rapidly decreasing world market share (Britain's share fell from 25 per cent in 1969 to 5 per cent in 1979) (Solomon 1986).

The Department of Trade and Industry proved susceptible to the arguments for liberalising terminal equipment supply and 'non-voice markets'. Large international telecommunications users, dependent upon reliable and specialised facilities, strongly criticised the Post Office/BT's sluggish quality of service and the poor performance of its network. They lodged complaints about the Post Office's inability to meet business demand: bureaucratic maintenance and installation procedures, time-consuming approval of equipment, major shortcomings in its commercial and marketing capabilities and the absence of quantity discounts for its bulk users (Carter Committee Appendix 1977:395–426). At the end of the 1970s, the City of London saw its position as one of the few global financial centres deteriorating as a result of relatively expensive trading costs (Hall 1987). In order to prevent business draining away from London, the City commenced improving the efficiency of the stock market by modernising its communications and computer systems. However, the City's direct demand for fast, reliable and cheap datacommunications facilities and a first-class communications infrastructure could not be immediately met by BT/the Post Office. In September 1979, Sir Keith Joseph, the Secretary of State for Industry, announced a review of the public network monopoly and promised he would consider liberalising the provision of terminal equipment and data services. These plans were strongly opposed by the public sector unions. Almost a year later in July 1980, Joseph announced to Parliament the government's intention of relaxing BT's monopoly and restructuring the corporation. The Post Office would be divided into two separate public corporations (i.e. Royal Mail and British Telecom) and the markets for terminal equipment and value-added network services (VANS) would be liberalised. The government's intention to increase domestic competition and improve the choice and services available to customers, were laid down in the Telecommunications Bill, published in November 1980.

With respect to the liberalisation of value-added services the government commissioned Professor Michael Beesley, then Professor of Economics at the London Business School, to undertake a study into the economic implications of allowing complete freedom to offer services to third parties over BT's network. In January 1981, Beesley (1981a) reported to the Secretary of State for Industry that the liberalisation of the use of the national public network would actively stimulate innovation and expansion in the market for value-added network services (VANS) If the resale of BT's capacity by its customers were to be allowed, benefits would most likely outweigh the projected loss of revenue by BT. Competition based on reselling BT's leased circuits would facilitate new entries and fairness in the market place and force BT to bring its prices closer to costs. Beesley argued that there should be no restrictions on the freedom of suppliers to use BT's infrastructure in providing telecommunications services to third parties at a flat rate. As a compensation for its losses, BT should be free to engage in competition in the non-voice markets via a separate accounting subsidiary to compete with these VANS suppliers.

BT (1981) vehemently opposed Beesley's suggestion to resell capacity, because of the foreseen losses of net revenue which would result from tariff rebalancing and cream skimming (traffic diverted to the leased lines). In the end this would trouble BT's financial position, hampering its investments and capacity to cross-subsidise services. BT, defending the long-standing monopoly tradition of economies of scale, universal supply and uniform tariffing, suggested a modest liberalisation measure, limited to true value-added services and excluding the resale of leased lines. BT argued that it needed more time to prepare itself for market competition, than the immediate implementation of the liberalisation of network capacity, as suggested by Beesley. BT stated that it needed at least five years to modernise its network and adapt its tariff structures and quality of service to the changing market environment.

The resulting British Telecommunications Act, receiving Royal Assent in October 1981, formalised the separation between the telephone services and the postal services by establishing British Telecom as a public corporation separate from the Post Office. The statutory monopoly was replaced by an exclusive privilege for BT to run the UK network. The Secretary of State for Industry was required by the terms of the Act to consult with BT before licensing any new entrant. The 1981 Act granted the government powers to license competitors in the provision of telecommunications facilities. The government announced that three licences for operating and exploiting a public network would be issued: two nationwide licences to British Telecom and Mercury Communications and one local franchise to the City of Kingston upon Hull. The right to compete with BT in the market of long-distance services was given to the Mercury consortium, which was granted licences between 1982–4 to build and operate an alternative high-capacity trunk network parallel to BT's network, linking the major cities and business centres in the UK, and provide international services. BT sought to preempt Mercury's plans for a parallel network by sharply reducing its tariffs for long-distance calls between major cities. Through the creation of a competitor, the government intended to reform BT and make it more efficient and answerable to its customers. Another reason the government had for developing an effective duopoly was to forestall the potential powers of the unions by having an alternative network in case of an all-out strike (Harper 1995, 1997). The Thatcher government followed a so-called duopoly policy, whereby no fixed-link operators would be licensed before 1990 other than British Telecom and Mercury.

The 1981 Act brought to an end BT's monopoly on terminal equipment supply. The first telephone set, however, remained the prerogative of BT. Also the liberalisation of the market for value-added network services was implemented by the government. Beesley's recommendation to use the national network to provide services to third parties was not followed by the government. The simple resale of leased circuits was regarded as too radical and was rejected. To prevent operators of managed data services from bypassing BT's network, simple resale would not be allowed before 1989. The government argued that BT needed sufficient time to rebalance its tariffs and bring prices more in line with costs. Network competition was allowed: the Secretary of State was given powers to assign licences to run telecommunications systems to (potential) competitors of BT. Furthermore, the Act installed an independent structure to set standards and approve equipment, including new responsibilities for the British Standard Institute (BSI), the Secretary of State and the newly established British Approval Board for Telecommunications (BABT). The creation of this BSI/BABT regime was regarded as essential if private suppliers were to compete effectively with BT on equal terms. In the 1982 White Paper (DoI 1982b) the government announced its plans to privatise BT and reform the licensing arrangements in order to curtail BT's exclusive privileges. It also proposed the establishment of a regulatory agency, to be modelled after the Office of Fair Trading, that would ensure fair competition and quality of service. This Office of Telecommunications (Oftel) would have to balance the interests of service providers, customers, equipment suppliers and other stakeholders.

It was only after the summer of 1981 that the privatisation of BT (and other public utilities) was discovered as the way to increase the economic performance and customer service of these companies and to curtail public expenditure. A large majority of the chairmen of the nationalised industries demanded more money to finance their major modernisation programmes and the sums required exceeded the strict cash limits imposed by the Treasury. In the mid-1970s, BT management was confronted with the problem of financing the heavy investments required to upgrade its infrastructure and its service provision level. The government and BT considered the possibility of raising funds outside the PSBR by giving BT direct access to the private capital market. In 1981, BT management was allowed to finance its major investment programme through the introduction of government-backed 'Buzby-bonds' that would raise a possible £100–200 million on the stock market. This operation, however, did not turn out to be a success, lacking sufficient support from the Treasury and institutional investors. The Treasury still considered the alternative of seeking private sector capital to be made subject to tight governmental controls. Potential investors regarded the Buzby-operation as a high-risk enterprise with restricted profitability, because external borrowing would be limited and tariffs would be kept low. The Thatcher government realised that with regard to the financing of nationalised industries, some crucial decisions had to be made if public expenditure was to be controlled.

The alternative of a large-scale privatisation of public utilities was considered a rather easy solution to reduce public expenditure (not touching upon or even reducing the PSBR) and to rely upon the capital market for the financing of these industries. For BT, privatisation would imply that it was to be converted into a Companies Act company and that a majority of the state assets would be sold on the stock market. In his Parliamentary Statement of July 1982 the Secretary of

State for Industry strongly promoted the government's policy, to be based on privatising and liberalising telecommunications:

> It's the Government's aim to promote consumer choice. Wherever possible, we want industrial and commercial decisions to be determined by the market and not by the State. We believe consumer choice and the disciplines of the market lead to more stable prices, improved efficiency and a higher quality of service.
> (DoI 1982b:1)

The Thatcher government argued that privatisation would also give BT access to the capital market as a way of generating the necessary funds for the modernisation and expansion of its network.

The original suggestion to divest BT was withdrawn by the Thatcher government, partly because of the strong pleas of BT's management and its unions to keep the privatised company integrated. The government had become convinced about the political and financial risks of a break-up of BT and decided to preserve it as a single unit (Wood 1994). A far-reaching reform of the industry by a corporate divestiture would complicate the whole operation of denationalising BT, confronting the government with the problem of finding buyers for BT's unprofitable parts and reducing the proceeds of the sale substantially (Harper 1997). In order to make a speedy flotation possible, however, the government relied upon the support of BT management and financial investors (Kay 1984; Vickers & Yarrow 1986; Hawkings 1987). For that purpose BT's figures and balance sheet were given a special treatment: accounting procedures were restated, asset lives shortened, depreciation schemes revised and cost structures reallocated. The attitude of BT management towards privatisation was conditionally supportive: it wanted to keep the corporation integrated, to be allowed to expand into new markets and it preferred a business environment that would be only moderately liberalised and regulated. BT's Corporate Affairs Department successfully lobbied in Standing Committees, the Commons and the DTI in favour of abandoning organisational divestiture and keeping BT as an integrated unit (Newman 1986). In its claim for remaining integrated, BT management was supported by the unions' association BTUC (1983a), which had commissioned a study of the US experiences following the break-up of AT&T. This study concluded that the American strategy was not applicable in the UK. BT was in a significantly weaker position than AT&T which would hamper its capacity to cope effectively with increasing competition and a probable divestiture would also diminish BT's ability to compete in world markets.

The Labour party, supported by the unions, strongly opposed the neo-liberal policy measures of the Thatcher government. Labour followed a straightforward strategy of proposing to keep or restore the state-owned monopoly of BT on the basis of social fairness, national safety, industrial/economic policy and efficiency concerns. The decision to privatise BT at first caused an enormous upheaval among BT's workforce and its unions, leading to one of the most widespread campaigns against the privatisation programme of the Conservative government. A majority of BT's employees was involved in the industrial action campaign of 1982–83 ('Day of Action') to oppose the privatisation proposals and the permission given to Mercury to interconnect its future networks with the BT system. The unions, organised in the BTUC (1983b, 1984), declared that 'selling a public service for private profit' would have the following major consequences:

- socially desirable but uneconomic services provided by BT would be threatened;
- working conditions within BT would deteriorate and there would probably be huge job losses;
- BT's 'buy British' policy would be abandoned, putting domestic suppliers and jobs at risk.

The action campaign of the unions in 1982–83, however, proved ineffective. The biggest union (the NCU) was almost destroyed financially by the consequences of the engineers' strike, industrial relations within BT became more confrontational and the re-elected Thatcher government went ahead with the preparations for floating BT. The original goal, set by the Labour party and the unions, of delaying the 1981–82 legislation, meant to be passed before the first term of office of the Thatcher government ended, was nonetheless accomplished. In its 1983 manifesto the Labour party vehemently went against the free market policies in the telecommunications domain, proposed by the Thatcher government, and suggested a prolongation of a publicly owned infrastructure. Such a national broadband system under firm public control could be accomplished by the integration of cable and the telecommunications networks of BT and Mercury. The unions and the Labour party, however, could not counteract the privatisation and liberalisation proposals of the Tory government *per se*. A clear election victory for the Thatcher government in 1983 led to a reintroduction of the 1981 Bill to Parliament and, after the Telecommunications Bill was passed in December 1983, the campaign to oppose the reform of UK telecommunications was phased out.

Although some of the criticisms launched by the unions were incorporated, the contents of the 1981 Bill had not been amended substantially. The explicit incorporation of consumer interest in the new institutional structure was one of the minor successes of the unions' campaign. In the original 1981 Bill, the government had proposed to abolish the consumer watchdog POUNC without an alternative to further and protect consumer interests. In the 1983 Bill, however, the government recognised the notion of consumer concern and proposed replacing POUNC by Oftel, designed to oversee and regulate the industry. After the decisions to privatise BT and to float it as an integrated unit were taken, the unions targeted their industrial actions against the envisaged duopoly and the projected parallel network of Mercury. They argued that Mercury would take profitable activities away from BT, which would jeopardise the public service character of telecommunications and threaten job security. In September 1983, BT's engineers and their unions refused to allow Mercury to interconnect with the public network. Mercury brought an injunction against the POEU, claiming that this action was against the law. The union's refusal to interconnect proved fruitless, when Mercury was vindicated in its claim by the High Court's ruling of November 1983. In the face of a legal attack for further damages and legal costs the POEU had to comply with the ruling on interconnection (Moon *et al.* 1986).

In October 1982 Professor Stephen Littlechild, then Professor of Commerce at the University of Birmingham, was asked to study and evaluate proposals to regulate BT's profitability and the liberalised telecommunications sector. The regulatory regime to be implemented had to meet the following criteria: to prevent BT from abusing its dominant position, to minimise the regulatory burden, to provide incentives for efficiency and innovation, to ensure the maximisation of net proceeds from the sale, and to facilitate a successful operation of BT as a commercial enterprise after flotation. In his report, published in February 1983, Littlechild (1983) investigated five regulatory alternatives:

1 no explicit regulatory constraints;
2 the maximum rate of return scheme;
3 the output related profits levy option;
4 profit ceiling regulation; and
5 the local tariff reduction scheme.

After evaluating and comparing the various alternatives, Littlechild expressed a clear preference for the local tariff reduction scheme. This would both protect domestic and small business customers and encourage new entry and competition. In other words, BT should be regulated by a system of price control in which BT's tariff increases were pinned down below the inflation rate. The government accepted the Littlechild recommendation for price-cap regulation and incorporated it in the new regulatory framework.

After the 1983 election, the government reintroduced a similar Telecommunications Bill, herewith reaffirming its commitment to the liberalisation and privatisation of telecommunications in Britain. The Secretary of State for Information Technology, Sir Kenneth Baker, confirmed that by the end of 1984 BT's monopoly of supply and maintenance of the prime telephone set would end. The Secretary of State for Industry would be given the power to transfer all property rights and liabilities of British Telecom to a future privatised company; he would also become the ultimate licensing authority for telecommunications activities. With the new Telecommunications Act finally receiving Royal Assent in April 1984, a new regulatory structure was set up, BT's exclusive privilege to run telecommunications systems was abolished and the privatisation of British Telecom was prepared. In the 1984 Act a regulatory framework was established based on a system of licences for all telecommunications operators. BT was placed on an equal footing with its potential competitors in having to obtain a number of licences for each of its operations, granted by the Secretary of State for Industry. The most important was the twenty-five-year master licence, that came into effect in June 1984, to run a public telecommunications system. BT had some obligations under the licence, such as the provision of universal service, call boxes, emergency services and services to the disadvantaged, and it had to give rebates to low users. A renewable twenty-five-year licence was also given to Mercury to install and operate an alternative national telecommunications system. Mercury did not have to fulfil any universal service obligations; it was free to provide all types of fixed-link services.

In the new situation, BT and Mercury would have a duopoly with regard to the provision of basic national and international services. It was decided not to license other network operators until 1990. The protection of BT and Mercury was limited to the core telephone services and the provision of existing data services on the BT network and would last until 1990. The principal reason behind this duopoly policy was to protect the revenue base of BT and Mercury in particular, from the potential competition that was emerging from third service operators (cable and VAN operators). Originally the Mercury consortium consisted of Cable & Wireless (C&W), British Petroleum (both 40 per cent) and Barclays Bank (20 per cent): the first provided engineering capabilities, the second project management expertise and the third financial expertise. Then, after the management buy-out of 1984, Mercury became a 100 per cent subsidiary of C&W. Following its launch in the City of London in April 1983, Mercury decided to concentrate almost exclusively on the provision of long-distance communications with special services to corporate clients between the large business centres of the UK. Mercury's licence also permitted it to interconnect with the British Telecom network and enter the profitable market for international services.

Through the 1984 Act BT was transferred to the private sector as a public limited company and consequently its flotation was prepared. Public and political support for the privatisation of BT was shaky in June 1983, so the Thatcher government attempted to persuade potential financial investors by making BT attractive as a solid and highly profitable investment that would certainly give value for money (De Zoete & Bevan 1984; Scrimgeour Kemp-Gee & Co. 1984). BT was entrusted with considerable funds for new business ventures and acquisitions and given sufficient time to build up an enterprise culture, improve its efficiency, upgrade its network and experiment with a new regulatory regime. The government had insisted upon a light regulatory regime (with price-cap regulation instead of profit regulation) and a transitional period of modest competition from small operators and service providers, while prohibiting the simple resale of private circuits. Probably the most successful instrument in turning the privatisation into a political success was the 'sweetener' of underpricing: a low share price would certainly persuade sceptical financial investors from the City to participate in the programme and maximise the number of shareholders in privatised businesses (Hawkings 1987). As a consequence, the flotation was a success: the largest share issue at that time raised almost £4bn, increasing the number of shareholders to 2.3m in 1984 (the number of shareholders fell to 1.3m in 1988). On the day of the flotation, the value of BT's underpriced shares went up by 85 per cent, offering the new shareholders substantial windfall profits. Although the unions were against share ownership by BT employees, the large majority of BT's workforce became a shareholder in the company: 96 per cent accepted the free share offer, about 80 per cent bought a matching share offer and a third participated in a company share-save scheme (Smith & Terry 1993).

Until recently, the Labour party and the unions have argued in favour of restoring the public monopoly on telecommunications (Labour Party 1983; NCU 1985; POEU 1985; STE 1986). They supported the return of BT to public ownership with defined social obligations for the national infrastructure and the relocation of BT and the Post Office into one publicly owned corporation. Labour and the unions argued furthermore for the simultaneous incorporation of the 'cream skimmer' Mercury into BT. Re-establishing public ownership could be accomplished by transforming the existing BT shares into non-voting shares and buying them back at the original issue price, while purchasing sufficient shares to give the government a majority. Labour and the unions argued that renationalisation would ensure public telecommunications services and facilitate a broad national strategy to construct one national broadband network. The availability of a national champion in the domestic industry would enable the government to achieve industrial policy objectives (e.g. innovation, employment, training and education).

Over the years, Labour and the unions have become more modest and have adopted a more pragmatic line towards share ownership, profit-sharing schemes for personnel and redundancy settlements. They have narrowed their focus, concentrating on the interests of workers and ordinary consumers by raising issues like job protection, training, universal service and improving regulation. Labour and the unions have started to recognise the market mechanism as the dominant force in the allocation of economic activities. A set of provocative proposals launched by Garnham (1990) in a discussion paper published by Labour's Fabian Society, aimed at a major restructuring of UK telecommunications, went largely unnoticed. Garnham recommended the creation of ten regional operating companies with a monopoly of local cable operation and the separation of customer service delivery from network provision; BT's remainder would be free to operate at a global level. Within Labour's ranks it has also been suggested that the UK network should be physically reunified and placed under the control of a reduced BT (Harper 1993, 1997). In this new setting BT would shed all its customer retail functions and become a regulated wholesaler of network capacity.

From managed to open competition?

In November 1990, after the seven-year commitment of the government not to license fixed-link operators other than BT and Mercury, the government officially started the Duopoly Review through the publication of the Consultative Document (DTI 1990). The government was inclined to terminate the duopoly policy and to encourage comments and recommendations on the shift from managed competition to open competition. This so-called 'Green Paper' suggested opening up markets further by licensing more players: class licences for self-provided circuits and two-way satellite services would be granted to new operators and mobile operators would be allowed to run fixed telecommunications services. The government adopted a position of technological neutrality, arguing that 'each technological solution should compete for its place in the market against all others' (DTI 1990:23). However, in order to make the policy of extending competition effective, the government had to prevent BT (and to some extent Mercury) from taking advantage of its market hegemony. Therefore the government suggested banning the designated public operators from providing mobile or cable services. The application of this so-called 'asymmetry rule' effectively prevented BT and Mercury/C&W from entering the Personal Communications Networks (PCN) and home entertainment services market for a ten-year period, while at the same time allowing new entrants to establish themselves in these markets. The government was furthermore inclined to consider applications for new licences to provide a full range of international services. The government also consulted the major stakeholders on specific policy issues, like number portability, volume discounts to certain customers by BT, increasing transparency and disclosure of financial information by the regulated industries.

Oftel (1991a) endorsed the broad outlines of the 1990 Green Paper, seeing them as further moves away from an administered market towards more general and effective competition. With respect to the promotion of international communications the position of Oftel (1991) was complicated and its recommendation contingent on the liberalisation strategies of other countries. It supported the need to restrict competition in the provision of international services, because a go-it-alone free trade strategy would be detrimental to UK interests, should other states fail to follow. BT's reaction to the government's intention to abandon the duopoly and to increase competition was ambiguous (BT 1990). BT responded positively to the notion of more domestic competition by not objecting to the liberalisation of private circuits, national fixed-link services, datacasting and satellite communications. However, BT disagreed with the government's asymmetry rules, excluding it from entering promising markets in order to protect new entrants from BT's dominant position. The involvement of BT and Mercury in cellular services was restricted to offering these services through autonomous subsidiaries only. Furthermore, the government barred BT and Mercury from providing entertainment services, while CATV companies were given local monopolies in the provision of cable television and were at the same time allowed to offer telecommunications services. According to BT, the creation of a genuinely competitive regime in UK telecommunications also meant allowing BT to participate in promising markets and technologies like broadband cable and radio:

> We therefore welcome further competition, but in an open market in which BT is allowed to compete on an equal basis with other companies. BT, naturally, accepts appropriate competition law rules governing market dominance; but the present regulatory rules go beyond that in a way that restricts BT's ability to respond to market needs in a commercial way.
>
> (BT 1990:1)

BT suggested that asymmetric regulation between BT/Mercury and the cable companies could be removed by the end of 1993. A three-year protection period would be adequate, taking into account that these cable operators collaborated closely with large foreign companies. While foreign telephone companies like US West, Bell South, Nynex, Bell Canada, Singapore Telecom were acquiring large stakes in British cable companies, BT was still excluded from the provision of entertainment and radio-based services and furthermore subject to universal service obligations (from which all the other national and foreign operators active in the domestic market were exempt). With respect to the reform proposals for international services, BT (1991c) fully accepted the need to reduce international call charges, but was very concerned about opening up the UK market unilaterally. Except in reciprocal circumstances, BT had strong reservations about unrestricted international competition, because this would not only damage the domestic activities and overseas aspirations of BT and Mercury, but also have a negative influence on the balance of payments.

Mercury (1991) supported the broad outlines of the government's Green Paper. It backed the government's plans to encourage the establishment of alternative local networks by removing the restrictions on cable operators to provide voice telephony. With respect to trunk services Mercury welcomed a modest increase of competition: it suggested that new carriers should be examined very carefully before being granted a licence to offer long-distance services. Mercury agreed with the government that a general distinction ought to be made in telecommunications regulation between dominant and non-dominant operators: the former would be prevented from providing radio-based and cable services and the latter would have the freedom to select and exploit the most profitable market segments such as fixed or radio-based services. Mercury, however, fully disagreed with the government's idea to classify both BT and Mercury as dominant operators and treat them similarly. For Mercury, the term 'dominant operator' applied exclusively to BT; in its view 'non-dominant operator' referred to all present and future operators, including Mercury itself, that were seeking to challenge BT's powerful position. Mercury argued that like any other non-dominant operator, it should be allowed to provide radio-based services by itself and construct a local network for the conveyance of both entertainment and telecommunications services. Like BT, Mercury was highly suspicious of the government's suggestion of licensing new operators for the provision of international services. Mercury believed that the existing level of competition in the international market was more than sufficient. A unilateral liberalisation of international communications could severely damage Mercury's economic position when foreign-based operators were allowed to take advantage of the open UK market, without British operators being accorded equal treatment abroad (Mercury 1991).

The user community supported the government proposals to license additional local services and remove all restrictions on satellite services, cable systems carrying telephony, and self-provided private circuits. TMA (1991:9), representing large business customers, expressed its support for the government's intention to open the market to new entrants, arguing that 'further competition is needed to achieve the objective of ensuring that users' needs are met by a wide range of services at prices that do not leave room for inefficiencies or excessive profits.' While supporting the government's suggestion to lift the ban on mobile operators providing fixed services, TMA went further by arguing for the removal of restrictions on the provision of mobile service providers by fixed service operators. TMA advocated the promotion of further competition in international services and suggested that BT should be allowed to offer selective volume discounts to large customers. To facilitate effective competition, TMA suggested that certain regulatory conditions had to be met, such as ensuring equal access to all trunk networks and full interconnectivity between all the infrastructures, establishing network-independent numbering, reviewing the price-control system, advancing more effective quality controls and disclosure of strategic information by the incumbent operator.

TUA (1990, 1991), representing residential users and small businesses, argued that the duopoly policy did not go far enough and had failed to bring effective competition to the market place. BT still had a virtual monopoly on the local loop as Mercury was not seeking national competition with BT and building up a large customer base, but instead looked for profitable niche markets. Although it did not rule out the possibility of establishing competition through the licensing of a third national operator, TUA preferred the active encouragement of entry into local loop provision. Regional fixed and mobile operators would use new and more cost-effective technologies, like cable tv, microwave, satellite links, which would be attractive to business users as a complementary or alternative infrastructure for both national and international services. TUA drew attention to the promotion and guarantee of equal access between network providers and the introduction of flexible pricing formulae, that would allow for a choice between higher rental and lower call charges, and lower rental and higher call charges packages.

Both the Consumer Association (CA 1991) and the National Consumer Council (NCC 1991) argued that the extension of competition in telecommunications was in the interest of residential consumers because choice would increase and prices would be significantly lower. The introduction of competition gave consumers clear benefits with regard to the supply of peripheral equipment, value-added services, trunk and international calls. The CA and NCC generally speaking agreed with the government's principle of technological neutrality and its consequent intention to liberalise the 'local loop' by introducing competition between the fixed infrastructure and radio technologies. While establishing effective competition, strong regulatory measures were still required to protect the interests of residential consumers *vis-à-vis* dominant suppliers and business users. The NCC suggested strengthening price controls on connection charges, rentals, local calls and enquiries and argued for more openness in the regulatory regime and for a fuller disclosure of information by the regulated parties. The two consumer organisations stated furthermore that the representation of (residential) consumers in the UK telecommunications framework needed to be strengthened. Oftel intended to combine the role of regulating the industry with that of representing the interests of consumers. The CA (1991) and the NCC (1991) made it clear that a regulatory body and a consumer representation agency were not identical and argued for the setting up of a properly equipped national telecommunications consumer council, that would act (semi-)independently from Oftel. This separately organised consumers' structure would be in charge of consumer advocacy and general policy representation of consumers' views.

The telecommunications unions, organised in the BTUC (1990, 1991), expressed a critical opinion of the Green Paper and the following White Paper. BTUC argued for a better balance between the interest of British companies, large users and average consumers. It set forth three objectives for telecommunications policy that were insufficiently met in the Duopoly Review: securing high-quality telecommunications for the maximum number of people at the lowest cost, proving the best infrastructure for British business, and ensuring the economic success of British telecommunications. BTUC criticised the neglect of the interests of the established British telecommunications companies in the national and international economy. Further evolution of liberalisation would pose a serious blow to the established industrial framework. The increase of network and services competition, the entry of new specialised operators, a possible divestiture of BT and tightening regulatory constraints, would especially harm the financial position of BT, its network and quality of service. This free-for-all competition would challenge the profitability of the fibre-optic networks of both BT and C&W/Mercury and would instigate a shift in these companies to short-term commercial returns and hereby sacrifice the longer-term goals of infrastructure modernisation, training, R&D and quality of service. Therefore BTUC asked to secure the threatened position of BT by giving it more freedom to enter the cable market and more pricing flexibility.

The level of competition in UK telecommunications was regarded by the unions as sufficient. No further licences for fixed networks were required: BT's single national grid would ensure that the universal service and social obligations would be met. Another related element of criticism referred to the underestimation of the rapid globalisation of the telecommunications industry and its consequences for the key British companies. BTUC stated that the government promoted the interests of major foreign companies at the expense of British business. BT and Mercury/C&W were prevented from entering the markets of PCN and home entertainment services for a fixed period while new entrants, most of them American companies, were given the opportunity to establish themselves in the local communications market. BTUC suggested that the government needed to further both the national and international interests of British business by abolishing asymmetric regulation in the cable and mobile markets and demanding reciprocity of market access at the international level. BTUC also observed deficiencies in the existing regulatory structure and in Oftel's functioning, lacking sufficient accountability to Parliament, the Department of Trade and Industry and the general public. The Association of the Electronics, Telecommunications and Business Equipment Industries (EEA) by and large spelled out the same arguments in support of the *status quo* as BT and the unions. Endorsing the main outlines of the Green Paper, the EEA argued that the position of both BT and Mercury/C&W needed to be protected and strengthened. Consequently, it suggested that duopoly on long-distance services should be maintained for at least another five years. The duopolists should be able to introduce and operate two-way broadband facilities.

The following White Paper contained some of the resolutions of the government in the light of the observations from the various stakeholders in the telecommunications domain (DTI 1991). It suggested a termination of the duopoly policy in favour of a competitive market environment, in which only minimal restrictions were imposed. This shift from managed competition to open competition would increase consumer choice, promote free enterprise and bring down prices. The 1991 White Paper suggested the opening up of the telecommunications market to new entrants and the progressive introduction of equal access and a further streamlining of the interconnection between the various networks. On the issue of the new entry into international service provision, the government reversed its position from the one put forward in the 1990 Green Paper. Despite its intention to terminate BT and Mercury's exclusive rights in international services, the government proved susceptible to the risk of following a go-it-alone strategy of free-for-all competition in the still strongly cartelised market for international communications. Instead the government followed a different path: it decided not to issue new licences in the short term and to put a regulatory restraint on BT's market power in international communications. From 1991 onwards, international services would be included in the price control agreements. The government proposed to facilitate competition between the telecommunications and the cable infrastructures. In the new situation, BT and Mercury were prohibited to provide entertainment services in their own right for at least ten years and cable operators were allowed to provide voice telephony without intervention from BT or Mercury. Furthermore, mobile operators were allowed to operate fixed telecommunications services using their radio networks. The government, however, kept the restriction on fixed operators with regards to the provision of mobile or multi-point services under their main licences. The proposals of the White Paper were finalised in September 1991.

After the adoption of the 1991 White Paper, the British telecommunications industry moved away from managed competition through a duopoly structure with restricted entry to an open market environment under detailed regulatory oversight. Officially there were no limits to the level of domestic competition: anyone with sufficient resources could at any time apply with a fair chance of success for a restricted number of licences. After 1991 BT could expect increasing competition in the local loop from cable operators and PCN operators and in the long-distance market from cable companies, public utilities and other new operators. The new regulatory system could be regarded as a revision of the duopoly regime, in which incumbent operators and new market players were sheltered in order to prepare themselves for competition, by shifting the protection to a new generation of operators and keeping the already established parties out of particular markets.

In May 1994 the Major government announced in a consultative document that new and high-capacity computer and telecommunications infrastructures would be utilised in improving and developing digital-based public services (CCTA 1994). Examples of applications with a widespread coverage to be distributed over these *information superhighways* could include electronic messaging between government agencies and citizens, information access, electronic trading, tele-reservation, distance learning etc. In a follow-up policy document, published in November 1994, the DTI (1994) argued that the UK needed a more coherent policy on national broadband infrastructure and had to develop new ways of providing services to citizens and linking public institutions with each other. The 1991 framework, which banned BT from using its network to carry entertainment services, would not be touched upon until the late 1990s. The government did consider, however, a progressive lifting of the ban over an eight-year period on the grounds that otherwise the UK would lag behind the US and Japan in the development of interactive superhighway services.

Since the liberalisation of international simple resale in 1994, companies such as AT&T and Sprint are permitted to lease lines from BT and C&W and international calls along the entire route can be handled by a single company. In December 1996, the UK liberalised international facilities a year in advance of the EU deadline. Office of Telecommunications (Oftel) and the Department of Trade and Industry (DTI) abandoned BT and Mercury's duopoly on international voice traffic by issuing forty-four licences to domestic and foreign carriers (e.g. Vodafone, Telia, Telstra). The approval of international competition allows competitors to build and operate their own networks rather than rely on leasing lines from BT and C&W. The government's decision to sanction international competition is an effort to establish the UK as Europe's intercontinental hub to route traffic to the USA.

4.5 The liberalisation of the UK telecommunications market

The years between 1981 and 1996 saw a progressive liberalisation of the various telecommunications markets. At that time the UK was still the only country in Europe without a state monopoly on basic transport services, but instead it had a duopoly shared between BT and Mercury. The markets for terminal equipment and value-added network services were almost fully liberalised with hardly any entry barriers. Since 1981 the provision, installation and maintenance of all terminal equipment is open to competition. BT's monopoly on the supply and maintenance of the first telephone connected to the network was removed in December 1984. Before the standard-setting institutions BABT/BSI were effectively institutionalised, type approval and testing remained BT's prerogative for a transitional period. This led to strong complaints of selective treatment and abuse of power by BT: rival suppliers were confronted with lengthy and excessive delays in the testing and approval of their equipment.

In 1982 Mercury received a twenty-five-year licence to run long-distance communications systems (reconfirmed under the new 1984 legislation). In 1985, Mercury installed a fully operational digital system, made up of fibre-optic cables and microwave, linking the major cities of the UK. The establishment of a second network operator was opposed by BT, who claimed that this would result in wasteful duplication and cream-skimming of its most profitable markets. At first Mercury was seriously impeded by the absence of any statutory right to interconnect its alternative network to the BT infrastructure. Instead Mercury had to rely upon bilateral negotiations with BT management over the terms of interconnection. A provisional agreement between the two operators was reached in 1982. Initially, BT engineers refused to link Mercury to the public infrastructure. In November 1983 the non-collaborative action on interconnectivity by the BT engineers and the POEU collapsed in the face of a court injunction. Finally, in 1985, Mercury received Oftel's support on its claim for full access to the BT network through Oftel's determination (Oftel 1985b). This decision was aimed at stimulating effective penetration by Mercury and had been based on the principle of 'any to any' (any customer of one network can access any customer of the other), freedom of choice (customers free to choose in routing messages) and 'equal charges for equal access' or cost-based interconnection charges. The introduction of Mercury was also slowed down by local authorities, supporting the Labour party and the public sector unions. Mercury successfully diversified into new businesses: in 1987 it was allowed to launch international services (relying upon the overseas links of its parent C&W), public call box services, followed in 1989 by a licence to offer personal communications services (in partnership with US West). Although BT's overall presence in the domestic telephone market is still dominant (about 90 per cent), Mercury has become successful in the business market. It has concentrated its marketing efforts on the most lucrative markets in and around the City of London, where it has gained a substantial market share of more than 25 per cent.

The market for value-added networks was opened in 1982 when the Department of Trade and Industry issued licences for value-added network services, group use of leased lines and the running of closed-user networks, eventually followed by a new class licence for value-added and data services in 1987. The Department of Trade and Industry licensed several hundred suppliers, who started to provide a wide range of specialised tele-services. In 1988, six licences were issued to provide specialised satellite services: British Aerospace, BSB, Maxwell, Uplink, SIS and the US-based company EDS (a subsidiary of GM). One year later, the UK government was the first in Europe to allow the simple resale of network capacity and leased lines from the designated operators in domestic telecommunications to third parties. The DTI issued licences to private network operators such as British Rail, Racal and Istel (in 1989 acquired by AT&T) to establish virtual network services. These liberalisation measures stimulated the utilisation of BT's network by other firms to provide value-added services.

In 1991 the duopoly expired and the government conducted a review of its telecommunications policy. The outcome of this research was the promotion of further liberalisation in the local and trunk services. The new policies favoured virtually free entry to the domestic telecommunications market, including cable, satellite communications and resale of capacity. Utility companies were allowed to utilise their corporate high capacity fixed networks for the provision of trunk network services. Since 1992 the DTI has granted licences to several new public telecommunications operators. Among others, Scottish Hydro-Electric, Scottish Power, Energis (a subsidiary of the National Grid) and Torch Ltd (a joint venture of Yorkshire Electricity and Kingston upon Hull Communications) were licensed to provide nationwide services. Ionica (owned by Yorkshire Electricity and Telecom Finland) was allowed to build radio-based networks for the local loop. WorldCom and Esprit Telecom were authorised to install terrestrial or fixed telephone links throughout the UK and to provide international services over links leased from other operators. Two fibre-optic networks – MFS and COLT – which are modelled on competitive access providers (CAPs) in the US, now operate in the City of London. The licensing of foreign entrants, such as the US-based MFS, Telia from Sweden, Telstra of Australia and the US-owned COLT, further challenged the privileged position of BT and Mercury in the City market for domestic and international business information services. Having already established full competition in equipment, mobile services, data services, network provision and voice telephony, international voice telephony was opened up to competition in summer 1996 by the issuing of forty-four licences to companies including Vodafone, Orange, Telia, Telstra, etc.

In its Cable Programme, launched in 1982–83, the Thatcher government sought to expand the installation of broadband systems in order to pull through advanced tele-services and to promote the domestic cable industry (HO & DoI 1983). The government made clear that cable television systems and a future nationwide broadband infrastructure, as an alternative local network to BT's network, would increase the possibilities of competition in the provision of communications services and equipment. It attempted to balance free entry to private companies to develop cable systems and allow only a restricted involvement of Mercury and BT in that market. BT and Mercury would not be given the exclusive right to run cable systems, nor would their participation in every cable company be mandatory. So cable operators were allowed to provide broadcasting and tele-services, but voice telephony and data services remained BT's and Mercury's exclusive responsibility. In order to protect their financial base, the two nationwide public operators would retain their exclusive right both to link local systems and to provide voice telephony on local cable systems. BT and Mercury were allowed to supply cable services, only through a subsidiary with a separate accounting system and in collaboration with a licensed cable company.

The government's strategy towards cable development and the concomitant revitalisation of the electronics industry was based on a market-led approach, financed by private investment and with a minimum of government intervention. The two sponsoring ministries, the Home Office and the Department of Industry (HO & DoI 1983) attempted to find a balance between deregulation and the existence of local monopolies through proposing 'regulation with a light touch', in which the licensing of cable services providers and the supervision of the industry was to be administered by a new independent regulatory body, called the Cable Authority. The Cable Programme was finalised with the coming into force of the Cable and Broadcasting Act of 1984. This deregulated the cable sector and actually established the Cable Authority as the body responsible for the promotion of cable, licensing, standardisation and compatibility matters and general oversight of the industry. In 1991, on the basis of the new Broadcasting Act, the Cable

Authority was integrated, together with the Independent Broadcasting Authority, in the newly established Independent Television Commission (ITC).

Before 1990 only a small number of cable franchises were assigned to CATV operators. Although the results of the pilot experiments were disappointing and the extent of cabling was still only limited, the government gave support to cable development, because it offered the prospect of rivalry between BT and Mercury as local network operators and cable operators in the near future. The decision taken in 1990 to permit cable companies to offer voice telephony services in a limited geographical area and the possibility that these cable networks could become interconnected to create a national or regional parallel network, was directly aimed at challenging BT's dominant position in the entire telecommunications market in the mid-term (Cornford & Gillespie 1993).

From 1991 onwards, the number of cable franchises granted by Oftel and ITC increased significantly: more than 150 CATV companies have been licensed to provide broadcast services and telephone services. As a consequence, North American cable and telecommunications companies, such as TeleWest (US West) and Nynex, have shown a big interest in developing the UK cable market with domestic partners, thereby stimulating service innovation, investment in new networks and price reductions in local access markets. Cable subscribers are being offered a full range of TV channels, telephone and value-added services including ISDN and rapid access to the Internet. In 1995 customers of the cable TV operators received discounts of up to 15 per cent on telephone services. This prompted BT to reduce its prices and introduce new services (such as video-on-demand using existing telephone lines) to counter the threat to its market share by cable TV companies offering lower prices. In June 1995, BT announced a reduction in call charges of £310m, the equivalent to 5 per cent on the average phone bill. A year later, BT was losing a net 50,000 customers a month attracted by the low prices for telephone calls offered by cable operators.

Since 1996 a process of concentration has heralded the return to duopoly or oligopoly in UK telecommunications markets. In October 1996, C&W announced the formation of Cable and Wireless Communications (CWC), a major new national telecommunications company designed to obtain economies of scale and scope by operating across local, long-distance and international markets. CWC was formed out of a merger of Mercury, its UK long-distance subsidiary, and three cable TV companies (Nynex CableComms, Bell CableMedia and Videotron). CWC will offer many new services such as asynchronous transfer mode (ATM) and existing services, such as integrated services digital network (ISDN), from a single source. However, Mercury's One-2-One GSM service, provided in partnership with US West, is not yet included in the deal. The emergence of CWC as a major operator able to challenge BT's dominant position in the UK has encouraged other UK long-distance operators (e.g. Energis and Racal-BR Telecommunications) and cable TV companies to enter into partnerships to offer a similar portfolio of services.

In the mobile services market the government also followed a duopoly strategy, allowing two competing players to develop a specific market exclusively and to recover the investments by prohibiting new entry for a certain period. In May 1983 two licences for the operation and exploitation of cellular radio systems were issued to Vodafone (owned by Racal and Millicom) and Cellnet (owned by Securicor and BT). The two licensed mobile operators were not allowed to provide value-added services or enter manufacturing and were required to operate their cellular activities at arm's length from their parent companies. Cellnet and Vodafone were granted a twenty-five-year licence to operate a cellular radio network (including GSM provision) with full interconnection rights to the networks of BT and Mercury. The cellular market really came into effect in 1985 when the services were actually launched, followed by a fierce competition between the two systems with Vodafone now having a slightly bigger market share and higher revenues than Cellnet. The competition both in retailing services and between network duopolists resulted in an intensive rivalry and a rapid market expansion (Müller & Toker 1994). In 1991, three licences for personal communications networks (PCN) were issued. After a consolidation phase, two groups remained and started to roll out and operate their network in 1993–4: Microtel (Hutchison and British Aerospace), later renamed Orange, and Mercury One-2-One (C&W and US West). The expectation of PCN technology was that it would not only stimulate competition with Cellnet and Vodafone, but would also promote competition in the mid-term with the fixed network services, in particular the provision of the local loop to the final customer (with BT and the cable operators as major rivals). Cellular prices have fallen dramatically and the number of cellular subscribers has increased significantly over the years, as a result of intense competition in the UK cellular market.

The equipment market in the UK was also opened up to competition. The closed telephone ring of suppliers was replaced by a cross-border merger, a foreign take-over and new entrants. These far-reaching developments indicate that the British equipment manufacturing industry was in a deep crisis, in which domestic development, production and sales were more and more replaced by competitive tendering, including new entry and imports (Morgan *et al.* 1989). There has always existed a large degree of mutual adjustment between BT and its preferred suppliers, but their relationships became particularly close when in 1977 the Post Office, GEC, Plessey and STC decided to collaborate on the development of System X. BT, irritated by the disappointing results and delays in the project caused by software problems, expressed its desire to reform the collaborative arrangements by transferring all the System X activities to a single company. Eventually, after STC backed out, a compromise was found in which Plessey and GEC agreed that Plessey would be the leading contractor and GEC the subcontractor.

At the same time, BT considered offering a part of its switching business to an international tender. In March 1985, BT finally abandoned System X and purchased the rivalling System Y from the new supplier, Thorn Ericsson. Besides the security of delivery, the major reason behind BT's strategy of second-sourcing was to put the System X manufacturers GEC and Plessey under competitive pressure to accelerate the upgrading of the network and cut costs in order to respond effectively to Mercury's threat. GEC and Plessey argued that by choosing a second telephone switch, reliant upon Swedish technology and components manufactured in the UK, BT was abusing its power and reducing the domestic and international market prospects for British switching technology. The decision to use System Y next to System X would, according to GEC and Plessey, clearly damage the interests of the national equipment manufacturing industry and would lead to a large number of job losses (Cawson *et al.* 1990:99–100). The regulatory body Oftel (1985a) responded favourably to BT's argument that second-sourcing would encourage competition in the domestic equipment manufacturing industry. It suggested a compromise in which BT was not allowed to purchase System Y for a transitional period of three years, thus giving the System X manufacturers GEC and Plessey sufficient time to adjust to a more competitive market and to redevelop their productive process. On the one hand, Oftel to some extent acknowledged the protectionist arguments of the UK telecommunications manufacturing industry, but it was on the other hand critical about its longer-term prospects in a globalising business environment: raise the exports of System X or collaborate with foreign partners.

BT's acquisition of a 51 per cent share in Mitel, the Canadian PABX manufacturer, was originally intended as a means to build a powerful presence in the North American market. BT argued that majority control of Mitel was vital to its corporate strategy of becoming a major equipment manufacturer. However, in its growing ambition to become an international vertically integrated company, BT glossed over the domestic implications of the link up. In the UK market, where Mitel had a 15–20 per cent share, the deal posed a considerable threat to the established producers. Notably, GEC and Plessey strongly opposed the BT-Mitel venture, claiming it was a severe distortion of competition in the light of BT's dominant position in purchasing and supplying equipment and exclusively stipulating network interface standards. The Secretary of State referred the case to the Monopolies and Mergers Commission (MMC). The MMC (1986) was highly critical of BT's take-over of Mitel, referring to BT strengthening its market power and distorting competition between suppliers in the domestic PABX market. The acquisition could nevertheless be allowed, provided some restrictive conditions were met: BT and Mitel had to keep their business activities separate and the use and supply of Mitel equipment in the UK should be limited. To some extent guided by the optimistic prospects of a successful entrance of BT into the American market, the Secretary of State in 1986 decided to endorse the MMC recommendations and allowed the Mitel take-over on the condition that a separate subsidiary was created for Mitel and its share in the UK market was limited to 25 per cent. BT's emphasis on vertical integration, however, proved misplaced: in 1990 BT sold its stake in Mitel, then a loss-making equipment manufacturer, for a quarter of the original price.

The process of concentration in UK telecommunications manufacturing continued in December 1985, when GEC made a bid to acquire Plessey. The envisaged merger was opposed by the MMC because it would be against the public interest and national defence interests. The MMC proved susceptible to the Ministry of Defence's argument, that a GEC-Plessey merger would seriously reduce competition in the supply of defence electronic equipment (Morgan *et al.* 1989). In 1987, GEC and Plessey suddenly announced that they would merge their manufacturing activities to achieve the economies of scale required for international competition. This joint venture, called GEC Plessey Telecommunications (GPT), was established in April 1988. Two years later, GEC, in close collaboration with Siemens, succeeded in gaining almost exclusive control over Plessey through a full take-over. Today 40 per cent of GPT is owned by Siemens of Germany.

The evolution of BT's third preferred supplier STC is somehow cumbersome: STC was excluded by BT from the digital exchange market in favour of Ericsson, which was allowed to enter the market for digital public switches. In 1984 STC linked up with ICL, in 1990 it was incorporated within the merger of ICL and Fujitsu and in 1991 was acquired by the Canadian/US manufacturer Northern Telecom (NT). In the early 1990s, two policy experts associated with the Labour Party, Harper (1993) and Garnham (1990) suggested that BT should seriously consider buying out Siemens in GPT in order to keep up the UK's competitive advantage in the entire telecommunications industry, including both services/network provision and equipment manufacturing.

4.6 The privatisation of British Telecom

In August 1984, BT was converted from a statutory corporation into a public limited company. BT was duly privatised in November of that year: its flotation put 50.2 per cent of the corporation's shares in 'public' hands, raising £3.9bn. BT's shares were highly undervalued, as the offer was 3.2 times oversubscribed. In addition to 49.8 per cent of the issued ordinary shares, the government possessed a special 'golden' share and preference shares. Although the government did not intend to use its right as a major shareholder to intervene in the management decisions of BT, the potential influence of the government upon some aspects was nevertheless substantial. For example the rights of the special share allowed the government to block changes in certain Articles of the company, especially in limiting large shareholdings to 15 per cent, ensuring the British nationality of the CEO, the creation of new classes of voting shares and the appointment of two non-executive directors to the BT Board. It was clear that the government's golden share protected BT from the threat of a hostile take-over (De Zoete & Bevan 1985).

The privatisation of BT put the government in a difficult position, because on the one hand it was ideologically in favour of increased competition and more efficiency in British telecommunications while on the other hand relying on BT for information and co-operation to ensure a successful sale of the majority of BT's shares (Veljanovski 1987). The critics argued that if BT was sold as one corporate entity the government was simply replacing a public monopoly with a private monopoly. In 1982–3, the government put forward the plan to float BT as a single entity rather than splitting it up into smaller independent units. This was an implicit recognition that BT would dominate the British market for years. BT's future role lay in becoming the flagship of Britain's IT industry. To achieve this, a strong and integrated BT was essential. In 1991, another 28.4 per cent of BT's shares were sold on the stock market, raising £5bn. In the summer of 1993 BT's privatisation continued when the government's remaining share of 21.8 per cent was sold on the stock exchange. BT's flotation was accomplished when the UK government sold its golden share in BT to smoothen the full merger with the US firm MCI (which eventually came to nothing).

Although the government still gave clear priority to BT at the expense of promoting market forces, it nevertheless expressed some concern with regard to encouraging new entry and establishing effective competition in telecommunications. In 1984, for example, the government rejected a far-reaching joint venture of BT and IBM to provide managed data services. At the basis of the refusal lay a mixture of competition policy and industrial policy: whether to see a powerful and integrated BT as a platform to develop an internationally successful electronic and IT industry or to use competition in network provision and apparatus supply to serve the interests of domestic users and the information services industry (Garnham 1985). Competing VAN operators believed that the BT-IBM alliance, called JOVE, would seriously threaten the interests of the domestic computer industry, notably ICL, by JOVE's reliance on IBM standards. The competitors feared a 'lock-in' situation that would force them to use IBM's SNA architecture. There was also much concern that a venture between two leading companies in the related telecommunications and computing market would surely dominate the value-added service market. After the blocking of the JOVE venture by the DTI, BT and IBM established a more moderate venture which applied successfully for a licence for the provision of data network management services. The major competitor in the VAN market, ICL, teamed up with GEIS in 1987 to establish International Network Services (INS).

The restructuring of British Telecom has been directed towards stimulating its market responsiveness and economic performance. BT's corporate culture successfully developed from a former state bureaucracy to a market-led, vertically integrated multi-product firm. The centralised functional form was replaced by a more decentralised divisional form, that has moved to neatly defined geographical markets and moreover to BT's various customer groups. The organisational changes included the external recruitment of senior management, the creation of independent and financially separate divisions and a decentralisation of decision making with the establishment of independent profit centres and geographical areas. The process started in 1982 with an internal reorganisation of the company. BT was divided into four divisions: BT Inland, BT International, BT Enterprises, and Development and Procurement, headed by the British Telecom Board. In 1983 the previous Inland division was split into the Local Communications Services (LCS) and National Networks (NN). The former was in charge of operating and exploiting the local telephone network for residential customers, while the latter was responsible for the operation and exploitation of the trunk network and the special business services.

The reason for this separation was twofold: much of BT's manpower was concentrated in the Inland division and BT, challenged by competition from Mercury, was more or less forced to make a distinction between the local loop and trunk communications. The LCS division was the major branch of BT in terms of revenues, assets and manpower (about 80 per cent of the total). Given the claims about lack of efficiency (mainly caused by outdated network equipment and overmanning), financial controls, flexibility and customer responsiveness, a further reorganisation of the LCS was regarded as necessary (De Zoete & Bevan 1984; Scrimgeour *et al.* 1984). In 1984, BT announced that its structure of ten regions and sixty-one areas would be replaced by a new structure of five territorial directorates and thirty-one districts. At local level the old telephone areas were replaced by twenty-seven geographical districts, that would be managed as semi-autonomous profit centres with increased decision-making powers for District Managers.

In 1987, the LCS and NN divisions merged again to create the division UK Communications, operating BT's local and long-distance networks and supplying peripheral equipment. BT International (BTI) was in charge of operating and exploiting international communications and business systems. For BT, as one of the largest carriers of international traffic, this branch was the most profitable and fastest growing part of the corporation. BT Enterprises (BTE), set up in 1981, focused upon newly emerging markets and comprised the following major businesses: Consumer Products (CPD), Merlin (business products and systems), Value-Added Systems and Services (VASS) (Yellow Pages, Spectrum/Dialcom services, Prestel), Mobile Systems and Services (MSS), and BT Vision (visual and broadcast services). The Engineering and Procurement division is responsible for R&D and the purchase of system equipment. Two supportive divisions have to be mentioned: BT International Products, which is responsible for developing, producing and marketing telematic products internationally, and BT Overseas, which sells BT's knowledge and expertise abroad.

In order to reduce organisational complexity and management costs BT was once again restructured in 1990–1. This new 'efficiency drive', named Project Sovereign, aimed at removing management layers and dismantling BT's established structure of relative geographical autonomy. The established divisional pattern, based on responsiveness to local customer needs, was replaced by a new structure, based on two business units. The first would focus on large individual businesses (notably large multinational companies) and the second would deal with residential customers and small and medium-sized businesses. The Business Communications division (e.g. Syncordia, GNS) would handle business customers by providing managed network services and systems integration services for the complex needs of leading-edge customers; in the UK BCD is structured on the basis of five geographical sectors. The Worldwide Networks division (combining BT's UK and International Networks) is responsible for domestic and international network management; in the UK market the division is made up of three geographical areas. The Personal Communications division provides services to residential consumers through nine regional PC Customer Service Zones.

BT has become more efficient with operations continuing to grow and new products and services being marketed successfully. BT has become much more sensitive both to its residential customers and to its business customers, through the implementation of an advanced computerised Customer Services System (CSS) (the 'Front Office'), which integrated customer billing, stores, fault diagnosis and repair, clerical work as well as centralised control over work in a single automated system. In 1989, BT introduced guaranteed contract terms for installing, repairing and maintaining its services and compensating customers financially if these terms were not met. Through this Customer Service Guarantee Scheme, BT (1991a) hoped to improve its relationship with clients by committing itself to keeping up a complete set of service standards and providing guarantee and compensation schemes for poor service. It also introduced a rebate scheme for low users, called Supportline. The combined effects of the introduction and implementation of new technologies (digitisation of the network and automated office systems), tight price controls, fierce competition in both domestic and international communications and the recession that manifested itself in the late 1980s, have fostered a continuous drive within the BT organisation to increase efficiency and keep up profitability levels. These objectives were achieved by cutting labour and overhead costs, short-term contracting and putting out ancillary services like the cabling and maintenance activities. BT's staffing levels were reduced significantly from 253,000 in 1981 to 130,000 in 1996 (BT Annual Report 1996). The temporary increase in manpower levels between 1987–90 was instigated by BT's programme to upgrade its performance through the modernisation of the infrastructure. This involved replacing analogue with digital switches and enhancing the speed and quality of service to its customers. A drastic reduction in staffing levels of almost 100,000 between 1990–4 was the result (see Table 4.1). The cost of financing these massive redundancy schemes, together with tighter price controls and the depreciation of loss-making equipment ventures, led to an incidental drop in BT's profits in 1993.

The effect of privatisation and deregulation on BT's system of industrial relations is rather ambiguous, characterised by drastic pay-level increases for every employee on the one hand and numerous redundancies, a decentralisation of collective bargaining, flexible pay terms and a less consensualist management style on the other. The labour of senior and middle management has been effectively challenged by BT's countermove from standard pay terms to favourable performance-related pay schedules and personal contracts (STE 1989). Since privatisation BT employees have done well, with pay increases well above the rate of inflation: wages have gone up 54.6 per cent for engineers, 66.9 per cent for general management and 344 per cent for the CEO (NCU 1992a). The intention of BT management was to achieve higher levels of productivity and profitability by increasing market shares and trimming operating costs (shedding labour and trimming overheads). The traditional co-operative system of industrial relations was replaced by bilateral and more flexible arrangements at the district and division levels of the corporation. Today, representatives of BT and the unions meet and settle labour agreements in seventeen decentralised units. Centralised pay bargaining still exists, but lower management tiers now handle some labour conditions (grade evaluation, training, hours of service).

Over the years, BT management has become more hostile towards the unions, providing them with less information about investment plans and strategic issues and gradually abandoning existing consultation and arbitration arrangements that had kept the industry virtually strike free. Although BT's strategy with respect to industrial relations was not going as far as the derecognition policy followed by C&W/Mercury and still resembles to some extent the consultative patterns of the past, there have been some attempts to divide, curtail and marginalise the unions (NCU 1992; Smith & Terry 1993). Notwithstanding the various attempts of BT management to split, circumvent and cut back union powers, the unions have remained relatively strong in terms of membership and density (a majority of BT's workforce is still unionised) and have consequently continued to be involved in collective bargaining on pay awards at national level. Traditionally, BT recognised four unions in collective bargaining: the National Communications Union (NCU) (a merger of the Post Office Engineering Union (POEU) and the Civil Public Servants Association (CPSA)), the Union of Communications Workers (UCW), the Society of Telecom Executives (STE) and the Communications Managers' Association (CMA) with the British Telecommunications Unions Committee (BTUC) functioning as the unions' federation. The unions became further isolated by BT management's reluctance to recognise trade unions in most of its new subsidiaries and acquisitions and by the extensive use of short-term contracts and contracting out of labour. The power of the unions was dispersed even further by the 'divide and rule' strategy of BT management, aimed at outplaying the individual unions against one another. It rejected collective negotiations and insisted on splitting collective bargaining into bilateral negotiations with individual unions. BT's union avoidance approach was reinforced by a lack of co-ordination within and between the various recognised unions. The larger NCU was still divided between engineers and clerical workers and major disagreements between the NCU and the medium-sized UCW have prevented effective collaboration and consolidation between the two. Furthermore, BT management has attempted to bypass the established union channels by strengthening direct communications with its personnel. The individualisation of labour relations has manifested itself through the introduction of employee share ownership, team briefings, total quality management and flexible remuneration arrangements with a more personal appraisal and individual rewards.

The year 1987 turned out to be crucial for BT's corporate development for two reasons: the industrial dispute between BT management and its engineers, and the implementation of new digital switching technology and its subsequent teething problems. The already bitter social climate in the aftermath of the 1987 strike worsened even further when BT management refused to pay 'bonuses' on employee shares to its personnel, despite huge profits being announced at the end of the year (BT Annual Report 1987). Together with the disastrous breakdown of the BT network in the City on 23 October 1987, these incidents seriously called into question BT's performance and quality of service. There were severe complaints from the entire user community and Oftel (1987). These pressures triggered BT management to initiate an ambitious campaign to improve services

Table 4.1 British Telecom: key data

BT	*'81*	*'82*	*'83*	*'84*	*'85*	*'86*	*'87*	*'88*	*'89*	*'90*	*'91*	*'92*	*'93*	*'94*	*'95*	*'96*
Workforce (x 1000)	253	252	246	241	235	236	234	237	244	246	227	211	171	156	138	131
Turnover (£b)	4.6	5.8	6.4	6.9	7.7	8.3	9.3	10.2	11.1	12.3	13.2	13.3	13.2	13.7	13.9	14.4
Profit before tax (£b)	0.6	0.9	1.0	1.0	1.8	1.5	2.1	2.3	2.4	2.7	3.1	3.1	2.0	2.8	2.7	3.0
Profits as a percentage of turnover	12.5	16.2	16.1	14.4	23.3	17.8	22.3	22.5	21.6	22.0	23.5	23.3	15.2	20.4	19.4	20.8

Source: BT *Annual Reports* (1981–96)

to customers, including better access to enquiry services and an increase in the number of public telephone boxes. After the network had been upgraded successfully and the industrial dispute had been settled, BT management responded with the implementation of a large-scale Total Quality Management programme. The greater emphasis on productivity and quality was also accomplished through Project Sovereign, designed to make BT's divisions more customer-focused, removing layers of management and reducing its workforce substantially. As a consequence of this almost total restructuring, characterised by a large number of redundancies and major threats to job security, complaints about a lack of control over work and deskilling of labour, and a devolution of concerted industrial action, had a depressing effect on the morale of the BT workforce (Smith & Terry 1993).

After the failure of System X, the newly privatised BT steadily loosened its cosy relationship with the domestic manufacturers GEC, Plessey and STC. BT reduced its dependence on the domestic industry by adding the semi-foreign manufacturer Ericsson/Thorn-EMI to its list of suppliers. With the acquisition of a 51 per cent share in the Canadian PABX manufacturer Mitel Corporation in 1986 BT attempted to establish itself as a supplier of exchange equipment. Further overseas acquisitions included cellular telephony and global value-added networks with the procurement of ITT Dialcom in 1986, the acquisition of a 20 per cent stake in McCaw Cellular (the leading US cellular service provider), and its purchase of the leading US-based value-added service provider Tymnet from McDonnell Douglas. Tymnet was integrated in BT/GNS Global Network Services and was later sold to MCI, as part of the formation of the Concert alliance. BT expanded its service to large users by establishing Atlanta-based Syncordia, which was later also incorporated into the BT/MCI joint venture Concert, providing network management tools for multinational customers. BT has set up partnerships in the major European countries: Telenordia in Sweden (with Telenor and Telecom Danmark), Telfort in the Netherlands (with NS Dutch Railways), Cegetel/BT in France (with CGE-SFR and SNCF, the French Railways company), Albacom in Italy (with Fininvest/Mediaset and Banco Nazionale del Lavoro), VIAG Interkom in Germany, and Airtel (with Airtouch and Banco Santander) in Spain. At the national level, BT participates in the mobile operator Cellnet (a 60 per cent stake) and has aligned itself with the commercial broadcaster BSkyB. BT also holds minority shares in the satellite consortia, Eutelsat, Inmarsat and Intelsat (see Table 4.2).

In 1989 BT reversed its corporate strategy, so far based on diversification and vertical integration, into an approach that concentrated explicitly on the core business of telecommunications. This global networking strategy is designed:

> . . . to provide world-class telecommunications and information products and services, and to develop and exploit our networks, at home and overseas, so that we can: meet the requirements of our customers, sustain growth in the earnings of the group on behalf of our shareholders, and make a fitting contribution to the community in which we conduct our business.
>
> (e.g. BT Annual Report 1994)

Table 4.2 British Telecom: major partnerships

Subsidiaries/participations of BT	*Activity*	*Interest of BT in allotted capital and associated partners (%)*	*Country of registration or location of headquarters*
Albacom	communication related services and products provider	35	Italy
BT Australasia Pty	communication related services and products provider	100	Australia
BT France	communication related services and products provider	100	France
BT Hong Kong Limited	communication related services and products provider	100	Hong Kong (China)
BT North America Inc	communication related services and products provider	100	USA
BT Telecommunicaciones SA	communication related services and products provider	100	Spain
Call Connections Ltd	cellular services provider	60	UK
Cellnet Solutions/Telecom Securicor Cellular Radio Ltd	mobile services operator and provider	60	UK
Concert Communications	telecommunication services and network systems provider	75	USA
Manx Telecom Ltd	telecom services supplier	100	Isle of Man
Syntegra BV	systems integration and application development	100	Netherlands
Westminster Cable Company	cable television service provider	100	UK

Yellow Pages Sales Ltd	Yellow Pages sales contractor	100	UK
Bharti Cellular Ltd	Mobile cellular operator	23	India
Clear Communications Ltd	telecommunications services supplier	25	New Zealand
Telfort BV	communication related services and products provider	50	Netherlands
Telenordia	communication related services and products provider	33	Sweden
VIAG Interkom	communication related services and products provider	100	Germany
Springboard Internet Services Ltd	Internet service provider	50	UK
MCI	large national and international operator	20	USA
Eutelsat	European satellite consortium	18	France
Inmarsat	maritime satellite consortium	10	UK
Intelsat	international satellite consortium	7	USA

Source: BT *Annual Report* (1996)

BT sought to divest itself of its equipment manufacturing activities and other activities, regarding them as loss making or lying outside its core business of global networking. It sold its Consumer Electronics subsidiary to STC/Northern Telecom in 1991 and two years later its 20 per cent share in McCaw to its large competitor AT&T (finalised in 1995), as well as disposing of its interests in the loss-making Mitel subsidiary. Today BT focuses on the major business customers in the three major world telecommunications markets, namely North America, Western Europe and the Asia/Pacific Regions. Important ventures within BT's new strategy of furnishing network services to international business, labelled 'Going Further Staying Closer', are the US-based Syncordia and Global Network Services (GNS) (BT 1991b). The former provides multinational customers with facility management and specialised outsourcing services for their telecommunications networks and the second offers advanced dataprocessing services and value-added networks.

BT's internationalisation strategy showed mixed results: initially, it found a strong North American partner, while acquisitions in Europe were still small, and coverage in the Pacific Rim was weak. As part of its aggressive strategy to become one of the leading telecommunications companies in the world, BT decided early in 1993 to penetrate the large US market by providing intercontinental services. Without collaboration from any of the American operators (AT&T, MCI and Sprint), BT intended to build a virtual private communications network in the US as another step in the construction of a global switched-voice network, connecting all the major business centres across the world. This project, entitled Cyclone, would, together with GNS and Syncordia, complete the portfolio of BT's services and facilities to multinational businesses. BT's ambitious worldwide project proved, however, more complicated than originally foreseen. BT relied on the consent of national governments to liberalise their protected markets, while at the same time facing demands for reciprocity in the opening up of the UK market. The realisation of BT's virtual network in the US ran into problems when BT applied for a US domestic and international licence from the FCC. After being put under strong pressure from AT&T to block BT's application for US operating licences, the FCC delayed the authorisation procedure. AT&T responded by applying reciprocally for a licence in the UK to offer big companies international communications services. BT was more or less forced to revise its ambitious plans to penetrate the highly competitive US market on its own by choosing a more collaborative path and teaming up with a US-based partner.

In the summer of 1993 BT joined forces with the second long-distance and international operator in the US market, MCI. BT obtained a 20 per cent stake in MCI ($4.3b) and jointly they embarked upon a $1b global services venture, known as Concert. This Anglo-American venture offered communications facilities for multinational businesses, developed global products and services, created integrated service management, end-to-end connectivity and established a global managed network platform for advanced business services. The two companies agreed on a geographical division of labour: MCI would become the distributor of all these Concert services in the USA, BT would cover Europe and Asia. After a year of investigations and negotiations the Concert joint venture was approved by the American FCC and the European Commission in June 1994. At more or less the same time, the Anglo-American market access dispute was solved: BT was granted an American international simple resale (ISR) licence, which would enable it to offer services to the UK from the US, while AT&T was offered a similar licence in the UK. In 1997, BT was willing to spend a total amount of almost $23bn for the remainder of the shares in order to have full control over MCI. After complaints about overvaluation from the company's shareholders, BT management announced an expenditure of $19bn to gain full control over MCI. The full acquisition of MCI was intended to take place with a bigger link-up with Spain's Telefonica and Telecom Portugal. The deal did not go ahead, because BT's bid was overshadowed by that of two American companies, GTE and Worldcom (to whom the company was sold in the end). An attempt to merge the United Kingdom's two largest operators, BT and C&W, which would have created the world's third largest operator, valued at £34bn, was abandoned in May 1996 following difficulties in overcoming regulatory, political and business obstacles.

BT's contenders

Until 1990, Mercury Communications Ltd was exclusively licensed to compete with BT. Mercury has become commercially successful in the markets of high-volume customers and cross-border communications. It ignored the local services market of residential users and, instead, exploited the lucrative business and international markets. Mercury followed a marketing strategy aimed at a competitive pricing (10 to 20 per cent below BT), high quality of service and product innovation (e.g. fully digital networks, itemised billing, call routing services). Furthermore, in 1993, Mercury launched PCN services in collaboration with US West. In the same year BCE, the leading Canadian telecommunications company, acquired a 20 per cent stake in Mercury, reducing C&W's stake to 80 per cent. In 1994 Mercury employed about 10,000 people and had a customer base of more than 50,000 corporate and professional customers, mostly located in the larger urbanised areas. In its management philosophy, Mercury strongly emphasised the increase of internal flexibility by the creation of a union-free workforce, quality circles and a cafeteria pay and reward system (Bradley 1992). Like its parent C&W, Mercury followed a derecognition policy towards the unions, setting individual work objectives instead of bargaining collectively. The implementation of performance-related contracts, employee participation schemes, total quality management and worker–management joint decision-making committees were illustrations of the more general trend towards the individualisation of industrial relations. In December 1994, Mercury announced that it was giving up its aspiration to become a local access provider: it discontinued its local loop and pay phone services and narrowed its focus to domestic subscribers, through third parties (i.e. cable operators or BT). As a consequence of this rationalisation measure, a quarter of Mercury's workforce was made redundant.

Mercury is a subsidiary of Cable and Wireless, one of the first state enterprises to be privatised by the Thatcher government in 1981. Three years later it acquired

79 per cent of the shares of the highly profitable Hong Kong Telephone Company (HKT). The company employs about 40,000 people: about 40 per cent of its workforce is Hong Kong based, 25 per cent UK based, and 20 per cent is located in the Caribbean. In 1993, C&W associated itself with Bell Canada through the latter's acquisition of a 20 per cent stake in Mercury for £982m and teamed up with West in the Mercury One-2-One joint venture. C&W also established international partnerships with Bell Canada Enterprises and VEBA (Germany). At the end of 1996, C&W, after having acquired the cable operators Bell Cablemedia, Videotron and Nynex, merged its subsidiaries, Mercury and the cable companies into a new company, C&W Communications.

The corporate strategy of C&W is aimed at strengthening its position as a major force in worldwide telecommunications through the establishment of a federation of telecommunications companies consisting of a global super-carrier and local niche operators. This worldwide strategy is made up of three related objectives: the provision of global end-to-end services for business customers, the expanding of basic telecommunications services (particularly in less developed countries) and the building up of a solid base in mobile communications around the world. Through its existing subsidiaries Mercury and Hong Kong Telecom, C&W already had a strong position in the UK/Europe and the Pacific Rim, but it also sought to penetrate the Japanese and Australian markets. C&W has taken an 18 per cent minority stake in the Japanese IDC company, the licensed international value-added services provider, and committed itself to a 21 per cent investment in the Australian second fixed/GSM operator, Optus. C&W has also built up a base in North America by acquiring the small US domestic and international carrier TRT/FTC in 1991, thereby creating C&W Inc. US. Over the years, C&W has built up a strong presence in developing countries through substantial equity stakes in dominant local carriers in the Middle East, Caribbean and the Asia-Pacific regions. C&W has also become strongly involved in the operation and exploitation of digital mobile services. In addition, C&W penetrated national markets in Europe by acquiring minority stakes in the Swedish mobile service provider Comvik/Tele 2, the German cellular service provider Mannesmann/D 2, and the French mobile operator Bouygues Telecom.

Another powerful player emerging from the liberalised UK telecommunications market is Vodafone. Originally a subsidiary of Racal Electronics, Vodafone launched its cellular system in the UK in 1985, and successfully built up an overseas presence. Vodafone has become a strong niche player in the cellular market as operator, service provider and mobile equipment manufacturer (through its subsidiary Orbitel). In 1993 it demerged from Racal and became fully independent. The company has minority shares in the domestic service providers for cellular networks, Martin Dawes and GMC Talkland, and minority commitments in foreign mobile operators: E-Plus Mobilfunk (Germany), Pacific Link Communications (Hong Kong), Panavox/Panafon (Greece), Vodam/Vodac (South Africa), SFR (France), Nordic Tel (Sweden), Dansk Mobiltelefon (Denmark), and Libertel (the Netherlands).

4.7 Regulatory reform of UK telecommunications policy

In July 1984, BT was finally relieved of its administrative responsibilities and ceased to fall under public law: its former policy supportive functions were transferred to the Department of Trade and Industry and its regulatory tasks to the Office of Telecommunications (Oftel). This body, set up under the 1984 Telecommunications Act, was a non-ministerial government department set up to insulate regulatory activities from short-term political pressures and government interference. Oftel's status as an independent specialised agency acting relatively independently from parliamentary and ministerial controls, allows for informal and secretive negotiations about the enforcement and modification of licences between the regulator and its industry. Oftel's staff are civil servants and its budget is provided by Parliament, but the cost is met almost entirely from licence fees which in the case of the larger franchised operators are broadly related to the size of the turnover of the licensed businesses. Oftel is headed by a Director General of Telecommunications (DGT), who is appointed for a fixed term, is independent from ministerial control and cannot be dismissed under normal circumstances. The DGT has the exclusive responsibility for the enforcement of licences with strong powers to ensure that the conditions of the licences are complied with. He alone can inaugurate an amendment of a licence, either by agreement with the licensee, or, failing to reach such an agreement, by making a reference to the MMC to review the proposed licence amendment. Officially, Oftel is responsible for:

- ensuring that all reasonable demands for telecommunications services throughout the UK are met (this requires the continued provision of emergency services, public call box services, and services in rural areas);
- ensuring that BT and other licensees comply with their licence conditions (i.e. meet statutory obligations and charge a fair price for their services where possible);
- promoting the interests of consumers and purchasers (with regard to price, quality and variety of services offered), efficiency, effective competition, and the development of a technological and economic base for the telecommunications equipment and information services industry (R&D, strengthening the ability of the domestic industry to compete at home and overseas and attracting foreign investment).

In contrast with the American regulatory agencies that are known for their open access conditions, Oftel at first exhibited a non-participatory model, facilitating informal negotiations between regulator and industry. It was not obliged to hold public hearings or be open to public scrutiny (Heald 1988, Graham & Prosser 1991; Veljanovski 1991a). Two other shortcomings have been identified in Oftel's regulatory structure: the near absence of appeal procedures and the concentration of power in the person of the DGT (Foster 1992; Baldwin 1995; Harper 1995). If agreement cannot be reached between the regulator and the regulated industry over changes to a licence or a price control, the issue may be referred to the MMC. This is an unattractive option for both parties involved in a dispute, however, given the fact that cases referred to the MMC are often time consuming and costly. The DGT, appointed by government and supported by a staff and an office, is personally responsible for the regulatory duties, as laid down in the law, to supervise parts of the industry's activities and apply competition legislation to the telecommunications industry.

From the early 1990s onwards, Oftel has acquired more transparency and openness in its decision making. Especially during the Duopoly Review, Oftel's role in the policy process was very active. It organised industry-wide consultations about future policy directions, explained its particular measures and decisions in public and published the relevant information for key regulatory issues and the DGT's advice to the Secretary of State. The small size of the regulator and its structural dependence for strategic information on the large enterprise to be regulated has thwarted its development towards becoming a powerful watchdog. Over the years, however, Oftel has had increasing influence in the shaping of the telecommunications market in Britain. This might be the result of political entrepreneurship from the DGT and the agency itself, but it is also the effect of an interdependent regulatory framework in which Oftel is loosely coupled to other institutions like the Department of Trade and Industry and the generic competition authorities (the OFT and the MMC), which gives the system flexibility and dynamism. With a staff of only approximately 160, Oftel is relatively small compared with the large regulatory agencies in the US. The British government rejected the US model of rate-of-return regulation. Such a regulatory regime, aiming at controlling profits, would lead to overcapitalisation (the gold-plating syndrome) and its effect upon encouraging innovation and competition was regarded as insufficient. Instead the government adopted a regulatory system based on price controls, as proposed by Littlechild (1983), that inserted particular conditions in BT's licence to prevent the prices of BT's monopoly services increasing faster than the annual inflation rate. The government argued that price-cap regulation integrated adequate safeguards to protect customers from the abuse of monopoly power with strong incentives to increase efficiency, tariff rebalancing and effective competition. The 1984 licence required BT to ensure that the charges for a basket of services, namely local and trunk calls, residential and business line rentals, did not rise faster than 3 percentage points less than the annual inflation rate. The RPI-3 regulatory scheme originally covered about 55 per cent of BT's activities. Before 1991 there was no price control on international services, customer premises equipment and value-added services; recently international services have been included in the basket. Later, Oftel set a more stringent 'cap' on BT's regulated prices by changing it to RPI-4.5 in 1989, revising it to RPI-6.25 in 1991 (also including international call prices) and to RPI-7.5 from August 1993 onwards. BT has received more freedom in its pricing policy with the packages concession to offer volume discounts on call charges and alternative tariffs to large users.

Compared to the shortcomings of traditional rate-of-return regulation, this RPI-X system too had its own serious drawbacks. As Garnham (1990:10) has argued: 'While prices are controlled, other parameters move out of control'. As a consequence of this regulatory system BT's profits went up rapidly, allowing BT to channel extra funds into new business ventures and overseas acquisitions instead of investing these enormous revenues in upgrading the domestic network, seriously constraining BT's quality of service at home. In 1992, when the company made huge profits, Oftel told BT to invest in the construction of a universal digital broadband infrastructure to serve the interests of the British economy as a whole. Another unexpected consequence of price-control regulation is that it encourages an organisation to strive for efficiency and phase out cross-subsidies; hence it puts a premium on cutting overhead and labour costs, resulting in a high number of lay-offs. The price-capping system has triggered a process of drastic tariff rebalancing by bringing tariffs more closely in line with costs and shifting charges by cutting prices in those areas where competition between BT and Mercury is present and raising them where exclusive service provision by BT prevailed. Tariff rebalancing has caused sharp price increases for residential customers (e.g. line rentals and connection charges), decreases in residential usage charges, and enormous price cuts for large business users, accomplished through bulk discounts and other tariff packages (OECD 1995, 1997).

Before workable competition could be established in a deregulated and privatised sector, the government was convinced that regulation could prevent the worst excesses of monopoly and protect emerging competition. Oftel became responsible for managing the transition from a monopolistic to a more open environment by encouraging genuine competition, promoting the development of innovative systems and protecting consumers. Through its detailed investigations into the telecommunications industry and its consequent decisions and determinations, Oftel actively promoted the goal of encouraging effective competition and increasing the availability of better and more policy information from and about the regulated parties. In 1984, when BT and IBM jointly applied for a licence to provide managed data network services (the JOVE initiative), Oftel advised negatively, on the grounds that such a joint venture would simply be too powerful. Oftel gave a positive recommendation on the BT/Mitel case. In the emerging market of cellular radio, Oftel stipulated that the fixed network operators, BT and Mercury, were not allowed to participate in their own right; a minority stake in the consortia to be established to run mobile networks was permitted. Another important contribution towards the development of a more open market was Oftel's determination to settle interconnectivity between the existing and the newly created networks. To provide its long-distance and international services and to enable fair competition between the two network operators, Mercury demanded full access to BT's local networks. Oftel gave a strong impetus to an effective duopoly with its ruling on interconnection between BT's and Mercury's networks. The disagreement between the two operators was finally settled early in 1986, when Oftel decided to give Mercury the right to interconnect with the BT system on favourable terms (i.e. at cost-based charges).

Oftel has become involved in encouraging and ensuring the quality of service provision. When BT stopped the systematic publication of quality-of-service statistics on the grounds of commercial confidentiality, Oftel sought to develop instruments of its own to monitor BT's quality-of-service provision. After the National Consumer Council (NCC 1987) made complaints about BT's disappointing performance in terms of faults, delays of new connections, unsatisfactory operator and enquiry services and inadequate regulatory regime, Oftel (1987) responded by pushing BT to accept higher levels of public accountability with regard to quality of service and internal management controls. BT promised to improve its service to customers: since 1987 it has published quality-of-service indicators every six months, installed a customer guarantee scheme (in case BT fails to meet certain service obligations) and introduced itemised billing on request. Mercury followed BT's example and also publishes bi-annual quality-of-service reports. The Competition and Service (Utilities) Act 1992 assigned additional powers to the public utility regulators, including DGT/Oftel, to set standards of public service provision (in terms of quality, choice, value and accountability), to determine compensation where those standards are not met and to set overall performance objectives. In the case of telecommunications, Oftel required the two 'designated' operators of relevant (basic) services, BT and Kingston upon Hull, to set adequate arrangements, dealing with the publication of information about performance and to install procedures dealing with complaints, dispute resolution and compensation.

After receiving complaints about the inequality of bargaining power and strategic information between BT and the competing operators, Oftel in 1992 demanded BT provide more detailed information about its interconnection agreements, cost allocation methodology, interconnection charges and revenues. Given the development towards 'multi-operator and multi-service' businesses, Oftel regarded the elements of interconnectivity and information symmetry as vital to its policies. In order to improve the equability and transparency of domestic services and to make the charges more efficient and sustainable, Oftel (1992, 1993a) urged BT to publish separate accounts for each of its businesses. For the development of standard interconnection charges and the establishment of a more transparent process for relating costs to charges, BT's collaboration was crucial for Oftel (BT 1993): any alternative going further than separate accounting would be strongly opposed by BT (e.g. implementing structural separation between the local and long-distance activities). Oftel (1994a, b) came up with a less radical solution by suggesting the implementation of a differentiated auditing regime for BT's Network (wholesale) and Retail and Access (connections and rentals) businesses. This would generate separate accounting for local, national and international calls, private circuits, public payphones, terminal equipment and non-regulated activities. BT would work towards fully separate accounts for its network, retail and access activities to be implemented before 1996. According to TMA (1993) and Mercury (1993), Oftel focused too narrowly on accounting separation, leaving more structural issues untouched, such as bottleneck and market dominance, the distinction between local loop operation, service provision and trunk network operation, and the unbundling of BT's network cost components.

An issue related to Oftel's work on accounting separation is the control BT still has over the interconnect and access charges. Under an amendment to BT's licence agreed in 1991, as part of the Duopoly Review, Oftel (1994b) issued a revision of the interconnect charges between BT and Mercury, requiring third operators to contribute to BT's access deficit (i.e. the compensation 'third operators' pay for BT's losses on its local networks). Under certain conditions (for instance, new operators with market share of less than 15 per cent), payment of these access deficit contributions (ADC) could be waived. The establishment of an ADC waiver regime would, according to Oftel (1994b), lower barriers to entry in the market. Until the post duopoly competitive market place was established this would require detailed regulatory intervention to persuade BT to disentangle the cost allocations between its access, network and retail businesses and to publish separate accounts for them. The introduction of accounting separation between BT's local and long-distance services would force BT's retail business to pay the same interconnect and access deficit charges as its competitors.

The increasingly complex regulatory regime was ill-equipped to deal with BT's anti-competitive behaviour or prevent BT from maintaining its dominant market position in areas such as interconnection, access and number portability. Since 1995, BT has faced an increasingly adversarial regulatory environment as Oftel (1995, 1996a, b) has introduced tougher regulation to promote competition. These efforts have included:

- reducing interconnection charges to carry and deliver calls through BT's network;
- controlling market abuse by amending BT's licence to grant Oftel greater anti-competition powers;
- imposing price caps on BT's services, where competition can effectively regulate prices (after 1997 Oftel seeks to cut back BT's regulated markets from 64 per cent to just under 26 per cent and reduce BT's market share from 80 per cent to about 60 per cent);
- ensuring and enforcing number portability between different networks and operators.

In December 1995, Oftel proposed introducing a clause in BT's licence which would require BT to refrain from anti-competitive behaviour. The amendment included a proposal that Oftel should acquire greater powers to act as an industry-specific competition authority. Oftel also has the right to recommend the structural separation, or break-up, of BT if the latest regulatory proposals prove unsuccessful. In BT's view, unfair competition from mainly foreign-owned cable TV companies and prohibitive price controls in the UK have encouraged its expansion abroad. A year later, Oftel (1997) issued a consultative document recognising that the transformation of BT into a global telecommunications company may mean that the company would be less willing to meet its UK licence obligations. The financial and managerial demands of BT's merger with MCI could jeopardise the public operator's ability to maintain and finance a high-quality UK network. BT's intention to become a major player in local markets worldwide brings larger financial commitments outside the UK and opportunities to achieve higher returns in overseas markets. The incentive to look abroad for higher levels of capital growth may become more attractive as domestic competition accelerates the erosion of the company's UK market share and reduces large profits made from international calls. The document maintained that it is Oftel's statutory duty to interfere with the regulated company's globalisation plans to safeguard future service levels in the UK.

4.8 Summary and concluding remarks

In the late 1970s, when demand from business customers for reliable, low-cost and high-capacity data services became manifest, the public telecommunications operator, the Post Office, was poorly equipped in terms of its strategic and organisational resources to respond effectively. The combination of four successive Tory governments, constantly arguing in favour of free trade and popular capitalism and against high levels of public expenditure, taxation and government intervention, and the economic and political weight of the financial and business services industry, has proved crucial to the reorganisation of UK telecommunications. Elaborating on the proposals of the Labour government (1974–9) for the reorganisation of the Post Office, the Thatcher government decided to separate telecommunications from postal services, give the newly established British Telecom (BT) a legal corporate status and sell off a majority of its shares. The markets for terminal equipment and enhanced services were fully liberalised. In the case of voice telephony and network provision, a market duopoly was established, whereby BT competed with the much smaller Mercury/Cable & Wireless. In the cellular market another duopoly was established, in which BT's subsidiary Cellnet competed with the new entrant Vodafone. A newly created administrative agency, Oftel, acting autonomously from the central government and British Telecom, was put in charge of supervising the telecommunications industry.

After 1991 when the duopoly path was abandoned, a fully privatised BT, Mercury/C&W and Vodafone faced increasing competition from a variety of newly licensed operators, such as utilities, cable companies, cellular service providers, foreign carriers and others. After the review of this duopoly structure, the entry of new telecommunications operators was encouraged by allowing cable companies, mobile service providers, utilities etc. to penetrate the markets for local access networks and international services. The two liberalised telephone administrations, BT and C&W, have transformed themselves successfully from rigid and hierarchical organisations into highly commercial international operators, providing a variety of telecommunications services all over the world. These major international carriers have grown in foreign markets by making acquisitions or taking out equity stakes in telecommunications operators, equipment, value-added services and entertainment companies.

The real winners from the privatisation and liberalisation of UK telecommunications have been the large business users, the shareholders (including the state), senior management of BT and the new market entrants. One important limitation on the introduction of genuine competition into the domestic market was the integrated transfer of BT from public to private ownership, without the kind of organisational divestiture that had taken place in the USA. The decision to privatise BT as an integrated national unit was supported by the government, the merchant banks and BT management. The first two parties agreed to maximise the proceeds from the sale of BT shares. BT management was in favour of keeping its dominant position in the industry by minimising competitive threats to the company and installing a relatively light regulatory regime that would give BT as much discretion as possible. Other measures limiting competition were the decision not to license any public network operators other than BT and Mercury and the decision not to allow simple resale, at least until 1990.

The liberalisation, privatisation and (re)regulation measures have turned out to be beneficial to large business users, BT and Mercury/C&W, and disadvantageous to households who are paying more for their local telephone calls. Demonopolisation has encouraged BT, Mercury and new operators to improve performance and increase efficiency and customer service; the efficiency gains, however, have not been equally distributed. While large and international business users and residential customers in the urbanised South clearly benefited from competition, the large majority of people saw no substantial decrease in their local charges. Although its network has been curtailed and it has been subject to price control where the provision of basic services is concerned, BT has developed into the most profitable European telecommunications operator. Mercury/C&W has effectively penetrated the business market for international and domestic trunk services, and effectively cut back BT's market shares in those markets.

Britain was the first country in Europe and so far the only one with such a drastic restructuring programme. The level of foreign presence in UK telecommunications is higher than anywhere else in Europe, where there is still a reluctance towards foreign (either EU or non-EU) penetration. There is, however, another side to the coin. The drastic deregulation process allowed BT to diversify unsuccessfully into equipment manufacturing and successfully into cellular radio, value-added services and global end-to-end services. While other PTTs were still concerned about their domestic interests, BT and C&W were establishing joint ventures with foreign local carriers in order to provide an integrated set of telecommunications facilities in the USA, Europe and the Far East. In terms of economic regulation, Britain seems to have a competitive advantage over Continental Europe as well. The pro-competitive policies of the British government were part of a larger strategy to attract foreign investment and strengthen the domestic services industries in order to achieve market leadership in the emerging multi-media business. The British government claimed that a substantial part of transatlantic traffic was now routed via the UK and the City of London had effectively strengthened its competitive advantage as a major international centre for business services, facilitated by low-cost and high-quality communications packages.

5 The liberalisation, privatisation and regulatory reform of telecommunications in the Netherlands

In pursuit of a new consensus?

5.1 Introduction

This chapter provides the second case study of a European country adjusting its domestic industrial organisation to the structural changes in technology, markets and international governance in telecommunications. The Netherlands, a small and open economy, sought to pursue a strategy of promoting national and global liberalisation while retaining its unanimous and consensualist decision-making structures at home. In the Netherlands there has always been a tendency as well as a necessity to follow major international developments. In order to strengthen their economic position as a major European centre for international trade, transportation and financial services, the Dutch decided relatively early to evaluate their existing telecommunications policy. The radical US policy to deregulate domestic telecommunications and to break up the Bell system, and the British decision to privatise the public operator and to establish a duopoly, were regarded as too controversial to be implemented in the Netherlands. The Dutch strategy showed a more gradual approach to the introduction of liberalisation, privatisation and regulatory reform in telecommunications.

Before discussing and assessing the traditional structure and changes in the Dutch telecommunications industry, we will examine the political and economic contingencies at the national level that have had an impact on the policy process. The sector's wider political and economic setting and the structural adjustment policies between 1980–98, are introduced in sections 5.2 and 5.3 respectively. After this general introduction to the relevant institutional framework, the focus will shift to the industry level, where the traditional structure governing Dutch telecommunications and the development towards demonopolisation in the 1980s and 1990s will be discussed (section 5.4). The processes of liberalisation, privatisation and regulatory reform are discussed in more detail in sections 5.5, 5.6 and 5.7 respectively. The chapter will close with a brief summary and some concluding remarks on the major shifts in Dutch telecommunications policy (section 5.9).

5.2 The wider political-economic setting of Dutch telecommunications

Before discussing and assessing the established structure and structural changes in telecommunications policy, we will examine the wider political and economic setting of Dutch society. This institutional environment will be portrayed by a brief discussion of the following national variables: the production profile, the organisation of government, the system of interest intermediation, approaches to economic policy, and the Dutch economy's international dependence.

Production profile

The Dutch have a small open economy of about 15 million people with a strong mercantile tradition, depending highly on international trade (especially with its Community partners Germany, Belgium, the UK and France). The high levels of exports and imports (about 60 per cent of GDP) illustrate both its openness and sensitivity to political and economic developments abroad. The Dutch economy is strongly internationalised and concentrated, with a strong presence of incoming and outgoing multinational firms. The small size of the home market has stimulated Dutch companies to enter foreign markets and encouraged the development of big (Anglo-)Dutch multinational companies such as Royal Dutch Shell, Unilever, Reed-Elsevier, Philips, DSM, AKZO. These companies dominate employment, R&D expenditures and the Amsterdam Stock Exchange. On the other hand, the openness of the economy has facilitated the entrance of foreign-based firms which have gained a strong footing in the Dutch market. Foreign investors have been attracted by the country's ideal geographical location for distributive activities: it is a gateway to Europe. The production profile of the economy is centred around a small set of key sectors: transportation and logistics, agri-business, chemical and refining industries, information-related services (printing, publishing, banking and insurance) and electronics (e.g. the home base of Philips NV). The country is relatively weak in high-tech industries and labour-intensive manufacturing (e.g. office machinery, automobiles).

The share of public enterprise in the Netherlands, made up of state enterprises and state shares, has traditionally been small (De Ru 1985; Andeweg 1989, 1994). An economic reason for this can be found in the openness and small size of the Dutch economy, making it difficult to use public enterprise as an instrument to accomplish national economic objectives such as innovation and growth. An additional political explanation can be found in the moderate attitude of the Social Democrats towards nationalisation. Immediately after World War II they dropped nationalisation in favour of indicative planning and functional decentralisation (Andeweg 1989). Apart from some minority state shares in the steel and chemicals/mining sectors (Hoogovens and DSM), public enterprise in the Netherlands has been largely restricted to the traditional public utilities (PTT, energy, railways, airlines, etc.) and the Central Bank. As a rule government interference in the daily operations and long-term planning of public enterprise has been minor. The main exception was PTT (and to a certain extent NS Dutch Rail), where the government actively intervened in decisions on tariff setting, wages, labour conditions and investments.

Organisation of the state

The Netherlands could be typified as a decentralised unitary state, in which the government fosters self-regulation within the various groupings of society and facilitates nationwide collaboration among the 'pillars' of society. Daalder (1966) saw the Netherlands as a pacification democracy in which a basic acceptance of disagreement and diversity in the political community (i.e. respecting the various ideological traditions within society) is combined with a willingness to seek compromises between the political élites of the different societal groupings through slow negotiations and mutual concessions. Lijphart (1975) described policy making in the Netherlands as the 'politics of accommodation': depoliticisation (pragmatism), tolerance (agree to disagree), summit diplomacy, consensualism, secrecy, proportionality and a considerable degree of restraint on the part of Parliament in its dealings with the government. Since none of the political parties has ever enjoyed an absolute majority, coalition cabinets are inevitable. In the long run this political system has produced a large degree of continuity and coherence in public policy.

The Dutch style of policy making could be characterised as both corporatist, illuminating the participation of the major stakeholders and their systemic cooperation in the policy process, and technocratic, relying upon 'objective' investigations and prospective studies carried out by independent experts (Van Doorn 1981; Hemerijck 1994). The political system exhibits an interpenetration of the public and private sector through various tripartite and consultative platforms, which play a crucial role in forging nationwide compromises between central government and organised interests about social and economic policies (Van Putten 1982). There is also an important technocratic element in Dutch policy making, namely the influential role played by independent non-representational experts and (*ad hoc*) advisory committees. Whenever complex policy problems or politically sensitive issues feature on the agenda, the corporatist routines, based on interest articulation and organisational concertation, are often replaced by a 'depoliticised' approach, relying upon the involvement of technocrats and external advisory bodies. Problems are solved in a laborious collective search for a broad compromise that will be acceptable to all major stakeholders in the policy community. Although it may facilitate domestic stability and gradual adjustment, such an accommodationist regime, however, does have drawbacks, such as conflict avoidance and a lack of decisiveness at critical junctures. In those situations the policy process in the Netherlands seems to follow the law of least societal resistance, manifested by the time-consuming character of the search for inclusive compromises between public and private actors and the risk-averse outcomes of collective bargaining (Geelhoed 1990). In case of major international and/or domestic challenges disrupting well-established patterns, the Dutch negotiated style has proven relatively ineffective and inefficient (Windmuller *et al.* 1987; Geelhoed 1989).

System of interest intermediation

The Dutch have a social market economy, characterised by a large redistributional role for the government (i.e. welfare state), an active involvement of organised business and labour in the formulation and implementation of socio-economic policies, and a 'power sharing' tradition, stressing the interdependence and interpenetration of public and private decision making. The origins of this negotiated or corporatist framework lie in the post-war settlement between government, business and trade unions when class-wide consensus about industrialisation, wage moderation and export promotion was considered essential to rebuilding the country. The Dutch corporatist framework is made up of the following institutions (Windmuller *et al.* 1987):

- the Socio-Economic Council (SER): a tripartite body with far-reaching advisory and executive powers in economic and industrial policy making;
- the Central Planning Bureau (CPB): an advisory body preparing economic forecasts on the government's instruction and carrying out prospective studies;
- a Statutory Industrial Organisation (PBO) made up of product and industry councils with far-reaching competencies (only effectively implemented in the agri-business);
- the Foundation of Labour SvdA, a consultative body set up by the central employers and labour organisations to oversee, negotiate and co-ordinate wage controls and employment conditions; and
- a unitary conception of the company, manifested by collaborative relationships between management and labour through works councils (OR), the board of directors and an independent supervisory board, with a collective responsibility for the well-being of the corporation.

This corporatist institutional matrix was relatively effective in a stable business environment of steady economic growth, facilitating gradual industrial adjustments and a high level of societal stability. The Dutch system favours the active involvement of labour and employers' organisations in public policy making. This made strong centralised associations of both workers and employers a prerequisite. There are two major federations of employers' organisations, the Federation of Dutch Industries (VNO) and the Dutch Christian Employers Union (NCW), representing the large majority of employers. With some smaller employers' associations, VNO and NCW work closely together in co-ordinating the broad lines of socio-economic policies. The large home-based firms have a disproportionate influence on the internal functioning and the external strategy of these employers' federations. The role of financial institutions in the system of interest intermediation can be typified by a commercial banking tradition, specialising in short-term credit provision and trade finance. The involvement of Dutch banks in corporate decision making is modest and their task limited to supporting the international commercial activities of domestic businesses.

The employee front is concentrated in one big trade union federation, the Dutch Federation of Trade Unions (FNV) and two smaller ones, the Christian Trade Union Federation (CNV) and the Federation of White-collar Employees (MHP). FNV is by far the biggest employee organisation with 60 per cent of all union membership, leaving 20 and 7 per cent to CNV and MHP respectively. Although the total unionisation rate in the Netherlands is relatively low and falling (from 41 per cent in 1977 to less than 25 per cent in 1990, although it increased to almost 30 per cent in 1996), its influence in the settlement of socio-economic matters must not be underestimated (Windmuller *et al.* 1987; Visser 1990; Visser & Hemerijck 1997). Because the three union federations are relatively centralised and exercise remarkable control over their member unions (e.g. an extremely low level of industrial militancy), together with organised business they are an acceptable partner for government in macro-economic policy making. Both the unions and the employers' associations focus on interests beyond mere wages and labour conditions, they consider all social and economic aspects of the national economy their domain (price levels, unemployment, training, innovation etc.).

Economic culture and ideology

With a small and open economy, the Dutch lack the political and economic instruments to control or even influence major industrial developments. Given the unfeasibility of protectionism and long-term economic planning, small countries have to rely on export-oriented and free trade strategies, supported by collaboration between national producers and compensation schemes between the ruling political and socio-economic coalitions at home (Katzenstein 1985). In the Netherlands these conditions have fostered a business culture characterised by concerted action on the part of the major interest associations in the formulation and implementation of economic policy, together with active export promotion and trade liberalisation abroad. The traditional encouragement of the accommodation of government, organised business and labour has facilitated the development of collaborative arrangements in the national economy, like a tolerant anti-trust policy and elaborate protective measures to restrict the influence of foreign and/or conflict-seeking shareholders on management decision making.

De Jong (1990, 1992) comparing the Dutch competition policy with that of other European countries has described the Netherlands as the 'cartel paradise of Europe'. The 1956 Competition Act, hardly amended over the years, left more room for explicit and implicit collusion between firms and market dominance than the anti-trust legislation of the European Community. Unlike EC law, Dutch competition policy did not prohibit anti-competitive practices, but aimed at limiting abuse of economic dominance and collusion in the market place. Furthermore, legislation on the scrutiny and control of mergers was absent and prosecution for abuse of monopoly power was rare. Competition legislation in the Netherlands was based on a case-by-case judicial review and the enforcement by law of privately conducted cartel agreements. The reason for this clear preference for business self-regulation and industrial co-operation over competition policy can be found in the negotiated character of the Dutch economy, with its strong international economic dependence requiring the collaboration at home between business, state and labour but also within the rankings of organised business.

As a consequence of the attacks by the European Commission on the *cartel-friendliness* of their economy in the early 1990s, the Dutch authorities suddenly realised that a complete revision of the Competition Act was needed to align it to the prohibitive EC rulings. From 1992 onwards a series of anti-trust measures, referring to prohibitions on horizontal price agreements, market sharing agreements and transparent public procurement, have been effectively implemented. In May 1996 a Bill for a new Competition Act was put forward in Parliament. The material provisions of this new Competition Act have been brought into line with Articles 85 and 86 of the EC Treaty. Furthermore, the Bill introduced a fully fledged framework of merger control provisions, which is also heavily inspired by the EC Merger Control Regulation. Compliance with the new competition rules will be enforced by a newly established anti-trust authority, called NMA, which will act independently but will still be answerable to the Minister of Economic Affairs. The Competition Act, which also gives the courts competence in appeals against decisions taken on the basis of the new Act, came into force in 1997.

Positioning in the international political-economic system

The strong dependence of the Netherlands on the world economy has fostered a pro-European attitude and a major interest in the further integration of national markets into a free trade regime. The Dutch have always been strong advocates of European integration, being one of the six founding members of the EC and playing a relatively important role in the establishment of the Common Market and the move towards closer economic, monetary and political union. The Dutch have developed a strong commercial interest in the EC with a vast majority of its exports and imports being exchanged with their EC partners. The economic gains from European integration have been high: agriculture, transportation and financial services and communications have clearly benefited from EC policies (Wolters & Coffey 1990). Dutch policy makers, however, have seriously underestimated the consequences of increasing European integration on domestic socio-economic policies (Van Schendelen 1991, 1993b). The rapid implementation of the Single Market Programme by 1993 and the concomitant irreversible shift of decision-making powers towards the European level took most Dutch ministries and key industries by surprise. Nowadays, public and private policy makers have become more aware of the significance of the draft phase of negotiations and the need to anticipate the final phase of decision making. In the Netherlands, the harmonisation of national policies to form one common policy and the enactment of guidelines, regulations and directives in order to establish the Common Market were widely regarded as major consequences of European integration. The extent to which the Internal Market Programme also instigated a process of international deregulation and policy competition between the national institutional frameworks was, however, underestimated. The emphasis of the Dutch political debate had been too much on the process towards European policy convergence, overlooking the fact that not only European businesses, but also governments are competing against each other for capital, labour and technology (WRR 1993). In such a market environment member states could optimise their comparative advantage by encouraging entrepreneurship, improving the quality of the industrial and educational infrastructure, and strengthening other competitive factor conditions.

5.3 The politics of economic adjustment in the Netherlands

Between 1945–73 a broad consensus existed in the Netherlands in favour of Keynesian demand management. Government, business and labour had committed themselves to the national goals of full employment, steady economic growth and the development of the welfare state. Important instruments to accomplish these objectives were legally enforced price controls and a centrally guided wage policy to keep the domestic production costs significantly below those of foreign competitors and advance the country's trading position. This successful reconstruction and industrialisation strategy in the post-war period has become known as '*le Miracle Hollandais*' (Van Den Brink 1984). When economic growth started to stagnate in the early 1970s, the thus far collaborative relationships between state, business and labour and within the ranks of organised labour and business became more polarised. The delicate balance between economic adjustment and political stability was effectively challenged. The sharp increase of wages and the drop of profits in the late 1960s encouraged business to modernise traditional industries (i.e. invest in labour-saving techniques) or shut down plants and move economic activities abroad. As a consequence the established corporatist framework was replaced by 'many little corporatisms' in which negotiations between organised interests and state agencies were decentralised and increasingly located at industry and firm level (Visser 1990).

In the 1970s there was a move from low to high politicisation of business in the Netherlands (Van Schendelen 1987). The centre left Den Uyl cabinet (1973–7), with its slogan *a fair distribution of knowledge, power and income,* put forward an ambitious programme to increase government spending on social welfare and social security, housing and education and to curtail the power of business by proposing measures to increase workers' involvement in corporate decision making. The country, already suffering from decreasing economic growth, was hit hard by the oil crises of 1973 and 1978. The situation became even worse as a consequence of the so-called Dutch disease: the windfall profits of the production and export of natural gas were not used to strengthen the country's competitiveness and economic structure, but were utilised for public consumption and the expansion of the welfare system. The bonanza of gas revenues made the inflation rate and the value of the guilder go up, eventually leading to a deteriorating competitive position in the world market. At the end of the 1970s the Dutch economy was confronted with increasing international competition and a shrinking economy, characterised by lagging investments, de-industrialisation, decreasing exports, increasing imports and a huge budget deficit (WRR 1980). An urgent need was felt to trim the welfare state with its high taxation and social security payments and to promote economic growth. The centre right Van Agt I cabinet (1977–81) started to abandon Keynesian policies in favour of neo-liberal alternatives by promoting (minor) cutbacks in public spending and a more positive appraisal of market principles. The implementation of these austerity measures did not go far enough to cope with the economic situation, however. In 1983–4 the unemployment rate rose to a record level of about 15 per cent and public spending to 70 per cent of GDP (Windmuller *et al.* 1987). The decision of the centre right Lubbers I cabinet to impose a 3.5 per cent cut on civil servants' salaries for 1984 triggered the biggest-ever civil servants' strike (including PTT staff) since the great railway strike of 1903 (Van Den Besselaar & Visser 1993). Confronted with a negative public opinion and after losing numerous law suits, the public sector unions accepted a 3 per cent salary cut.

The established corporatist framework hampered the formation of an adequate policy response to the new international and economic conditions. After a short interval of the centre left Van Agt II cabinet (1981–2), the centre right Lubbers cabinets (1981–5 and 1985–8) pursued no-nonsense polices aimed at restoring investment and economic growth, reducing unemployment and maintaining price levels. Only then did the Dutch economy begin to show signs of economic recovery. Another reason for the country's economic revival was the general upswing in the world economy throughout the 1980s and 1990s. Between 1991–6, the Dutch policy of wage moderation and welfare state reform without big bang showed a higher than average economic performance, with strong job growth and a fall in unemployment to almost half the current level in the EU (Visser & Hemerijck 1996). The major shift in economic policy took place between 1981–3 when the foundation was laid for the revival of Dutch industry and trade. The effective remedies consisted of a mix of cutbacks in public spending, wage moderation, a market orientation towards the labour market and business in general and the active promotion of innovation policy. The steady flow of large government subsidies for ailing industries (and other domains such as culture and housing) was cut dramatically: partly as a deliberate consequence of balancing the budget by the centre right Lubbers cabinet but also due to misuse of public funds (e.g. shipbuilding). This package of measures proved effective over the years: it stimulated economic growth and cut back unemployment as well as the budget deficit (with inflation remaining low).

The austerity policy of the Lubbers government was reinforced by the agreement between business and labour in the Foundation of Labour SvdA (instigated under pressure from the government) regarding continued wage moderation and the voluntary decentralisation of collective bargaining. From that moment onwards, organised business gradually took over the initiative from labour in setting wages and other employment conditions. In 1981–2 the Lubbers I cabinet implemented an ambitious programme, called the Major Operations, to facilitate structural reform of public administration and a retreat of the state. The Major Operations consisted of the following sub-programmes: a reconsideration of public expenditures, a reorganisation and decentralisation of government administration, deregulation, and debureaucratisation (cuts in civil service salaries and/or staff). Privatisation was another instrument in redefining the boundaries between the public and private sectors. The Dutch privatisation programme had no explicit ideological motivation: it was considered just another pragmatic instrument to economise on the state budget and to strengthen the market sector. Officially, the objectives were to reorganise the public bureaucracy and improve efficiency within government, to reduce public expenditure and the budget deficit and to stimulate entrepreneurship and market responsiveness to economic revitalisation (Andeweg 1989, 1994; De Ru 1992). The budgetary motive to trim the public sector was prevalent in the privatisation programme and the Treasury was the driving force behind its implementation. The final results were somehow disappointing: there were hardly any state enterprises that could be privatised and the operation was carried out gradually without a clear mission. Although characterised by a large non-profit sector, the Dutch economy has only a small number of state enterprises that could possibly be privatised. Leaving aside the hiving-off of some smaller state-owned services, only two larger companies were privatised between 1981 and 1994, namely KPN/PTT and the Postbank. The Dutch way of privatisation is characterised by a two-stage approach, emphasising an incremental adjustment of the public enterprise to the new market conditions and avoiding the radical option of selling a majority of the state assets immediately. As a first step, state enterprises are restructured and brought under private law, without touching upon ownership and market conditions. Next, the independent state-owned company is given some breathing space to become a market-driven corporation, before (some of) its shares are sold off by the government (its single shareholder). The employment levels of civil servants and the interests of the newly established corporation are protected as much as possible to overcome the employees' resistance to change. At the same time the former civil servants and the autonomous state-owned company are prepared for the market environment, and the legislative framework is developed and implemented. This transitional period of *pseudo-privatisation* helps to improve the value of a private share offering by enabling the operator to become more efficient and allowing it time and space to restructure its business activities effectively, sheltered from disruptive market forces (De Ru 1992). This is what happened with the chemical company DSM, the Postbank, and KPN/PTT. The recent restructuring of the public utilities has been dominated by the reorganisation of its sectoral champions, namely KPN/PTT, NS Dutch Rail and the power utilities, who have been given more managerial and financial autonomy through their conversion into state-owned companies with a limited liability status. As a consequence, regulatory issues to be negotiated and settled at the industry level, like universal service requirements, consumer protection, interconnectivity and fair competition, played a minor role in the political process.

Privatisation met only minor political opposition and legal-constitutional problems. The pragmatic aspects of the privatisation proposals to a large extent prevented political opposition, which came only from the small radical left-wing parties. The Christian Democrats, the Liberal Conservatives and the centre left D66 were cautiously in favour and the main opposition party of the 1980s, the Social Democratic PvdA, was in principle not against the proposals. The privatisation programme was, however, opposed by the civil servants' unions. They saw the privatisation programme as just an excuse for restructuring an inefficient and ineffective civil service, and feared losses of jobs, income and pension rights (Andeweg 1994). The public sector unions (with a unionisation rate of about 50 per cent) argued that alternatives like decentralisation and industrial democracy were more appropriate to making the civil service and public management more flexible (Soeterbroek & Walravens 1985). When their rank and file did not wholeheartedly support these views, the public sector unions became more moderate and co-operative. If particular labour conditions, such as job security, payments and social security (such as the breach in pension schemes) were safeguarded, the civil servants themselves were not necessarily against privatisation. Furthermore, the two major labour union federations, FNV and CNV, did not support their civil service unions very strongly. Privatisation did not have an influence on their total unionisation rate. It only implied a (possible) transfer of members from public sector to private sector unions. The employers' associations regarded the sale of government shares and the contracting out of public services as part of an overall market-oriented economic policy aimed at stimulating the business spirit and encouraging market forces.

The Dutch path throughout the 1980s was aimed at creating more flexibility in the national economy by non-consensual programmes, like an austerity policy, decentralised bargaining, deregulation, administrative reform and the privatisation of the Postbank and KPN/PTT. After seven years of centre right government, in 1989 the pendulum once more swung to the other side of the political spectrum. The centre left Lubbers III cabinet (1989–94) went ahead with the difficult task of trimming the welfare state and establishing a smaller and more effective public sector. The levels of unemployment, government expenditure and the budget deficit were reduced, but they were still comparatively high (e.g. unemployment at about 8 per cent and public expenditure at 51 per cent of GDP in 1991) (WRR 1993). The installation of the 'purple' Kok cabinet in the summer of 1994, consisting of Liberal Conservatives, Liberal Democrats and Social Democrats, has not led to any big changes in public policy. Emphasis is still on balancing public expenditure to reduce social security costs and promote public investments and increasing participation levels in the labour market by the implementation of deregulatory measures. One of the main action lines set out by the Kok cabinet has been the introduction and further elaboration of the so-called 'Competition, Deregulation and Quality of Legislation' project, aimed at stimulating the functioning of the market, reconsidering existing rules as regards their effect on the dynamics of the economy and examining the quality of the various legal provisions in the light of changing demands. The results have been various; for instance, while the extension of shopping hours was a clear success, deregulating water distribution is still pending, the opening up of the public transport market was found unfeasible and was postponed. Although less dominant and persistent than before, corporatist structures and tripartite agreements have not disappeared over the years. To this day, organised business and labour still attempt to reach common views on more narrowly defined subjects such as pensions, indexation, shorter hours of work etc. Nevertheless, the corporatist framework may still be appropriate for adequate and responsive strategies on certain aspects of macro-economic policy and for the implementation of EC policies.

5.4 Demonopolisation of telecommunications in the Netherlands

Introduction

Until 1989, the Dutch PTT was part of the government administration as a corporate unit of the Department of Transport and Public Works, called Staatsbedrijf der Posterijen, Telegrafie en Telefonie. Traditionally the PTT Directorate was subordinate to the Minister of Transport and Public Works (T&PW) and subject to parliamentary supervision. The PTT administration was assigned a *de facto* public monopoly on the control of the postal and telecommunications infrastructure (including the responsibility for broadcasting transmission) and the operation of message transportation services. The Director General of PTT (DG) was granted certain administrative powers related to the control of the infrastructure, such as granting licences for private networks and cable systems, allocating frequencies, standardising and approving terminal equipment. The PTT administration was also involved on behalf of the Minister in policy formation and implementation. Due to a lack of expertise on the part of the government, postal and telecommunications policies were often prepared and ghost-written by PTT officials. The Advisory PTT Council (PTT-Raad) served as a consultative platform, through which the PTT and its major stakeholders could discuss the social, political and economic aspects of postal and telecommunications policy. Besides facilitating the exchange of information and views among the parties involved, this neo-corporatist body, whose secretarial functions were carried out by the PTT administration, was also required to give recommendations to the Minister of T&PW. The system of industrial relations within the PTT consisted of a Departmental Works Council at the central level and 200 sub-departmental committees at the lower ranks (Van Den Besselaar & Visser 1993). The prominent labour union, representing PTT staff (with a unionisation rate of nearly 50 per cent), was the centre left AbvaKabo, a public sector union loosely connected to the Social Democratic party. Besides the operation of postal and telecommunications services, the PTT administration was also responsible for providing money services to the general public (notably giro and savings) through its branches of the Postal Savings Bank RPS and the Postal Giro Bank PCGD. In 1986 the money services were integrated into the state-owned Postbank company and hived-off from the PTT administration.

The PTT administration was subject to close political intervention with respect to its day-to-day operations and strategic investments, which were often 'intensive, complicated, rather incoherent and very detailed' (Commissie Steenbergen 1985a:25). All the financial requirements of the PTT had to be met through the National Budget: the allocation of its resources was decided on a year-to-year basis by the government and agreed upon by Parliament. Given the fact that its profits and corporate financing were government-controlled, PTT therefore had to rely on internal financing and the cross-subsidisation between the profitable telecommunications function and the loss-making postal function. The PTT administration was regularly used to balance the budget. As a consequence, its long-term strategy was seriously hampered by short-term budgetary concerns, often leading to its charges being raised with the explicit purpose of increasing the financial contribution to the Treasury and postponing investments. The Home Office, in charge of stipulating the recruitment and employment conditions within the civil service, set clear limits on PTT's personnel policy. The task of rate setting also fell under the government's jurisdiction. Tariff changes were proposed by the Ministry of Transport & Public Works (T&PW), in close collaboration with the Ministry of Economic Affairs and the Treasury, and finally agreed upon by Parliament. In addition to being responsible for tariff policy, the Ministry of Economic Affairs used the PTT administration to further industrial policy objectives, notably the promotion of innovation and regional economic development. The Ministry for Welfare, Health and Culture had to be consulted in all matters related to cable television and media policy.

Traditionally the market for public switching equipment has been relatively closed: with the major supplier being Philips, the country's largest industrial employer and one of the world's largest diversified electronics firms, and the foreign companies ITT/NSEM (USA), Ericsson (Sweden) and Siemens (Germany) being additional suppliers. With respect to switching equipment Philips, forced by its relatively small share in the world market and lack of technological competitiveness, decided to change its corporate strategy by abandoning the national champion alternative for a more collaborative strategy. In 1983, Philips Telecommunications decided to catch up by joining forces with AT&T through the creation of APT. This link up with an American company was going against the ideals of establishing a common European technological base and a Common Market, in which Philips was deeply involved at the time. For AT&T the agreement was a bridgehead on the highly protected European market, quickly followed by partnerships with Olivetti, Italtel and Telefonica.

At the end of the 1970s the PTT monopoly attracted more and more criticism. Both government and business were convinced that a thorough reconsideration of the established regime and the effects of technological, economic and international developments in telecommunications was required. It also became clear that the PTT could only fulfil its proper function in this changing business environment if it gained more corporate autonomy. The 1979 *ad hoc* Advisory Committee Rathenau, dealing with the socio-economic consequences of micro-electronics, conceived of the PTT as a key actor in the development of dataprocessing, office automation and home information services. In the Committee's view, the PTT also played a crucial role where the formulation and implementation of communications policies were concerned. The Rathenau Committee argued that a major investigation was needed to determine whether the position of the PTT deserved reconsideration in the light of the new information and communications technologies (Adviesgroep Rathenau 1979).

The PTT monopoly was not only questioned by independent experts but by big business as well. The increasing demand of business for data services and leased circuits in the 1970s challenged the reliability, flexibility and capacity of the then nearly saturated telephone network. In order to meet business demand, an upgrade and expansion of the infrastructure was needed, but the PTT found itself facing with government restrictions on access to capital and qualified personnel. The banking, insurance, computing and publishing sectors were far from satisfied with the PTT framework and the provision of telecommunications services and equipment. These information-related industries complained about detailed political interference in the PTT's long-term investments, services and tariffs, the restricted or deficient supply of terminal equipment, leased lines and data services, restrictions in the use of other facilities than the ones supplied by the PTT, lack of market responsiveness and absence of user consultations, and conflict of interest between the public utility, regulatory, policy supportive and commercial functions of the PTT.

The Swarttouw Committee: setting the agenda for structural reform (1981–4)

After telecommunications and micro-electronics were also mentioned as priority sectors in the Report on Reindustrialisation (WRR 1980), the government responded in the summer of 1981 by installing the Swarttouw Committee to investigate and put forward institutional changes in the telecommunications domain (Commissie Swarttouw 1982). The government had become convinced that an advanced communications system was necessary to support the vital trade, distribution and financial interests of the Dutch economy. The Swarttouw Committee, named after its chairman (President of Fokker Aerospace Industries), consisted of six members, representing the industrial, technological and business user aspects, with the PTT providing the secretariat. The Committee consulted representatives of the PTT, the Advisory PTT Council, the domestic equipment industry, and the computing and information-related industries; labour unions and consumer organisations were not consulted. In its 1982 report the Committee stated that structural reform of Dutch telecommunications was necessary (Commissie Swarttouw 1982). Although it argued in favour of preserving the PTT's monopoly on the infrastructure and traditional telecommunications services, the Committee concluded that the market for terminal equipment and value-added services could be liberalised and direct political influence on the PTT should be diminished. The provision of the infrastructure, the telephone, telex and datatraffic, together with the tasks of standard setting, certification and type approval, should remain the exclusive jurisdiction of the PTT. The markets for terminal equipment and tele-services should be liberalised and developed into a level playing field in which the PTT would compete with others on equal conditions. The Committee stated furthermore that the delineation of the public network and the competitive equipment and tele-services market needed further specification. Notwithstanding the fact that the PTT as a public utility could play a stimulating and leading role in industrial and innovation policy, the Committee argued that the PTT needed more autonomy with respect to its investment policy, rate setting and service provision. To that end the Committee emphasised the need for a transformation of the PTT administration from a government department into a state-owned joint-stock company and the installation of a separate supervisory council, in which the major organised interests within society were represented. With respect to the PTT's internal organisation, the Committee stated that the money services, telecommunications and postal functions could remain incorporated within the administration for synergetic reasons, on condition that there would be a clear-cut separation between these three activities.

The Swarttouw Report was welcomed with moderate enthusiasm. The PTT, the Advisory PTT Council and the representatives of the information industry welcomed the general suggestions made to increase the PTT's corporate responsibility, liberalise some sections of the market and revise the institutional framework. PTT management and the Advisory PTT Council saw their enduring pleas for more autonomy from the government answered, while at the same time preserving the postal and the telecommunications functions within the corporation. The possibility of obtaining a legal entity status appealed to the PTT's growing commercial ambitions. The creation of a Supervisory Council that would act on behalf of the government and organised business and replace direct political scrutiny, reflected the administrative ambitions of the Advisory PTT Council to incorporate the body in its wider corporatist framework. For the PTT, the Advisory PTT Council and the labour unions, the proposals accompanying the change in status from the public to the private sector were important in the medium term, but they needed to be supported by short-term measures aimed at enlarging the corporation's autonomy with respect to investments, tariff setting and joint ventures.

The proposal to liberalise the terminal equipment market (excluding 'conventional' telephone sets) was not fully supported. The PTT, the Advisory PTT Council and the unions argued that the PTT should have a say in which market players and products could be admitted without undermining the integrity and reliability of the network. The retention of 'conventional' terminal equipment in the public monopoly would also secure employment within the PTT administration and the domestic manufacturing industry. Although welcoming the general suggestion to distinguish between the PTT's public and commercial tasks, the information industry and large business users made it clear that the Committee's recommendations did not go far enough. They stressed the necessity for a clear-cut separation between the operational and the regulatory functions of the PTT and the establishment of an independent Telecommunications Chamber, that could deal with rule setting, supervision, tariff regulation, standardisation and arbitration. The leading business interest association, the Centre for Information Policy (CIB 1982), was not really convinced about the validity of the Committee's point that the telecommunications division could remain integrated within the PTT administration. The decision to hive-off the telecommunications branch into an independent subsidiary would improve customer flexibility and market transparency.

In 1984, nearly two years after the publication of the Swarttouw Report, the centre right Lubbers I cabinet endorsed the main recommendations of the Committee, although actual measures were not yet taken. This delay was caused by changes in the cabinet from centre right to centre left to centre right again and by opposition from the Social Democrats and the Christian Democrats, who considered the Swarttouw proposals too radical. The negative effects of a proposed liberalisation of the equipment market upon employment, the hiving-off of PTT from the government administration, and the creation of an independent supervisory council outside parliamentary scrutiny, caused serious political tensions. The Swarttouw Report aroused opposition from the domestic equipment industry and the trade unions, while the PTT and the Advisory PTT Council could not fully agree with the Swarttouw proposals. In 1982 the Social Democratic Minister for PTT Affairs responded by shelving the Report, encouraged by the protest of the domestic equipment industry and the trade unions' fears that the attack on the equipment monopoly would mean a loss of market share to Far East producers and eventual job redundancies within the PTT administration and the domestic electronics industry (AbvaKabo 1984). In November 1983, political opposition was still so strong that in a draft White Paper the existing structure of the PTT administration remained largely intact (apart from a reorganisation caused by the denationalisation of the PTT money services). Two months later the situation had improved as a result of the priority given by the centre right Lubbers I cabinet to the retreat of the state from the national economy. In order to improve the country's vital interests of commerce, transportation and finance, the government regarded an advanced communications infrastructure and a leading role for the PTT in the development of new services of vital importance. The cabinet argued that Dutch telecommunications should follow developments in the US and UK and because of its strong international dependence structural adjustment had to be seriously considered.

The 1984 White Paper enforced the broad outlines of the Swarttouw Report by following its suggestion to restrict the PTT's future monopoly to the installation and maintenance of the public infrastructure, and the exclusive provision of telegraph, telephone, telex and dataprocessing services. The government opened up the way to providing certain advanced services by private companies in competition with the PTT through the public network. The use of leased lines for communications purposes by parties other than the contracting party was not allowed. In one respect the 1984 White Paper diverged markedly from the Swarttouw Report: while the latter proposed only a partial termination of the PTT's monopoly on telecommunications equipment, the former recommended the liberalisation of all terminal equipment. There would be a liberalisation of the PTT's monopoly on the procurement of *advanced or non-traditional* terminal equipment, implying competition between the PTT and private industry. The supply of traditional equipment, such as telephones, private telephone exchanges and telex equipment, would remain the exclusive domain of the PTT. The White Paper argued that an adequate provision of telecommunications facilities could only be assured when the PTT had joint responsibility over the infrastructure and the equipment directly attached to it. The main reason for maintaining the PTT's monopoly on traditional equipment seemed to be defensive, however. Termination would have serious effects on employment in the domestic equipment industry and the PTT administration. Furthermore, the White Paper announced the establishment of an advisory committee to investigate the possibility of a future integration of the cable networks and the local PTT controlled telecommunications infrastructure. The government also indicated that a follow-up to the Swarttouw study was needed to examine the status and the structure of the PTT to cope with the present and future developments in (inter-)national telecommunications.

The information, computing and services industry was somewhat disappointed by the government's official reaction to the Swarttouw Report. First, the White Paper was published almost three years after the Report; second, it did not put forward any new measures (only a new examination); and third, the differences made between the provision of basic transport services and value-added services on the one hand and the PTT's operational and regulatory and policy supportive functions on the other, were not clear. The policy changes proposed in the 1984 White Paper were considered to be directed towards protecting the established interests of the PTT and the domestic equipment industry. The representatives of the information industry stated that telecommunications policy, besides safeguarding public sector interests, should include new goals such as the active promotion of market entry and fair competition. The organised information industry argued that the delineation between a public monopoly and market competition should be based on a definition that was understandable, technologically independent and in line with international developments. Therefore the PTT's monopoly should be restricted to the provision of basic services, and its regulatory and supervisory activities should be transferred to a government agency. A clear liberalisation of the market for terminal equipment and value-added services would give a strong impetus for innovation and entrepreneurship in tele-matics. According to the information industry representatives, the commercial development of new services required not only the introduction of market competition, but a solid infrastructure that could be provided by both PTT and the cable companies. The possibility of utilising cable networks for service provision, however, was insufficiently recognised by the government.

The Steenbergen Committee: outlining structural reform (1984–5)

The next step was the installation of another committee in June 1984 to evaluate the present responsibilities and functioning of the PTT administration and to propose adjustments in the status, structure, task and supervision of the PTT to improve its flexibility, decisiveness and entrepreneurship (Commissie Steenbergen 1985a). In preparing its recommendations, the Committee worked together with the international management consultancy firm McKinsey & Co. The Committee's Report, published in July 1985, stated that a future telecommunications policy should balance monopoly and competition and elaborate upon the dual role of the PTT (Commissie Steenbergen 1985a). As a public utility it should provide the network and be able to play a leading role in the development of new services. As a private enterprise PTT would be allowed to offer terminal equipment and tele-services in open competition with third parties. The Steenbergen Committee was susceptible to the complaints made by business, consumers and PTT itself about the political and administrative restrictions confronting PTT and the domestic telecommunications sector. These constraints limited PTT's manœuvrability with respect to infrastructural investments, the promotion of new services and the development of new markets, and impeded the overall flexibility and innovativeness of Dutch telecommunications. The Committee separated the PTT's traditional activities into four separate functions:

1 the public utility function: a responsibility for the construction and maintenance of the infrastructure and basic services;
2 an entrepreneurial role: the provision of value-added network services and terminal equipment in an open market environment;
3 the regulatory task related to standard setting and approval, rule setting and supervision;
4 the policy supportive function: its influential recommendations on government policy with respect to post and telecommunications.

The Steenbergen Committee stated that the PTT's monopoly should be restricted to the installation, exploitation and maintenance of the national infrastructure and the supply of the basic services of telephony, telegraphy, telex, and data traffic, together with an obligation to provide leased lines. The present public monopoly on the provision of terminal equipment and tele-services should come to an end, allowing competition and permitting PTT to participate on equal conditions. With this liberalisation proposal the Committee bypassed the modest liberalisation measure put forward in the 1984 White Paper. The Committee expressed a liberal attitude towards the use of leased circuits by corporate users and new service providers. It argued that a ban on the resale of capacity and third-party use could impede the development of new services. In view of this, the Committee proposed that the provision of leased lines be included in the universal service obligation: the PTT should be forced to provide leased circuits on demand. Furthermore the resale of transmission capacity and the interconnection with other leased circuits to the public switched network should be allowed, on condition that these circuits be used for providing tele-services. The Steenbergen Committee suggested that a special agency (to be established in the future) would issue permits towards that end.

The Committee argued for a new governance structure in which the PTT would be granted an exclusive concession for the operation of the public infrastructure and the provision of the basic post and telecommunications services. In this concession-based regime a line could be drawn between the mandatory provision of reserved services and services offered in an open market. The concession could contain certain guidelines and preconditions with respect to universal service provision, quality of service, tariffing, consultation etc., to be acted upon by the concession holder in the operation of the licensed services. The Committee stated that the PTT's traditional involvement in regulation, licensing, type approval, standardisation and policy formulation needed to be separated from its operational function of providing services. The functions of rule setting, licensing and supervision would be taken away from the PTT altogether and assigned to a special agency in charge of post and telecommunications regulation attached to the Ministry of T&PW. This RVPT agency would regulate the PTT's provision of public services (i.e. tariffs and quality of service) and would furthermore be entrusted with the tasks of granting concessions and licences for network operation, recognition of testing institutions, type approval and certification and frequency allocation. The future institutional framework would include two other institutions, that would replace the Advisory PTT Council. The RATB, a committee of independent experts, would advise the Minister on post and telecommunications issues and longer-term planning with regards to the public infrastructure and the reserved services. The Consultative Body PTT would serve as a platform for the PTT and its major stakeholders to discuss future investment and marketing decisions with respect to the reserved infrastructure and basic services.

Regarding PTT's future legal status, the Committee proposed transforming it into a state-owned company, thereby lifting the political constraints of the civil servant statute and external borrowing prohibitions. The Committee showed that the PTT's profits and depreciations were increasingly transferred to the Treasury instead of remaining within the PTT administration to build a solid financial base. In order to finance the accelerating investments in network modernisation and the development of new services in the 1980s, the PTT required more control over its operating cash flow and direct access to outside debt funds. The Committee strongly recommended a legal entity status through a joint-stock company: this would enable the PTT to carry out its corporate and public functions more freely and prevent the government from interfering directly with the daily operations and hampering corporate planning (Commissie Steenbergen 1985a, b; McKinsey 1985). There would still be indirect governmental control because all the shares would remain in the hands of the state. Besides improving its financial flexibility and long-term return on investment, the legal-corporate structure of a public limited company for PTT would imply a replacement of the civil servants' statute, thus facilitating market-oriented wage and employment conditions. The Committee proposed that the future PTT holding should develop into a decentralised holding with three subsidiaries, one dedicated to mail and two dedicated to telecommunications. This meant that PTT was forced to split its telecommunications function into a *public utility* company exclusively in charge of the infrastructure and the provision of reserved services, and a *commercial* company active in the competitive markets of terminal equipment and tele-services. The public and private activities of the telecom branch were separated to secure fair competition by preventing a cross-subsidisation of its competitive services. The Committee did not see any reason to separate the utility and entrepreneurial function of the postal branch.

In December 1985, the cabinet endorsed the broad outlines of the Steenbergen Report, which fitted perfectly into its official privatisation and deregulation policy. The government argued in favour of an orchestrated and simultaneous implementation of the policy measures aimed at the structural reform of the PTT, the liberalisation of the markets for terminal equipment and tele-services and the enactment of a new institutional framework by 1989. The 1985 White Paper was directed towards encouraging innovation and competition in the provision of telecommunications facilities, while at the same time securing the nationwide and uniform provision of basic communications services. The government proposed opening up the markets of tele-services and equipment to competition and attempted to solve the discussion about the delineation of PTT's monopoly in terms of traditional and non-traditional terminal equipment (initiated by the 1984 White Paper). It adopted the Steenbergen recommendations and thereby overruled its previous decision. The government precluded the resale of transport capacity. To increase the efficiency and market responsiveness of the PTT administration, the government supported the change from a state department into a holding company with plc status (NV) and two separate post and telecommunications subsidiaries with Ltd corporate status (BV), all subject to civil law. Like any other private enterprise, PTT would have to aim for sufficient levels of return on investment, and was allowed to create subsidiaries and enter into joint ventures with other companies. This future PTT with the state as its sole shareholder, should be given permission to negotiate loans on the capital market, to enter into joint ventures and develop a market-based wage and working conditions policy.

The majority of stakeholders reacted positively to the Steenbergen Report and the subsequent 1985 White Paper. The large users, computing firms, service providers and consumer representatives saw their pleas for a more transparent market answered. A more flexible PTT, operating at arm's length from the government, could more easily respond to market demand. The demands for a functional separation of PTT's public utility and the commercial services, and the transfer of its regulatory and licensing activities to an administrative body, were met in the White Paper. The domestic equipment industry was one of the few parties, along with PTT and its unions, to criticise the reform proposals as being too radical. The manufacturing association asked for a phased introduction of the restructuring measures over five years. This would allow the equipment industry sufficient time to adjust itself to the new market conditions. Such a step-by-step implementation would also keep up employment levels. Other objections came from PTT, workers' representatives and the public sector unions. PTT found some major and minor shortcomings in the Steenbergen Report and the consequent White Paper, but endorsed the majority of the policy recommendations (PTT 1985a, b). PTT management supported the Committee's proposals for the concession-based system based on the exclusive provision of the infrastructure and basic services, and the liberalisation of the equipment and the tele-services market, as long as there remained a protected core telecommunications monopoly. It also strongly supported the proposal that the postal and telecommunications functions remain integrated within the future corporation, to be granted a legal entity status, operating autonomously from the government. The PTT strongly advocated cutting back political intervention and actively promoted a loosening of external borrowing and corporate employment restrictions. The PTT was in favour of granting the future corporation plc status, and agreed with the plan to create a new framework for consultation, advice, regulation and supervision. As far as the advisory body RATB and the Consultative Body PTT were concerned, the PTT was unable to discover a significant difference in terms of status and jurisdiction with the Advisory PTT Council.

Although broadly endorsing the Steenbergen recommendations, the PTT put forward some preconditions and criticised some of the proposals. First, the PTT stated that it needed sufficient time to prepare and adjust its internal organisation and corporate strategy to the new business environment. In order to start on an equal footing with its major competitors, it advocated prompt political decision making and a simultaneous implementation of all the measures. Second, the PTT detected a pro-computing bias in the Steenbergen Report, and an underestimation of the structural conflict in terms of regulation and standardisation between the information/computer industry with its global free trade ideology and the telecommunications sector with its nationwide public service tradition. The Steenbergen Report's insistence on allowing the resale of leased lines aroused in PTT a fear of the dangers of bypassing and cream skimming. Private companies could provide services selectively and at a lower price than the PTT, thereby leaving it with high-cost low-revenue customers. Third, the PTT strongly opposed the proposed split of its telecommunications function into separate public utility and commercial subsidiaries. Such an organisational separation was considered undesirable and detrimental as it would obstruct the exploitation of economies of scale and scope between the provision of the basic and the enhanced services. A divestiture of the telecommunications division would also imply another major internal restructuring, while that division was already in the process of reorganising its functional structure to increase market responsiveness and relocating its headquarters to Groningen as part of the government's regional economic policy.

The PTT workforce, organised in Departmental Committees and public sector unions, agreed with corporate management to keep both the telecommunications and postal services and the public and commercial activities integrated within the telecommunications function (AbvaKabo 1984). Both PTT management and the unions were convinced of the necessity for the PTT to develop its own corporate investment and employment policy, otherwise the company would lose its competitiveness and consequently market shares and jobs. The labour representatives were disappointed at the Steenbergen Report's views on the consequences of automation and digitisation, and the proposed far-reaching institutional changes for the PTT workforce. Unions and management were not really convinced that the plc structure, put forward by the Steenbergen Committee, was the appropriate status in the future. The public sector unions therefore commissioned an in-depth comparative study on the privatisation measure and the *sui generis* alternative. This study largely subscribed to the Steenbergen proposals: a joint-stock company, subject to company law, would have clear advantages over the *sui generis* construction (i.e. a para-state organisation) subject to public law in terms of corporate flexibility (access to the capital market, employment conditions and allowance for joint ventures) and legal-administrative transparency (limits to political interference, less complex and more compliant to cope with changes) (Berenschot Consultancy 1985). When the public sector unions realised that there was no real alternative to the clear-cut privatisation of PTT, the employee representatives moderated their criticism and joined the discussion on the reorganisation and the implementation of the new PTT structure. Since the question whether PTT ought to have a public or a private status had become less important, two things remained important for the organised PTT workforce: it strongly favoured an integrated organisation that would remain 100 per cent state owned.

Implementing structural reform (1986–9)

After the two Committees had put forward their proposals and the two White Papers were published, it was time for Parliament to prepare legislation. In April 1986 the Standing Parliamentary Committee on Transport and Public Works endorsed the outlines of the 1985 White Paper and authorised draft legislation. In December 1987 a package of four bills was introduced in Parliament:

1. the PTT Personnel Bill, including rules regarding the future legal status of PTT personnel;
2. the Telecommunications Bill, replacing the 1904 Telegraph and Telephone Act;
3. the PTT Authorisation Bill, regulating the new role for PTT as a limited liability company and granting the concessions under the new post and telecommunications legislation;
4. the Post Office Bill, revising and updating the 1954 Post Office Act.

This legislative package, to be implemented in January 1989, was discussed early in 1988 and was widely accepted by Parliament in June and enforced in October of the same year. The majority of the Steenbergen recommendations were followed by the government, only a few were not implemented. The Committee's proposal for a formal division of PTT Telecom into a public utility company and a commercial enterprise to avoid cross-subsidisation was not fully adopted by the government. The major political parties were divided: the major opposition party, the Social Democrats, was against, the Christian Democrats (in government) preferred to give PTT time to restructure itself, whereas the Liberal Conservatives (also in government) were in favour of a separation of PTT Telecom. The cabinet was put under pressure by the information and computing industry, supported by the Ministry of Economic Affairs, in favour of a divestiture so as to secure fair competition and market transparency. Finally a political compromise was reached in which the centre right cabinet did not rule out the legal separation of the telecommunications division, but its final decision was postponed and would be taken within five years after 1989. This would give PTT time to prepare its internal organisation for a more liberalised environment. Before the separation into two subsidiaries was decided upon and carried out by government, cross-subsidisation had to be precluded to make fair competition possible. In order to make the information flows and economic transactions between PTT's public and private activities more transparent, appropriate safeguards like separate accounting and reporting systems would be created.

The government furthermore changed the statute of the interdepartmental Advisory Council on Post and Telecommunications (RATB) and the regulatory/supervisory body (RVPT) acting autonomously from the Ministry, turning them into clear departmental bodies. The government decided furthermore that the new administrative structure would not be implemented before 1988, as part of the package deal to restructure the entire telecommunications sector. The information industry criticised the government for overlooking the conflicting interests of the Ministry as both regulator and sole shareholder of the PTT and underestimating the administrative 'teething problems' that would occur in the transition of a public monopoly to a regulated market. Representatives of the information industry strongly pushed for an independent regulatory framework and an interdepartmental advisory body on media and telecommunications policy, entrusted with a clear mandate and installed as soon as possible to accommodate the far-reaching changes to come. The government, however, gave clear priority to reforming the PTT organisation in favour of the restructuring of its external environment. The liberalisation of the equipment and service markets and the creation of a new regulatory, advisory and consultative structure were due to be implemented after 1988. The government argued that the new style PTT should start from a position of strength and that for the transitional period 1985–8 the tight political controls on the PTT with respect to its financial management and working conditions could be loosened provisionally. A near complete internal reorganisation of the telecommunications entity between 1983–5 persuaded the government to postpone the decision to separate the public and commercial activities until 1994. The limited size of the regulatory and supervisory body (HDTP) would prevent any regulatory interference and detailed investigations. The PTT (1985a, b) argued that the regulatory body had to be kept small (both in the number of people employed and in expertise and archives/data gathering) to co-operate more effectively (wherever necessary the PTT could provide technical support).

PTT prepared itself for the new situation by changing its management structure from that of a government department. It organised six directorates (Postal, Telecommunications and Money Services, Technical Affairs, Personnel, Financial-Economic Affairs and a Central Board) into an autonomous commercial corporation owned by the state, to be named KPN Royal PTT Netherlands. Direct management would shift from the Minister (supported by Parliament and the DG PTT) to the Executive Board of Directors of PTT, supported by a Supervisory Board of non-executives (with a minority of its members appointed by the government). Some of the members of the Executive Board were already appointed by government in 1987 and began to act in a shadow capacity behind the DG and the Central Board to prepare the corporatisation and eventual privatisation of the PTT administration. The majority of the newly installed Executive Board had experience in top positions in the private sector. Only a few of the then PTT directors were allowed to continue in an executive role in the new-style corporation, while others switched over to fulfil an advisory role or left the company. The distance between the state and PTT increased in the new setting, but governmental influence was still prevalent by way of the appointment of commissioners, supervision over the provision of network capacity and basic services, and the regulation of tariffs. The government also had an important say in the extent and timing of PTT's flotation (five years after putting in place the 1989 framework).

As early as 1986 management and employee representatives of the PTT started the official negotiations about the employment effects of the shift from a hierarchical administration towards a market-driven organisation. Initially the civil servants' unions rejected the plan to restructure the PTT, because they feared job losses and a deterioration of employment conditions. They regarded privatisation as an easy measure to combat mismanagement caused by lagging investments and constrained corporate planning (Soeterbroek & Walravens 1985). The PTT unions, however, found themselves alone in their opposition to the privatisation plans after PTT management had abandoned the *sui generis* construction. When the majority of PTT employees also supported the privatisation alternative, because such a switch would imply a prolongation or clear improvement of their salaries, secondary benefits and/or pension schemes, the unions decided to change their position and accepted the privatisation proposals (Van den Besselaar & Visser 1993). More or less convinced about the necessity to restructure the company and the consensus regarding the PTT's corporate autonomy from the government, the unions gave up opposition and opted for an active involvement in the policy process by participating in the negotiations about the employment effects. In May 1987 an agreement was reached between management and the unions to make every effort to keep employment levels up and prevent any layoffs within the company for the transitional period 1987–91 (PTT 1987). The government furthermore stipulated that PTT management had to make sure that in 1989 none of its employees would be worse off in terms of income. For a transitional period of eight years the PTT had to maintain and gradually phase out the pre-1989 income levels for the lower ranks in its workforce by reducing compensation schemes. The higher ranks saw an immediate increase in salaries. An agreement was reached between PTT management and labour representatives about the replacement of the civil servants' statute by a single collective labour agreement to be decided at the holding level, including both the capital-intensive telecommunications and the labour-intensive postal subsidiaries. In December 1988 the first collective labour agreement between management and the unions was reached. For the first time in its corporate history PTT staff were no longer paid as civil servants and tied to the pay scales of the government administration, but subject to a corporate labour agreement in line with the private communications industry.

A second round of structural reform: 1989–94

The Dutch legislative framework, established in 1989, was based on a system of privileges and obligations for KPN. This was to secure its exclusiveness in the provision of basic services and to protect the public interest by forcing KPN to safeguard the provision of these services. The Telecommunications Act was designed to serve the following purposes: a) to promote the development of the public infrastructure and basic services in a socially and economically responsible way; b) to increase KPN's operational, financial and strategic scope; and c) to increase business opportunities for other companies in the equipment and teleservices markets. After the 1989 framework was installed, the pace of liberalisation and privatisation was slow. The government paid lip service to the principle of increased competition and sometimes used delaying tactics, in order, among other reasons, to prepare the PTT company for its eventual privatisation and to protect the newly established regime from a hard confrontation with international market forces and new technologies.

The present legislation did not make clear how the government proposed effectively to regulate such a dominant market player as KPN/PTT. The proposal for a legal separation of PTT was originally included in the 1989 Act to serve the policy objective of fair competition in the domestic market by making cross-subsidisation between the utility and the commercial divisions virtually impossible. The advocates of separation maintained that the established structure enabled PTT Telecom to subsidise its commercial activities in the non-licensed markets (notably terminal equipment and foreign investments) through its exclusively licensed services. The government and KPN agreed in 1992 that a divestiture of PTT Telecom was politically undesirable and would be counterproductive for the domestic telecommunications sector, given the relatively small size of the Dutch operator and the intensity of international competition. A structural separation of PTT Telecom would not only hamper its foreign expansion, it would complicate the flotation process, generating less revenue for the government. PTT Telecom should be allowed to build up a strong presence in the international market through internal growth and strategic partnerships on the basis of a strong presence in its home market. PTT Telecom argued that, in view of increasing competition and the size of its foreign competitors, such a legal separation would hamper its universal public service obligation in the domestic market and adversely affect its position abroad.

The government also contributed willingly or unwillingly to a lack of transparency in the Dutch telecommunications field. Besides acting in a supervisory capacity with regard to the single operator and the industry, the government, as sole shareholder, had a direct interest in the regulated company. As legal experts have pointed out, the 1989 Act incorporated four partly conflicting goals (De Ru 1988; Dommering 1988): protecting the PTT's monopoly position in view of the supply of public services and the infrastructure; boosting the PTT's international position; enlarging the possibilities of third parties to provide services and equipment in order to stimulate innovation; and generating substantial proceeds for the Treasury. The socio-economic mission of PTT was hampered by the government's demand for high dividends and the (potentially) high revenues to be generated by a flotation of the PTT, envisaged in the near future. The government clearly attached more importance to the PTT's contributions to the national budget than to lower consumer prices. PTT management was clearly in favour of privatisation, provided that the company would remain integrated and retain some of its monopolistic privileges. It consolidated its dominant position in the domestic market segments and embarked upon an international expansion strategy through the active penetration of new geographical and/or product markets and the build up of strategic alliances. Another argument for dropping the legal separation of the PTT had to do with the fact that the divestiture of the PTT's public and commercial functions into two separate companies presupposed a formal and stable boundary between the regulated and non-regulated markets. In a business environment where rapid technological developments and international deregulation already challenged some of the dedicated services, the stipulation of such a public–private borderline seemed to be very difficult to accomplish. The Minister considered the monitoring and supervision of fair competition in Dutch telecommunications to be the responsibility of the European Commission, thereby circumventing any national responsibilities in applying Dutch competition policy to these matters.

Between 1989–94, Dutch telecommunications policy was brought into line with the enlarged scope of European Community legislation by (partially) opening up the markets of satellite, mobile and datacommunications to competition. The liberalisation process in the Netherlands received a new impetus in the summer of 1992, when McKinsey (1993) was commissioned to investigate the needs of the business user community with respect to the availability, quality and pricing of telecommunications services. The McKinsey Report contained a positive evaluation of PTT Telecom's residential telephone services, but criticised the operator's business services. Both PTT Telecom's domestic and international telephone services were rated as more than adequate and comparatively cheap. McKinsey made clear that service provision to large users was below average, mentioning complaints about high tariffs, poor quality of service, and shortcomings in the range of services provided (e.g. low degree of network digitisation, restricted supply of leased lines with large bandwidth, intelligent network services, cellular services etc.). The consultancy firm was also asked by the Ministry to examine the then existing telecommunications policy and to make recommendations about a future organisation of the market. These policy alternatives should be based on increasing service provision to business users, strengthening KPN's position in the international market and ensuring equal access and universal service. McKinsey (1993) proposed four scenarios for the Dutch telecommunications market in the medium term:

1 liberalisation of public telephony while preserving the exclusive rights of PTT with regard to the infrastructure: the provision of telephone services on the basis of the public network (leased lines);

2 the prolongation of the exclusive rights of PTT on the network and the reserved services, together with government regulation with regard to the introduction of new services, quality of service, tariffing and infrastructural development;
3 limited network competition: PTT has an exclusive service obligation for voice telephony and alternative operators are allowed to provide data services on the basis of leased circuits;
4 free-for-all competition on the former dedicated markets of network provision and voice telephony.

After less than four years, it became clear that the focus in Dutch policy making had to be shifted from giving priority to the general socio-economic interest of telecommunications, as expressed in the 1989 Act, to focusing on the interests of large business users and the promotion of competition. On the basis of the evidence provided by McKinsey, the Minister argued for the option of *managed competition* between 1993 and 1998, trying to find a balance between state-protected monopoly and free-for-all liberalisation. The public monopoly of the past would be phased out and replaced by an oligopolistic telecommunications market, characterised by new entries and competition between PTT and mobile operators, cable operators, service providers, etc. The basic assumption of the 1989 Act, namely the principle of *unity of control*, was no longer valid. The notion that network management had to be vested in one single entity had effectively been challenged by techno-economic developments and the increasing influence of EC legislation upon domestic policy (V&W 1993). The PTT controlled monopoly was increasingly put under pressure by the European Commission, targeting both the liberalisation of the European telecommunications market and the instigation of the relatively detailed Community regulations to ensure open network provision and fair competition, that were imposed on the national levels of policy making.

The Dutch government proposed introducing network competition by allowing 'private' companies with wayleaves to sell excess network capacity to third parties (V&W 1993). By offering a premium for a nationwide licence to provide an alternative fixed network, the Minister intended to stimulate cable operators, Dutch Rail NS and power utilities to integrate their networks and merge their businesses into a new public operator. If this plan were to be approved and implemented, the Netherlands would become one of the few European countries to allow network competition. From 1995 onwards, competition between PTT and a single contender in the provision of leased lines, data services and closed-user-group telephony, would then be permitted. In line with the Commission's plan for an Internal Services Market to be implemented by 1998, the government suggested that public voice telephony would continue to be provided exclusively by KPN in this transitional period. With these liberalisation measures, the Minister clearly responded to the demands of business customers and 'private' operators willing to become the second public operator. In its quest to increase network competition, the Dutch government itself was strongly involved in the development of this duopoly by protecting KPN and imposing the pre-selected collaboration of semi-public operators to create an alternative provider (instead of organising an open tender). At that time, the Dutch government still controlled KPN and NS Dutch Rail in the capacity of single shareholder and the cable/utility companies (the large majority owned by local/regional public authorities). The government argued that the corporate interests and long-term viability of PTT could be improved by gradually liberalising its critical markets of network provision and basic services, while establishing a transitional duopoly before genuine competition could eventually be introduced. The government assumed that phasing out PTT's exclusive rights before 1998 would have a positive effect on the valuation and flotation of KPN after 1993.

The reactions to the government's plan were relatively positive. CAPT (1993b, c) and NOTA (1993), two public bodies advising the government on telecommunications and technology policy respectively, welcomed full market liberalisation. They were, however, against posing limitations on the government's intentions to restrict network competition to a duopoly, and a single contender of KPN, consisting of Dutch public enterprises willing to enter telecommunications. Instead the two advisory bodies suggested a speeding up of decision making and demanded a more liberal approach, emphasising the active promotion of competition, asymmetric regulation that would curtail the incumbent and protect any prospective competitor(s), the establishment of an independent regulatory agency, that would ensure effective competition and quality of service provision. The CAPT body (1993a, b) argued that limiting the market to two players was not desirable. Besides nationwide competition between KPN and its contender, competition at the local level needed to be facilitated by granting licences to local and regional operators with only restricted service requirements and/or PTT being prohibited from entering cable distribution. Furthermore, in order to establish workable competition, CAPT demanded that overseas companies should be allowed to invest and participate in the alternative network venture. Given the huge costs of upgrading and integrating these networks, foreign capital and expertise (notably technological and marketing know how) were strongly needed.

Much like the government, organised business welcomed the promotion of competition as the appropriate instrument to improve PTT's quality of service provision and to increase the overall efficiency of Dutch services' industries. BTG (1993), representing the large business users, complained about the liberalisation plan not going far and fast enough, partly due to the government's slow response and its overly patronising attitude. BTG complained about the lack of urgency and feasibility of the plan to open up the telecommunications market gradually between 1995–8. Given the technological convergence caused by digitisation and the absence of an adequate definition of the 'closed-user-group' concept (i.e. separating private from public telephony), BTG demanded the liberalisation of the infrastructure and service provision as soon as possible (for instance by 1995). The employers' association RCO (1994) criticised the Ministry for underestimating the neglect of foreign participation in the Dutch market, the absence of closed-user-groups, provisions, and the underestimation of the necessity to design a new regulatory regime.

Encouraged by the government to become PTT Telecom's single competitor (owning about half of the cable networks), Dutch Rail decided to join forces and establish the Enertel/NS consortium with the power utilities having a 75 per cent and the railways a 25 per cent participation. Instigated by the government's condition to grant the licence if nationwide coverage would be achieved, the power utilities of the consortium started to acquire local cable networks and upgrade their infrastructure in order to make switching possible. The participation of the largest cable operator Casema in the alternative network venture would be crucial. Although Casema was still owned by KPN, the government announced that KPN should loosen its ties with its subsidiary in order to give nationwide network competition a chance. Before entering into genuine market competition, Enertel/NS sought temporary preferential treatment on the basis of the infant industry argument to become a full service company providing network capacity and fixed and mobile network services. Enertel/NS demanded from the government a permission for the provision of voice telephony, access to PTT Telecom's network on cost-related interconnect charges, asymmetric regulation (a temporary exemption from universal service requirements and ONP conditions). They also hoped to be allowed to collaborate closely with the alternative cellular operator(s) in order to achieve substantial economies of scale and scope.

Government and Parliament proved susceptible to some of the suggestions provided by the stakeholders in the policy process. The cabinet, for instance, would permit overseas investment in the prospective second operator if certain conditions were met. Foreign participation would be restricted to 25 per cent and the international operator would be selected on the basis of the reciprocity argument (i.e. the openness of the candidates' home markets). On the basis of these preconditions concerning any future strategic alliance, Enertel/NS chose the US-based BellSouth as their international partner. The evidence for establishing asymmetric regulation for a transitional period was also recognised by the government when it proposed to grant the alternative carrier a licence by the summer of 1995, even if it could not guarantee a nationwide coverage during the first years of operation. The government was also not necessarily against any business linkage between the prospective second GSM operator and the future alternative network provider. The pleas for speeding up the liberalisation process and installing an independent regulatory agency separate from the Ministry were considered understandable but were nevertheless pushed aside as not being urgent enough.

The last hurdle (1995–8)

In the period 1993–5, the government sought to prepare new legislation that would succeed the 1989 Act, based on the establishment of infrastructure duopoly between PTT Telecom and a second franchisee, to include all the alternative network operators in the country (e.g. energy, railways and cable companies). In order to be awarded the second franchise to compete at the different levels of the network with PTT, far-reaching collaboration and consolidation was required in this

Enertel/Telecom-2 joint venture. This fell apart because some of its demands could not be met by the Dutch government (e.g. GSM licence and asymmetric regulation and protection), but also because of internal disagreement. While Dutch Railways NS and the energy utilities without cable networks preferred the strategy of constructing and operating trunk networks, the cable network operators and the utilities with a stake in cable networks gave priority to the promotion of local/regional services.

In July 1996, with the adoption of the Interim Act, the fixed infrastructure was liberalised, permitting alternative networks to have a competitive edge over new service providers before full competition would develop (from July 1997). Furthermore, the Interim Act introduced 'permits', based on registering network/services, and 'licences', based on the licensing of network/services with special rights and obligations. Besides treating a permit holder and a licensee differently, the new licensing regime makes a distinction between a licence without territorial restrictions and one with territorial restrictions. When granted a licence after a competitive tender, the infrastructure licensee was entitled to provide all kinds of services on the basis of a system of rights and obligations (i.e. rights of way, interconnection rights and obligations, and ONP obligations). The permit holder was allowed to provide services after simple registration with no special rights and obligations. The distinction among licensees between those with and those without geographical restriction refers to national and regional operators. After a competitive tender, national infrastructure licences were granted to Telfort (BT and Dutch Railways NS) and Enertel (an association of regional utility companies); these two groups are now competing with the concession holder KPN/PTT Telecom in the market for network provision. Existing alternative infrastructure providers (e.g. cable network operators) are given preferential treatment in the allocation of regional licences. The licensing instrument has been used by the Minister to curtail the dominance in the domestic telecommunications market. She refused to grant a licence to an integrated KPN conglomerate and made it divest itself of the cable subsidiary KPN Kabel/Casema.

The final steps to create an open and transparent telecommunications market in the Netherlands are to permit (limited) entry for new players in the key voice telephony market and to promote and safeguard competition through an appropriate regulatory framework. In July 1997, the exclusive rights of KPN/PTT Telecom to provide voice services were withdrawn and replaced by a licensing regime in which new national and regional service providers compete with KPN/PTT Telecom. Also in 1997 an independent telecommunications authority was established that is responsible for supervision (e.g. licence compliance), conflict resolution (e.g. competition issues and interconnection disputes), quality of service and consumer protection. The post-1998 open market and regulatory framework will be governed through a new Revised Telecommunications Bill, replacing the 1989 Act and its subsequent amendments, which is due to be passed in Parliament and come into force in winter 1998.

5.5 The liberalisation of the Dutch telecommunications market

In the new telecommunications setting, introduced in 1989, PTT's public monopoly was reduced to a core monopoly and replaced by a concession-based system. PTT became the sole concession holder for the provision of the infrastructure and the reserved telephone, telex, datatransmission services on a universal basis, including international service and the provision of leased circuits to any interested party on request. Although special licences were needed to move beyond broadcasting, the operation of local cable networks by municipalities or private companies was tolerated. The 1989 legislation allowed for the establishment of alternative private networks should PTT fail to provide the facilities required or meet the minimum requirements. The interconnection between private and/or cable networks without using the PTT infrastructure was, however, strictly forbidden. All other telecommunications facilities, value-added network services and terminal equipment could be provided in a free market environment. No licensing or registration regime has been installed for the provision of these non-basic services. In addition to the liberalisation of the markets for telecommunications services (e.g. non-basic services) and terminal equipment, already enacted in 1989, the scope of KPN's exclusive concession was gradually reduced to the 'core' of providing voice telephony services. Between 1991 and 1997 the Dutch government liberalised the following markets: satellite communications (1991), databearer services (1993), mobile communications (1994) and the fixed infrastructure (i.e. network capacity) (1996). Voice telephony was eventually opened to competition in July 1997, just before the EU's 1998 liberalisation deadline. Early in 1998, competition in the cellular market duopoly (PTT Telecom and Libertel) was stimulated through the granting of a DCS 1800 and another GSM licence. After a brief discussion of the changes in the equipment market, we will discuss in more detail how PTT successfully attempted to expand its public monopoly in the newly emerging markets of tele-information services, cable networks and cellular services.

The liberalisation of the equipment market

Traditionally, the provision of telephone equipment has been a PTT monopoly. The terminal equipment PTT rented to its customers and the network switches were purchased from its preferred suppliers Philips, Ericsson and ITT/NSEM, with Philips being the dominant manufacturer. With respect to switching equipment AT&T, (after taking over the Philips subsidiary), held more than 50 per cent of the market, with Ericsson and Alcatel (having acquired ITT/NSEM) making up the other half (PA 1991–2). The market for switching equipment was highly sensitive, in which the political interests of supporting the domestic industry often prevailed over the efficiency arguments of foreign competitors and the corporate interests of PTT. An illustration of this is the announcement by the government in 1984–5, that three manufacturers would be selected to upgrade and digitise the public infrastructure. PTT had asked for only two suppliers of digital switching equipment, notably AT&T/Philips and Ericsson, but the government decided to add ITT/Alcatel to the list, fearing the closing down of its Dutch subsidiary. Ten years later, when PTT had finally achieved corporate autonomy, the Alcatel-ITT agreement was reversed: PTT Telecom decided to drop Alcatel as major supplier (with the exception of GSM equipment). In the post-1989 situation, PTT benefited from a clear bias in the liberalisation measures in the terminal equipment market in favour of the former monopolist by being allowed to remain actively involved in the testing and certification of terminal equipment. PTT's equipment was automatically approved, while its competitors had to rely on time-consuming procedures at the recognised testing agencies and long delays before certification was given. The departmental agency HDTP, the notifying body in telecommunications normalisation, had delegated the task of type approval to the designated bodies KEMA and the PTT subsidiary NKT/PTT Contest, while also recognising the EC certification agencies. Whether this situation with respect to PTT being player and referee at the same time will become more controllable with the 1995 agreement to merge NKT and KEMA into one certification agency, called Telefication, remains to be seen.

Network integration or network competition?

In the late 1970s the Dutch telecommunications system contained two separate local infrastructures: a PTT-controlled system and (until recently) an extensive, decentralised and poorly interconnected network of more than 1,000 cable systems, controlled by municipal authorities and operated by regional utilities or private cable companies. Due to an obscure formulation in the 1904 T&T Act, municipal cable networks were not part of the PTT monopoly (Hins 1991). The installation and exploitation of these local networks was solely allowed for broadcasting on the basis of an exclusive authorisation given by PTT. For new interactive services, like videotex, telemetry, alarm systems and the like, a special franchise was needed. Only a few of these licences were granted to cable companies (mainly for experimental purposes). PTT determined the technical conditions cable networks had to meet and exclusively provided the broadcasting signals for the distribution to customers as well as the interfaces between these local cable networks. Although the penetration rate in the Netherlands was already high in the early 1980s (about 90 per cent), there was little or no harmonisation of standards in the national cable system. Furthermore, the legislative regime prohibited any collaborative efforts among these local monopolists, and PTT was reluctant to furnish licences for the provision of non-broadcasting services distributed via cable. PTT regarded the cable system as belonging to its public monopoly. To that end, PTT established Casema, a private cable operator which became actively involved in the cable market.

Originally the two local networks were different in their technical specifications, communication patterns, regulation and use. However, as a consequence of technological developments (digitisation and the diffusion of optical fibres) and an increasing demand for tele-services, the evolution towards one integrated broadband communications network was regarded as inevitable (Commissie

Swarttouw 1982). After the 1984 White Paper had suggested further research on the subject of infrastructural co-ordination, the cabinet in 1985 announced the installation of an advisory committee dealing with the investigation of harmonising the cable and the local PTT networks. The Committee, named after its chairman Walter Zegveld (then Director of the Centre for Technology and Policy at the Netherlands' Organisation for Applied Scientific Research TNO/STB) was asked to investigate the possibility and desirability of integrating the two separate local infrastructures into a single network controlled by PTT and to give policy recommendations on its implementation (Commissie Zegveld 1986). The Committee included representatives from the Ministry of Economic Affairs, the Ministry of T&PW, information providers, equipment suppliers, cable operators, and the Union of Municipalities (VNG).

In 1986, the Zegveld Committee advised in favour of the future harmonisation of the two local communications infrastructures into one integrated broadband network to be controlled by PTT (Commissie Zegveld 1986). In its plea for network integration the Committee was led by the considerations of economies of scale and scope and by optimism concerning the large-scale diffusion of fibre optics and eventual technological convergence. The low costs of a universal network and high-capacity optical fibres would enable a cost-effective exploitation of cable television and facilitate the promotion of tele-services. The modernisation and enlargement of the two existing networks without at the same time integrating them would lead to notable losses of capital. The Committee concluded that PTT should be allowed to integrate all the cable networks into one fully controlled local infrastructure over a period of twenty years. The development towards an integrated broadband system and the transfer of the cable networks to PTT should be given a flexible legal basis, taking into account local conditions and possibilities. The timing of these take-overs could vary from one cable network to another, depending upon the necessity for and financing of infrastructural investments and the depreciation schemes of the cable network involved. In order to co-ordinate the replacement of the local infrastructure over different time spans and facilitate mutual adjustment between PTT and cable operators, the Committee demanded an institutional framework of various committees at local, regional and national levels with sufficient public and private support. The Committee also proposed that the government facilitate the collaboration between PTT and the group of cable operators by a financial contribution of NLG 600m and the creation of a broadly represented national Steering Committee.

The response to the Zegveld Report was mixed. PTT, supported by its subsidiary Casema, reacted positively to the recommendation for an integration of the two local infrastructures under its exclusive responsibility. On the other hand, PTT dismissed as too complex the Committee's proposal to install a framework to co-ordinate the integration process. PTT felt that bilateral negotiations with the cable operators would be more appropriate. The information providers asked for a clear separation between PTT's public utility and competitive functions and measures aimed at stimulating the market for tele-services and strengthening the position of the information providers. The think-tank of organised business

CIB-RCO (1987) criticised the Zegveld Report as being too absolute in its twenty-year integration path and thereby overlooking major uncertainties in the telecommunications environment in terms of technology and market demand and the envisaged privatisation of PTT.

The Report was not supported by the cable lobby of the municipal authorities and cable operators, who complained that it failed to give them sufficient commercial opportunities. According to the municipalities and cable operators, the Zegveld Report was too technocratic and too centralist to cope with the local variety and differentiation of cable networks, and it overestimated the cable operators readiness of cable operators to invest in their networks (VNG-VECAI). They recognised the necessity to integrate the integration of the local PTT infrastructure and the cable networks, to work out plans and scenarios for the future (together with PTT) and to initiate pilot experiments. A future PTT monopoly should be restricted to control over the infrastructure and not include the provision of services.

In 1988 the White Paper on the Zegveld recommendations was presented to Parliament. The government made it clear that the large-scale harmonisation of the local infrastructure required a balanced approach, going beyond the technical aspects. The cabinet emphasised that a duplication of the local infrastructure should be prevented and network integration ought to be achieved in terms of its technical aspects, economic control and administrative regulation. Like the Zegveld Committee, the cabinet expected that in the long term an integrated broadband network would take shape, in which the distinction between the PTT and the cable network would disappear. This future broadband network should be part of the exclusive concession given to the PTT. Although cable operators would lose their licence to operate a future cable network, the cabinet suggested that they could still provide broadcasting and broadcasting-related services. The cabinet argued that the time horizon for complete network integration of twenty years, as proposed by the Committee, could be reduced to twelve years. Rather than opting for an overarching bargaining framework for PTT, cable operators, municipalities and a national steering committee, the Cabinet argued for a more passive role for government by delegating essential issues such as achieving network integration, setting a timetable and agreeing about a take-over price to the PTT and the association of cable operators VECAI. The government also refused any financial participation in the process of network integration. The financial burden of the operation should be carried entirely by the local cable operators and PTT.

In 1989, PTT and VECAI started bilateral negotiations to work out a joint approach to the integration of local networks, dealing with such issues as the planning of the conveyance, compensation and the promotion of and control over the exploitation of new tele-services at local level. The cable operators, supported by the Liberal Conservatives in Parliament, found the authorisation regime of the PTT too restrictive for the development of new tele-services (VECAI 1990). The government decided to make the licensing of cable operators for the provision of new services dependent on the collaborative plans of PTT and VECAI. By linking the issue of franchising to laborious negotiations between the two stakeholders in the network market, the already restrictive cable regime with respect to the provision of tele-services became even more constrained. While PTT was allowed to provide all the various services, the cable operators were excluded from the supply of voice and data services and were only allowed to provide tele-services after being authorised by PTT. VECAI attempted to balance technical integration and cable liberalisation by promoting some kind of collaboration with PTT, while at the same time preventing a take-over of their local networks. The cable operators were, however, internally divided. The smaller municipal operators were in favour of selling or contracting out their networks to KPN/Casema, while the larger cable operators demanded a more liberal approach towards the operation of cable networks.

In 1991, the PTT-VECAI negotiations collapsed. VECAI found it impossible to accept the idea that network distribution and the exploitation of interactive broadband services would fall exclusively within the parameters of the PTT, making it virtually impossible for cable operators to move into new businesses. According to the cable operators PTT should be restricted to network provision for broadcasting purposes, while new interactive services should be left to others. PTT insisted on a local broadband communications network under its exclusive centralised control. Subsequently it has taken over several cable networks and entered into strategic partnerships with other cable operators and potential operators. Another factor that has surely contributed to the breakdown of negotiations is the increasing influence of EC legislation upon domestic telecommunications policies. The Dutch strategy of network integration could not help but go against the deregulation trajectory as developed by the European Commission.

Besides the fact that huge investments were needed to upgrade the infrastructure, the Dutch cable sector suffered from political intervention, internal fragmentation and a lack of market responsiveness. To strengthen its market position, VECAI (1990) demanded a liberalisation of the cable market and a cutting back of detailed political interference at both local and national level. In the spring of 1993, however, the government took the decision to liberalise the cable regime in the near future. From 1994 onwards, cable operators were permitted to collaborate and merge their networks and were allowed to offer interactive applications and data services (with the exception of network provision and voice telephony which would be liberalised later). The efforts of the cable companies to develop some form of regional collaboration and eventual network integration might seem promising, but there is still a long way to go before they will be able to compete effectively with PTT. A first trend towards concentration emerged in the Dutch cable sector between 1992–5, when several municipalities sold their cable networks to the power utilities, Casema and PTT Telecom. In 1994 Casema (with a market share of 17 per cent) acquired CAI-NKM (one of the larger cable companies with a market share of 5 per cent. Under domestic political pressure, both from the national and EU authorities to establish network and services competition, KPN suggested that it would reduce its majority share in Casema by establishing joint ventures with local and regional cable operators and service providers.

In 1993–4 the government announced that a temporary market duopoly between PTT and a prospective alternative network provider, consisting of the railway company, power utilities and cable operators, would be installed before full liberalisation of the Dutch market in 1998. The proposals to liberalise voice telephony and network provision in two stages, and to force the railway company, utilities and cable operators to merge into one competitive domestic carrier, were politically controversial. This transitional period of restricted competition would enable the alternative operators to pool their networks, operational capabilities, capital, technical expertise and marketing experience and catch up with their major contender. PTT, however, protected from immediate competition by strong (foreign) outsiders, would have more time to prepare itself for head-on competition and complete its corporate restructuring. The plans to install a managed duopoly for voice telephony and network provision and the alternative operator encouraged by the government were actually abandoned in the summer of 1995. The ambitions of the three constituents diverged between following a niche strategy by either becoming a local/regional full service provider, or a long-distance operator, or to compete head-on with PTT Telecom by offering local, long-distance and international services.

To enable the granting of infrastructure licences and liberalising voice telephony, the Interim Act and the Licensing Act were passed in the summer of 1996. This legislative package allowed for a managed and staged transition from monopoly to competition in network provision and voice telephony, endorsing new entry for Enertel (a group of energy and cable companies) and Telfort (BT and the railways company NS), and any other carrier meeting the qualifications. The government showed its determination to promote competition in fixed network and voice services in November 1996, when KPN's suggestion to transform Casema into a company in which the influence of shareholders is restricted and transferred to an independent board (*structuurvennootschap*) was not accepted, and the incumbent operator had to divest the majority of its CATV interests, including Casema.

The sale of the Amsterdam cable television network to the American-Dutch A-2000 consortium (including US West and United Philips Corporation) in 1995 was not just another step towards concentration, but also triggered a process towards foreign direct investment in the Dutch cable and telecommunications industry. Most of the eighty independent companies were then already sold off by the municipal owners and at present the sector is dominated by Enertel, a collaboration of the power companies and their shareholders, the regional authorities. One year later the Dutch Railways linked up with British Telecom and they set up the national operator Telfort, and MFS/WorldCom was granted a licence to build a telecommunications network for the business sector in Amsterdam. In the summer of 1996, Casema joined the Enertel federation by acquiring a 17 per cent share in the third Dutch fixed-line operator. One year later, KPN sold the Casema company to France Télécom for NLG 1.7bn, and the regional cable operator NUOM/Telekabel, one of the Enertel participants, merged with UPC.

The market for mobile communications

In the early 1990s the Dutch cellular market was relatively underdeveloped with a low market penetration, overcharged pricing, and faults in service provision. Like Belgium and Italy, the Netherlands have been slow in opening up the market for mobile communications. This was partly caused by a hesitant and protective government, insufficiently aware of the economic benefits of mobile communications and the positive effects of competition upon overall productivity and efficiency levels (PA 1993; Slaa 1993). PTT Telecom has also been relatively slow to realise the market potential of cellular technologies (ABN AMRO 1993; BZW 1993; McKinsey 1993). The reasons for this are manifold: poor marketing (one-sidedly concentrating on car telephony), poor quality of service, lack of price differentiation, and a go-it-alone strategy regarding sales and service provision. It was not until 1993–4 that PTT Telecom extended service provision to third parties, such as 'airtime resellers' Martin Dawes, Debitel and Talkline, established a network of retail outlets for the sale of cellular equipment and introduced a more flexible charging structure, based on varying combinations of subscription, usage and territory. Instigated by the liberalisation proposals of the European Commission and the fear of losing the country's competitiveness, the Minister and the relevant policy community were persuaded to replace the public monopoly on mobile communications by a more open mobile market in 1992. The political process, however, lost momentum and came to a complete standstill one year later. In 1995 a cellular duopoly was established, in which a second operator, Libertel, competed with PTT. So far decision making about the opening up of mobile communications (GSM) has been characterised by slow responsiveness and strategic neglect (by the government), tactical manœuvring (by PTT) and delays caused by interdepartmental clashes and legal-administrative dismissal.

Liberalisation on the part of the mobile communications market started in July 1992, when a provisional agreement was reached between the Ministries involved over an entrance fee for the two GSM providers of NLG 40m for the second licence. In the established legislative framework, cellular communications belonged to the reserved transport services and to grant additional licences in the provision of mobile services, an amendment to the 1989 Telecommunications Act was needed. Initially, PTT strongly opposed these new liberalisation proposals by arguing that the country was too small for more than one operator and that neither its investments in analogue mobile facilities nor the initial losses were fully recovered. While attempting to delay institutional change PTT at the same time started to upgrade its analogue mobile network by preparing the 'roll out' of its digital GSM network. At the end of 1992 the Minister drafted the amendment that would enable the opening up of the mobile communications market by suggesting the creation of a competitive licensing regime. Two licences with a duration of fifteen years would be issued: one would be automatically reserved for PTT Telecom (on the basis of its exclusive concession regarding the provision of reserved services), while the other would be awarded through a public tender procedure. After hearing about the promising prospects of mobile communications, the Treasury demanded a rise in the entrance fee to NLG 500m. The Treasury was keen to benefit from the windfall profits of selling mobile telecommunications licences in order to balance the budget. The Ministries of Economic Affairs and T&PW, backed by organised business, were strongly against such a high entrance fee. The two Ministries argued that this would be ineffective for an infant industry like mobile communications and would furthermore seriously damage the competitive position of the Dutch service industries in general.

After almost a year of departmental conflicts in the spring of 1993, agreement was reached between the Treasury and the two other Ministries involved to abolish the entrance fee proposal and to levy an additional 7.5 per cent profit tax from the two operators after they had installed their mobile networks properly. Although a broad consensus was developing about opening up mobile communications for competition, it also became clear that crucial issues like fair competition and interconnectivity still needed administrative settlement. Given the fact that the second GSM operator would be dependent on PTT Telecom for interconnections, prospective mobile operators feared that without a clear organisational separation of PTT Telecom's fixed network and mobile activities, a level playing field could not be established. Their proposal that PTT Telecom had to establish a legally separate company for operating its GSM licence was not taken over by the Minister, however. The proposal for a mobile communications duopoly suffered another setback when in the summer of 1993 the Council of State failed to endorse the legislative amendment. The Council of State argued that the proposal to impose an additional profit tax on the domestic GSM operators would go against Internal Market rules. The consultative body OPT (1993) argued that a duopoly model could easily instigate anti-competitive practices and collusion between the two operators. In case workable competition did not emerge, OPT suggested the possibility of allowing market entry for a third player. Like the Council of State, OPT was against the proposal of a 7.5 per cent levy on the profits of the two licence holders. Such a profit tax would be included in tariffing passed on to consumers and lead to higher charges, and it would also hamper a rapid development and diffusion of cellular technologies.

In 1994 the government decided to abandon the profit tax measure and to replace it by a virtually negligible fee that basically included administrative charges, to be paid by the two licensees. After several years delay, the new legislative amendment was discussed in Parliament and finally adopted in the summer of 1994. The assignment of the second licence was made official in spring 1995: the Libertel/MT-2 consortium (including ING Bank and Vodafone) was granted the second GSM licence, allowing it to start with the construction and operation of an alternative mobile network. This duopoly was only fully operational as early as the end of 1996. With PTT Telecom's present *de facto* monopoly on analogue mobile communications and the initial build-up period of an alternative digital cellular network which began in the summer of 1994, PTT has had substantial first-mover advantages and takes up a considerable market share, while serious competition may only occur in the medium term. Obviously, PTT will benefit from all mobile traffic including flows from its competitor, which is routed through its fixed infrastructure. With the envisaged arrival of a second fixed infrastructure provider, this one-sided dependency may eventually disappear. Furthermore, in February of 1998, the Dutch government auctioned two additional national cellular licences: the Federa consortium (Global One, Rabobank, ABN AMRO bank) paid NLG 600m for a DCS 1800 frequency and Telfort paid NLG 545m for an extended GSM frequency.

5.6 The privatisation of KPN/PTT Telecom

In 1989, the new public limited company KPN Royal PTT Netherlands became the largest private employer in the country with a workforce of 100,000, of whom about 30,000 were PTT Telecom employees. After gaining access to the capital market, the improvement of KPN's financial flexibility led to a steady increase of investments in the telecommunications and postal infrastructures to annual levels of about NLG 3.5bn. KPN transformed itself effectively from a government bureaucracy into a commercially driven company with a stronger focus on various product/market combinations. KPN has developed into a transnational multimedia conglomerate by safeguarding its dominant position on the Dutch information and communication markets, while at the same time building up a strong worldwide presence through establishing a chain of sales offices and linking up with international partners. In 1994, for instance, KPN had a turnover of about NLG 19bn with operating results at NLG 3.5bn with PTT Telecom accounting for 68 and 83 and PTT Post 31 and 16 per cent of KPN's total revenues and profits, respectively (KPN Annual Report 1994).

The public offering of a minority of KPN shares in 1994 was the final step in the privatisation process that has transformed the pseudo-privatised state-owned corporation into a normal joint stock company listed on the stock exchange. In order to become more efficient and effective in the market place and more attractive for potential financial investors, KPN/PTT started substantially trimming overhead and labour costs and entered into international strategic alliances. The government must have been pleased with the effect of KPN's internal efficiency and commercial successes, increasing the company's value at the moment of flotation. Before the initial public offering, financial investors estimated that KPN's value lay between NLG 12bn and 28bn. For instance, ABN AMRO (1993) and BZW (1993) mentioned KPN's market-responsive strategy, its flourishing results and the socio-economic geography of the Netherlands ('Entrepôt of Europe and Gateway to Europe'), all of which would eventually contribute to a successful privatisation. Referring to the government's decision to sell off a minority of the KPN shares, the valuation could be negatively influenced by the PTT Post subsidiary, that together with the commercially more attractive PTT Telecom branch, would be included in a flotation of the integrated holding company. Compared to the profits and revenue levels of PTT Telecom, PTT Post surely fell behind, although at the time it was one of the very few profitable postal administrations in Europe, relatively efficient and with high service levels. After its acquisition of TNT, the fourth largest logistics company, in 1996, PTT Post successfully built up a strong position in the international parcel markets and made clear that it was among the more aggressive and innovative companies in the field.

According to the 1989 legislation the government was allowed to sell off only a minority of the KPN shares (49 per cent). In the summer of 1993, the Minister announced that between 1994–8 a majority of the shares (varying between 51 per cent and 70 per cent) would eventually be sold on the Amsterdam stock exchange. The state would retain about 30 per cent in the newly privatised company. Notwithstanding the fact that the cabinet decided against retaining a majority interest, a set of protective measures has been built into the flotation procedure to leave the government's controlling interest untouched and to prevent a hostile take-over. The persistent influence of the government in a future privatised KPN is ensured through the issuing of certified shares ('prefs'), and the creation of a golden share, referring to the power to block changes in certain articles of association and granting the government the right to appoint three of the nine directors. In practice, this means that for critical policy issues, like the sale of a new tranche of the shares, the formation of strategic alliances (mergers or take-overs), large investments, and substantial tariff increases, KPN needs permission from central government. In June 1994, the government sold about 30 per cent of the KPN/PTT shares on the stock market. Shortly before the public offering, the pay out ratio was increased from 40 to 50 per cent to please potential investors and thereby promote the overall sale of shares. The centre left government furthermore facilitated the purchase of shares by KPN staff and small investors by underpricing them and giving particular 'sweeteners' (i.e. a discount) to private shareholders. The revenue of the sale was approximately NLG 6.8bn, with roughly a third of KPN's total workforce (i.e. 31,000) acquiring shares. In 1995, the second tranche of KPN shares was sold, reducing government ownership to 45 per cent and raising another NLG 5bn for the Treasury. The government agreed that the revenue generated by the flotation would be used to reduce the relatively high public deficit and to strengthen the country's economic structure through infrastructural investments. Like before, KPN proved a useful instrument to accomplish the government's fiscal objectives: once more the 'windfall profits' of the sale of KPN/PTT shares were used to balance the budget. In the course of 1997, KPN announced its intention to split up into independent companies by 1998. Such a demerger of postal and telecommunications interests would offer better strategic opportunities for both companies (e.g. allowing equity swaps for cementing alliances, different scales of investments) and would make the valuation of the shares more transparent. The Dutch government, which holds a golden share, would have to improve the plan for KPN's divestiture.

Changes in the internal organisation of PTT Telecom

In 1994, PTT Telecom had a turnover of about NLG 14bn, pre-tax results of NLG 3.5bn, and a rate of return of 13.5 (see Table 5.1). Despite competitive constraints in some of its markets, PTT Telecom's turnover and profit levels increased in most of them (with the exception of emerging markets and equipment sales that

Table 5.1 PTT Telecom Netherlands: key data

	1981	*1982*	*1983*	*1984*	*1985*	*1986*	*1987*	*1988*	*1989*	*1990*	*1991*	*1992*	*1993*	*1994*	*1995*	*1996*
Turnover (in NLG bn)	na	na	na	na	na	na	na	na	8.8	9.5	10.3	11.1	11.8	12.7	13.6	14.3
Operating results (NLG bn)	0.3	0.3	0.7	0.9	1.1	1.1	1.2	1.2	2.4	2.6	2.6	2.8	2.7	2.9	3.2	3.5
Rate of return	6.1*	6.6*	7.6*	8.6*	9.6*	9.7*	10.3*	9.5*	12.3	12.7	12.2	12.4	12.1	13.5	14.6	14.9
Workforce (x 1000)	29.6	29.8	29.5	29.7	30.6	31.7	31.8	31.4	31.5	31.8	34.5	34.9	34.4	33.9	32.3	32.3

* accounting system did not differentiate between postal and telecom services

Source: PTT/KPN *Annual Reports* (1981–96)

went down dramatically). The initial results of the internationalisation strategy of PTT Telecom were somewhat disappointing. The large majority of PTT Telecom's revenue (more than 80 per cent) is still generated by the provision of its exclusively licensed services. The results of the domestic and international participations of PTT Telecom were only modest: in its commercial activities in Central Europe, the Ukraine, Indonesia, and the cross-national Unisource joint venture, it faced high start-up costs, and as a result it took a relatively long time to recover investments. In its investments over the last five years, PTT Telecom has given priority to upgrading its infrastructure and developing new services: the roll out of cellular communications (GSM), the digitisation of the fixed infrastructure, optical fibres in the local network, and the promotion of a myriad new broadband and intelligent services.

After the denationalisation and liberalisation measures of 1989 were implemented, KPN had to adapt its internal organisation to the changed external environment and – like any other firm – it had to strive for profit-maximisation and return on investment. KPN staff lost its civil servant status and became subject to a single collective labour agreement that brought wages and working conditions more in line with the private sector. The higher ranks of PTT Telecom's management (about 200) were moved from collective pay terms to performance-related and personal contracts. As a consequence overall salary levels were reshuffled with wages in the lower ranks decreasing in relative terms and wages in the higher grades increasing (Van Den Besselaar & Visser 1993). A Central Corporate Works Council was installed, in which employee representatives and company management negotiate about operational and strategic issues. Under the new regime the trade unions are only represented at the holding level, they do not have access to the district level, subsidiaries and related companies. The transition from a state bureaucracy to a market-driven enterprise was not without its teething problems, caused by an inflexibility that still persisted within the new PTT organisation. Initially, the unions complained about the social climate deteriorating and the work load becoming too heavy. PTT Telecom aimed at creating sufficient internal flexibility by implementing reorganisation measures like decentralisation and divisionalisation, upgrading its personnel by job replacement and retraining programmes and the recruitment of highly qualified staff from outside. In 1990 PTT Telecom initiated a total quality management campaign to increase its quality of service to residential and business users by emphasising customer care, client satisfaction and account management (PTT Telecom 1993). PTT Telecom especially targeted business customers through creating decentralised regional units, and appointing account managers to settle customised contracts with large users (e.g. bulk discounts and tailored services).

After 1989, PTT Telecom's workforce initially rose from approximately 30,000 to 34,000 in 1992, and then fell to about 32,000 by 1996. This increase in staff levels made PTT decide to digitise its network, promote new telematic applications and hire new skills and capabilities (notably sales and marketing expertise). In 1992, PTT Telecom's management announced an efficiency and rationalisation programme, arguing that in the medium term growth could only be realised by consolidation and cost efficiency at home and expansion abroad. The programme outlined cutbacks of 10 per cent of its labour force and a reduction of overhead costs within two years. The shedding of labour (about 1,000 employees) would take place without compulsory layoffs. Although already one of Europe's most efficient telecommunications companies, in 1994 PTT Telecom announced a new efficiency drive to cut back a further 10 per cent of its total workforce, which meant 3,000 to 4,000 layoffs in the medium term. Facing the effects of increasing international and domestic competition and far-reaching automation of network switching, the unions were convinced that it was better to participate in the negotiations than to be left out. Although the potential for staff cuts may have been higher, the actual number of redundancies was restricted by the agreements made by management and the unions in 1988, the active involvement of the unions in reshuffling, and the stringent industrial relations legislation in the Netherlands.

After 1989, PTT Telecom separated its activities into five business units based on the following product market combinations:

1 network management: responsible for planning, installing, maintaining and operating the telecommunications infrastructure and the basic services, assigned to the company by law;
2 international telecommunications: in charge of handling cross-border services and co-ordinating the other international activities of PTT Telecom;
3 business market: telecommunications facilities and services for business users;
4 telematic systems and services: focusing on the new markets of value-added network services and tele-services: datatransport services (DN-1), videotex (Viditel), electronic mail, EDI services and other VAN services;
5 residential market: supplying a range of terminal equipment and services to households and small businesses.

After establishing the global carrier Unisource in 1992 (together with European partners), PTT Telecom started to transfer and integrate an increasing number of its business operations and international activities to Unisource. One year later, the Unisource initiative was extended to include the Swiss PTT, with ownership equally divided among the three participants. In the autumn of 1997, Unisource and AT&T aligned themselves through a small minority share in Telecom Italia while the Italian carrier announced it was to acquire 30 per cent of the shares of AT&T-Unisource. Early in 1998, however, Telecom Italia decided not to go ahead and loosened its ties with AT&T-Unisource, instead deciding to link up with Cable & Wireless.

In addition to the reorganisation of its business units, PTT Telecom decided that the lower district levels of the corporation needed restructuring. To remain close to the market and to cut operational management costs, the functional form of the thirteen districts, organised through a general board and four directorates for technical, commercial, financial and social issues, had to be replaced by a more decentralised grid structure, linking territorially defined basic units with expert-like support units. On behalf of the corporation's headquarters, the districts would remain responsible for the construction, operation and maintenance of the networks within particular territorial areas. In every district two or three regional business units would be established for sales and services, consisting of about 250 employees, responsible for carrying out the whole range of service provision in the business and residential markets. Support units, covering a larger territory of three or more districts, would offer specialised expertise to individual district units, like consultancy, engineering and (project) management support. In 1993 a beginning was made with the implementation of this new organisational structure. Almost five years later, a further decentralisation and flexibility strategy was implemented with the establishment of only three divisions with far-reaching responsibilities: infrastructure services, sales and innovation ventures.

Changes in the external organisation of PTT Telecom

With the telecommunications market losing more and more of its monopolistic structures, PTT Telecom has transformed itself from a centralised functional single-product organisation into a multi-divisional company that is far more responsive to the market than before. It has attempted to diversify into a multi-services provider and build up a customer orientation by aiming marketing efforts at profitable markets (data services and international telecommunications) and establishing price differentiation schemes. In its attempt to meet the demands of both residential and business customers PTT Telecom only partially succeeded: while the quality levels of its residential services were rated more than sufficient, its service provision to business users showed several shortcomings (McKinsey 1993). PTT Telecom began to use tariffs as a strategic instrument with the aim of being one of the cheapest operators, both for residential and business telephony services. The provision of the reserved services was still subject to a price-control regime: tariff increases must be below the consumer index. In order to match cost-based pricing with these tariff regulations, PTT Telecom followed a policy of shifting variable to fixed costs and of overall tariff rebalancing. After 1989, tariffs were restructured by increasing the rental (or subscription costs) of the telephone service, raising the charges of local calls (of residential consumers) and cutting the charges of business customers for international communications and datatraffic.

The erosion of its monopoly has prompted PTT to diversify into various segments of the electronic information market (see Table 5.2). PTT Telecom has moved into the expanding market of business communications by responding to the need of large diversified companies to manage their information flows and provide customised network services with 'one-stop-shopping' facilities. Another stronghold of PTT Telecom is in the market of value-added network services, electronic data interchange and dedicated wide-area networks. Here PTT has participations in ventures like Travel-net, Intis, PTT EDISon, PTT EDINova, Schiphol Telematics, Surfnet, Encompass Europe and Transopen. KPN/PTT Telecom is expanding its base from the telecommunications sector to the multimedia industries and from the provision of network capacity to advanced and value-added services. Besides having exclusive rights for the provision of infrastructure and basic services (through PTT Telecom), KPN had substantial control over the leading cable operator in the country, Casema. With the creation of the subsidiary KPN Kabel at the end of 1994, KPN has become active in cable development and network operations overseas. KPN/PTT Telecom has diversified into audio-visual production (the acquisition of Cinevideo), video conferencing and satellite news gathering services (Intrax venture), broadcasting (participation in Canal Plus Benelux/Nethold), and multifunctional chipcards (with the Postbank). Furthermore PTT is involved in the development of videotex and Internet services (through its participation in VTX: World Access/Planet Internet) and has gained substantial shares in type approval and consultancy (through Telefication and Intercai).

In the market for international business services, PTT Telecom has followed an expansionist strategy by entering new geographical markets and joining forces with foreign business partners. The reasons for establishing cross-national alliances were the limited size of the Dutch home market and the search for increasing economies of scale in its business operations and a desire to achieve a global coverage for its services. In 1989, the newly established PTT Telecom already had minority stakes in the satellite co-operatives Intelsat, Inmarsat and Eutelsat and in addition acquired a small stake in Infonet, the international services provider owned by a group of PTTs. A major international endeavour of PTT Telecom was its active participation, together with middle-sized national operators, to establish Unisource, a venture that is to become a true European long-distance operator. In the summer of 1994, Unisource became an equity partner in the Worldpartners Company, an AT&T-dominated global association of national carriers that delivered international data, voice and messaging services. At the end of 1996, AT&T and Unisource decided to deepen their collaboration in a joint venture in Europe through AT&T-Unisource Communications Services and the two carriers, together with KDD Japan and Singapore Telecom, are the key shareholders of WorldSource partners in the Pacific Rim.

Unisource was established in 1992 as a strategic alliance of PTT Telecom with Televerket/Telia (i.e. the Swedish PTT) with the purpose of providing managed network services to large international business customers. The Unisource venture did not emerge out of the blue, it was an extension of a collaboration between the Dutch and Swedish PTTs that had started earlier with the Vesatel initiative in 1991. The decision of the two carriers to provide international satellite services together through Vesatel proved to be a test case for the strategic international alliance of Unisource, that was established a year later (incorporating Vesatel). In order to develop international end-to-end services with one-stop-shopping facilities Unisource entered into a commercial agreement in 1992 with Sprint (the third US long-distance operator) to increase worldwide coverage. One year later, the Unisource initiative was extended to include the Swiss PTT, with ownership equally divided among the three participants. In 1997, Unisource and AT&T aligned themselves through a small minority share in Telecom Italia and the Italian carrier acquired 30 per cent of the shares of AT&T-Unisource. Furthermore, AT&T and Unisource have strategic partnerships with alternative carriers in Germany (Mannesmann-Arcor) and Siris (France).

The business portfolio of Unisource gradually expanded by the incorporation of the constituent partners' activities: electronic mail, EDI, virtual private networks, video conferencing, mobile services and dataprocessing services. PTT Telecom, like its partners, transferred its business unit Telematic Systems and Services and its chain of international sales offices to Unisource. In the near future Unisource will take over all its constituents' international licences activities, mobile operations and business development. Although Unisource established itself effectively as a European long-distance carrier, handling international voice and datatraffic, the company did experience teething problems. When Unisource started to integrate the various business operations of its members and reorganise them to eliminate overlapping activities, co-ordination problems between the constituent partners arose. For instance, 'turf battles' took place between PTT Telecom and Unisource about the precise delineation of responsibilities in the distribution of international business network services. The collaboration with Sprint turned out to be a technical and commercial disappointment. After problems emerged with regard to the interconnection of the two networks, due to differences in the basic technology used, Unisource decided to terminate the agreement with Sprint. Instead, Unisource linked up with SITA, the airlines telecommunications co-operative, and with KDD, the Japanese international operator, for service provision outside Europe. The collaboration between Unisource and the Spanish carrier Telefonica also came to nothing. Between 1995–7, Telefonica had a 25 per cent stake in Unisource, only to leave Unisource for BT and MCI's Concert. After Telefonica left, the remaining key constituents decided to pool their wholesale transmission services and merge their international carrier services.

To supplement its activities in the European market for business services, PTT Telecom also entered the newly emerging markets in Ireland, Central and Eastern Europe and Indonesia. With Swiss Telecom, PTT Telecom now participates in SPT Telecom of the Czech Republic (together 27 per cent) and, with Telia, PTT owns 20 per cent of the assets of Telecom Eireann (Ireland). PTT Telecom's strategy to penetrate the Hungarian telecommunications market was successful as well: the Pannon Mobile Services consortium, in which PTT Telecom, Unisource partner Telia, and other Scandinavian operators were represented, became the country's second mobile operator. PTT Telecom has also actively participated in overhauling the Ukrainian telecommunications infrastructure. Together with its partners AT&T, Bundespost Telekom, Telecom Denmark and the Ukrainian administration, PTT Telecom is involved in the construction, modernisation and operation of a fully fledged national and international network.

PTT Telecom has also attempted to acquire stakes in the national operators of newly industrialising countries in South East Asia. This strategy proved relatively successful in Indonesia, where PTT Telecom became actively involved in the modernisation and expansion of the existing telecommunications network in 1993 through a strategic alliance with Telkom, the country's national operator. To that end it decided to expand the collaboration with local Indonesian partners by acquiring a 30 per cent stake in PT Bakrie Electronic Company, a private operator, one year later and obtaining a stake in the local mobile services operator PT

Table 5.2 PTT Telecom Netherlands: major partnerships

	Country of registration	*Group interest in allotted capital (%)*	*Major business partners of PTT Telecom*
Telecom Eireann	Ireland	10	through Comsource, the joint venture with Telia
SPT Telecom	Czech Republic	14	through Telsource, the joint venture with Swiss PTT
Unisource NV	NL	33	Telia, Swiss PTT
Intrax	NL	50	NOB (Netherlands Broadcasting Facilities provider)
Cinevideo Enterprises BV	NL	40	–
Intis	NL	51	Harbour of Rotterdam, Cap Volmac (replacing BSO automation)
Intercai	NL	50	Getronics
KPN Multimedia	NL	50	PTT Post NL
SurfNet BV	NL	49	–
PT Ratelindo	NL	17	PT Bakrie Electronics Company (30% share for PTT Telecom)
UMC Ukrainian Mobile Communications	Ukraine	16	DBP Telekom (FRG), Telecom Denmark, Ukrainian government
Pannon GSM	Hungary	23	a.o. Telecom Denmark, Telia/ Televerket (Swedish Telecom)
Utel Ukrainian Telecom	Ukraine	10	DBP Telekom (FRG), AT&T, Ukrainian government
Infonet Services Corporation	USA	14	DBP Telekom, Belgacom, Swiss PTT, Telia, Transpac/France Telecom, Telefonica (E), KDD (Japan), Singapore Telecom, Telecom Australia, MCI (USA)
Eutelsat	France	7	various European public operators
Inmarsat	United Kingdom	4	various national operators
Jasz-Tel	Hungary	50	–
Telkomsel	Indonesia	17	Telkom (Indonesia)
Intelsat	USA	1	various national operators

* currently in the process of demerging from the KPN holding

Source: KPN *Annual Reports* (1989–96)

Telkomsel in 1996. Vision Networks, formerly known as KPN Kabel, PTT Telecom's sister, has also engaged in foreign expansion and acquired regional cable TV franchises in the UK (Telecential and ComTel/IVS), shares in Réseaux Câbles de France (the fourth largest cable operator in France), and minority shares in Kabelmedia (Germany), Interkabel Polska, and Intercable (Czech Republic).

5.7 Regulatory reform of Dutch telecommunications policy

The creation of a new regulatory regime

In the post-1989 telecommunications framework, PTT's operational and regulatory tasks were separated through the establishment of a concession-based regime. The government had granted the concession-holder KPN the exclusive right to install and operate the public infrastructure and basic services. At the same time KPN was forced by law to offer these reserved services nationwide and meet certain performance standards. Because of its obligation to provide public services at uniform charges, PTT needed some sort of protection to be able to carry out and finance its exclusively licensed activities and to protect it from the effects of (potential) competitors bypassing the network and cream-skimming the commercially attractive markets. The Ministry of Transport & Public Works (T&PW) was given the responsibility to regulate and supervise the concession holder and the postal and telecommunications industry. With the exception of the concessioned services which were subject to price controls, PTT was responsible for setting its own tariffs. For the provision of the reserved basic facilities a price-cap system was installed, that was based on the weighted average of tariff changes of the dedicated services. This system of price controls implied that the cumulative tariff increases had to be beneath the Consumer Price Index (CPI) (for telecommunications) and the wage index (for postal affairs). PTT had to submit a preliminary proposal for tariff changes to the Minister, who, after having ensured that the terms of the price-cap formula were respected, would enforce the plans. It was believed that price-cap regulation would secure the uniform pricing of the reserved services and protect domestic customers from excessive price increases. Thus far the process of tariff rebalancing has been carried out gradually by PTT Telecom with modest price increases in local calls and line rentals, all within the CPI-0 price cap, and by a continuous series of substantial price cuts in international communications. In April 1994, PTT Telecom made it clear that tariffing would be more market-based by increasing rentals by approximately 5 per cent, local calls by 20 per cent, and inquiry services by 400 per cent, while making substantial cuts in international communications. The reaction of consumers' representatives to the tariff rebalancing has been modest and balanced: they have accepted the price increases of the basic services while criticising the raise of rental charges (Consumentenbond 1993). The consumer association considered PTT Telecom's decision to raise rental charges unnecessary, because they were already among the highest in Western Europe. Furthermore, raising rental charges would be an easy way for PTT Telecom to secure its standing revenues without seeking real competition.

The 1989 Act required PTT to submit an audited annual report to the Minister on the results and development of the reserved services (i.e. publishing statistics on quality of service, network capacity, tariffs, financial results, etc.). PTT Telecom also had a statutory obligation to provide its major stakeholders with accurate planning indications for the medium term, concerning investments, prices, standards and policies regarding the future provision of the infrastructure and the dedicated facilities. In the value-added services and terminal equipment markets PTT was allowed to compete freely. Present legislation, originally, presupposed a clear separation between PTT Telecom's public and commercial activities through distinct reporting and accounting systems. Such a separation between the monopoly and market activities was seen as the first step to a *de facto* divestiture of PTT Telecom into two subsidiaries, originally proposed for 1994. The decision to separate the public and commercial tasks of PTT Telecom, however, was abandoned in 1992. The then Minister dropped the idea of an organisational divestiture, because this would hamper PTT's economic position in the relatively open home market and would be counterproductive to its internationalisation strategy.

The liberalisation and (pseudo-)privatisation measures were accompanied by the creation of a new framework for advice, consultation and supervision in telecommunications policy. In the new situation the political and administrative jurisdiction for both policy making and rule setting concerning the post and telecommunications sectors fell under the directorate HDTP of the Ministry of T&PW. Besides keeping responsibility for draft legislation, HDTP acted as a kind of regulator and supervisor for the postal and telecommunications sectors. The establishment of HDTP in 1988 implied that regulation and supervision were separated from the operation of the exclusively licensed functions, which until then had all been integrated within the PTT administration. HDTP was entrusted with the administrative task of surveying the reserved functions of KPN and controlling the charges of its reserved business activities. Furthermore it watched KPN closely, from cross-subsidising its reserved and commercial activities and became ultimately responsible for type approval of terminal equipment, frequency allocation, and licensing operators and cable systems. HDTP also acted as PTT's (sole) shareholder on behalf of the government.

The advisory and consultative structure of the Advisory PTT Council (PTT-Raad) was replaced by the Advisory Council for Post and Telecommunications (RAPT) and the Consultative Body PTT Telecom (Overlegorgaan PTT). RAPT was a statutory advisory committee made up of independent experts, making recommendations to the government on the broad lines of policy, technological developments and parliamentary proposals in the fields of post and telecommunications. It was also consulted by the Minister in the review of KPN's annual concession report. Compared to the former Advisory PTT Council, RAPT covered a larger area, but on the other hand it was mainly concerned with advice on broad policy lines. Having been granted a provisional legal status in 1989, RAPT was renamed the Commission for Advice on Postal and Telecommunications

Policy (CAPT) in 1992 and integrated into the departmental Advisory Council of Transport and Public Works. The purpose of the Consultative Body PTT was to act as intermediary between PTT and its stakeholders in the discussion of policy issues of a general nature and national significance, that had a direct relevance to the exclusively concessioned services assigned to PTT. According to the original plan the Consultative Body had to develop into a platform, composed on a representational basis, for consultations between PTT and its major stakeholders at least twice a year. In the Consultative Body were represented, apart from PTT, equipment suppliers, service providers, customers (households, small and medium-sized firms, large users and government), employers and employees. In 1992 a new body was established, the Consultative Body on Postal and Telecommunications policy (OPT), in which government and the major stakeholders of the telecommunications industry would periodically meet. The reason for the creation of this new departmental body was the absence of a consultative platform where the Ministry of T&PW could discuss operational aspects and policy issues with the public operator KPN/PTT and its stakeholders.

Shortcomings of the newly created institutional framework

Initially, the institutional framework for advice, consultation, supervision and regulation was created to guarantee a well-balanced relation between PTT's public function and the private interests of its major stakeholders and to increase the strategic flexibility of the Dutch telecommunications sector. The framework institutionalised the official exchange of information concerning such policy issues as price controls, quality of service and long-term developments in infrastructural planning between PTT and the Ministry, Parliament, equipment manufacturers, service providers, customers etc. The present institutional framework of Dutch telecommunications fell short in at least four ways. First, the policy information made public by PTT in its capacity of exclusive licensee was considered insufficient by its major stakeholders. In practice the lack of information provided by the regulated PTT obscured its public accountability and hindered parliamentary oversight of the industry. Second, the newly created market was not yet working perfectly. It lacked the proper conditions such as equal access and fair play. Third, it was thought that the new framework was inadequate for monitoring PTT's activities and promoting new entry, innovation and fair competition. And fourth, the newly established regime was unable to cope with the demands and requirements of the Internal Market and EC legislation.

The first criticism launched by the business community concerned PTT Telecom's secrecy about the prevalent conflicting loyalties between its commercial and utility functions and its reluctance to give policy information about its public activities (i.e. the operation of its reserved services and the long-term plans concerning the infrastructure). In the newly installed regulatory and supervisory framework PTT was obliged to report on investments and long-term planning directly related to its public activities and subjected to an auditing regime. Minister and Parliament were supposed to check these plans on the basis of criteria such as economies of scale, universal service, international competitiveness and innovation capacity. Especially in the beginning, PTT was not able to comply fully with this condition, and it appeared unable or reluctant to present adequate long-term strategic information with respect to its public functions. RAPT, the Minister and Parliament complained about the vagueness of the first long-term plans of the concession holder and demanded more detailed and concrete information about future investments to promote technological developments and tariff rebalancing. They argued that the plans contained little information and that it was impossible to maintain proper supervision without adequate information.

The second complaint referred to the fact that the separation of PTT's public utility and commercial activities was obscure and the relationship between PTT and its stakeholders (competitors and customers) in the post-monopoly market was not yet transparent. The dominant position held by PTT both in the market place and in the political arena impeded the equal access of new entrants to the markets of cable, information services and mobile communications, hampering the effective supervision of PTT as both concession holder and market player. The third criticism referred to the relatively weak position held by the administrative agencies *vis-à-vis* PTT. HDTP, RAPT and the Consultative Body PTT were originally designed to act as countervailing powers to PTT's monopoly. In practice, however, things were different. In terms of expertise and staff HDTP was too poorly equipped to supervise and guide the liberalisation process. HDTP's manœuvrability was restricted by the structural information asymmetry between the concession holder and its supervisor, and by its close ties with the government. The small size of the Directorate, ineffective recruitment of staff, and a relatively late creation (spring of 1988) have all contributed to the malfunctioning of HDTP both as a watchdog and as a policy unit in charge of stimulating innovation and market competition. The poor quality of the regulatory and supervisory functions was illustrated further by the difficult contacts between PTT as concession holder and its customers in the Consultative Body PTT. This administrative entity was originally established as a platform for regular consultations between PTT Telecom and its customers about the capacity, character and quality of the infrastructure. PTT Telecom's stakeholders were critical, both of the information provided by PTT and of the entity's functioning. The negotiations within the Body were open-ended and it did not have any compliance mechanisms that could force PTT to go into detail on its long-term infrastructural planning. PTT Telecom clearly thought little of the body as a formal consultative platform, having a preference for more informal and bilateral contacts with each of the individual parties involved. So the present institutional framework in Dutch telecommunications has generated an obscure market, with PTT acting as a kind of self-regulating conglomerate, dominating the policy process on the basis of its virtual information monopoly, recruitment of policy experts (i.e. a braindrain of civil servants) and the 'divide and rule' tactic of collaborating with third parties.

The reluctance of PTT to give information about its long-term strategy, and the relative inability of RAPT, HDTP and the Consultative Body PTT to do anything about it, made it clear that supervision over the concessionaire was insufficient. The

Consultative Body and the advisory body RAPT, both keeping a low profile, were not effective as strong countervailing powers to the established parties. Consequently, effective monitoring of the implementation of the administrative separation between the public function of PTT and its commercial activities in the liberalised markets of terminal equipment and advanced services was lacking. This triggered rumours about anti-competitive practices and collusion between the concession and the commercial functions of PTT Telecom (e.g. cross-subsidisation, artificial pricing, dirty tricks). The dominant political and economic position of PTT, reinforcing structural information asymmetry between PTT and its stakeholders, and lack of political-institutional oversight, hindering the monitoring of cross-subsidisation, gave support to those in favour of a formal separation of the concession and the commercial part of PTT Telecom. The large users, public utilities, and cable operators complained about PTT's market dominance and demanded a loosening of the restrictions to allow the resale of capacity to third parties and to further network competition. When they considered enforcing their argument by going to the courts, the responsible Minister responded that consultation and accommodation of the involved parties, PTT, private network operators, service providers and the large users, was a better way to handle this problem:

> In the meantime I want to impress all the parties involved to opt for a practical approach. Consultation and collaboration are more in line with the Dutch tradition than battling out this question by going to Court.
>
> (NRC Handelsblad 10 April 1991)

The fourth shortcoming of the established framework lay in its inflexibility to respond to the increasing jurisdiction of the European Community on domestic policies. The Internal Market Programme of the Commission, together with the ongoing international deregulation of telecommunications, forced Dutch policy makers to update their legislation on a relatively short notice. The effects of EC legislation were underestimated by the Dutch telecommunications community (V&W 1993). The Dutch governance regime, characterised by a retreat of the state, certain privileges for KPN as the national champion, and self-regulation, could not cope with the detailed regulations set by the European Commission to secure open network provision (ONP) and fair competition. The fact that emergent issues like interconnection, transparent tariffing, non-discrimination and fair play, needed detailed supervision, dispute settlement mechanisms or even asymmetric regulation to 'protect effective new entry', was not directly understood in the Netherlands. The EC proposals dealing with the mutual recognition of national licences in the European market were another argument in favour of a need for structural reform of the 1989 framework. The Dutch regime, based on an exclusive concession for the provision of dedicated services, needs to be replaced by a national licensing system and the mutual recognition of these licences throughout the European Community. The establishment of a national and EC-wide licensing regime will offer opportunities for Dutch operators to enter foreign markets (it will of course open up the domestic market to foreign competitors as well).

An independent regulatory agency?

The need for an independent regulatory watchdog and more market transparency was strongly felt by the new entrants in the market place, the large and residential user communities, and independent experts. In view of the limited capacity and insufficient expertise of HDTP and its incorporation within the government administration, the elements of appropriate expertise and administrative independence have not been sufficiently recognised in the Netherlands. Clear political leadership was lacking in the restructuring of Dutch telecommunications, due to the incumbent PTT squeezing market competition and government playing the conflicting roles of KPN's shareholder, policy maker and regulator of the industry at the same time. In addition the ONP directives of the European Community intended to ensure equal access to networks and services, required the establishment of an independent regulator with the authority to settle potential problems with interconnectivity, number planning, fair competition etc. As a consequence, by 1994, the government separated the role of policy maker and regulator, eventually establishing a (semi-)autonomous regulatory unit with sufficient means to police the telecommunications industry. The government only partially followed the pleas for an FCC/OFTEL-like regulatory body. At the end of 1994 the government announced that TND, a new regulatory unit, would be set up next to the policy-making and operational units in the HDTP department. This regulatory unit would temporarily be a semi-autonomous agency, with a budget of its own (but still operating within the Ministry), dealing with the overall supervision of the telecommunications sector, acting as referee between contending market players and the policing of market behaviour.

Initially, the Minister made it clear that such an independent regulatory authority would not be established before 1998, when all service monopolies would have been abolished. The reason for postponement was the legal-administrative obstacles over the definition of ministerial responsibility and the authority's legal status. The alternative of an independent regulatory authority to be installed as soon as possible was strongly pushed by a majority in Parliament who criticised the slow implementation of the open network provisions and lack of market transparency. After complaints put forward by service providers and large business customers about KPN/PTT Telecom overcharging for leased lines, the Minister responded in 1996 by forcing the incumbent to cut its charges for leased lines by about 25 per cent. The Minister also decided eventually to endorse the setting up of an independent telecommunications authority ahead of the 1998 deadline. In September 1997, TND was hived-off from government and established as an autonomous authority, called OPTA, in charge of resolving interconnection disputes, and in general promoting a more competitive environment. So, in the end, the Dutch government had to reconsider its initially modest and distant role *vis-à-vis* the dominant position of KPN/PTT in the telecommunications domain and to adopt a new regulatory approach that seeks to combine the promotion of competition and regulatory supervision to protect the interests of new entrants and consumers. Another sign of the more adversarial regulatory climate is that KPN/PTT has had to give up its dominant share in the Dutch cable television market in order to facilitate competition in the local loop market by demerging Casema. Furthermore, PTT Telecom's controversial plan to introduce new voice mail services and dial-up Internet access on a nationwide basis is currently being reviewed by the government. OPTA already vetoed PTT Telecom's offer of substantive bulk discounts to business users just before the liberalisation of the market in July 1997, and is now preoccupied with resolving interconnection disputes (e.g. between PTT and Enertel, PTT and Telfort).

5.8 Summary and concluding remarks

Before 1989 the Dutch PTT had a *de facto* monopoly on the provision of telecommunications (and postal) services. PTT was a state enterprise belonging to a government department, its employees were civil servants and its corporate decisions concerning tariffs, revenue and expenditure needed political approval. PTT needed large investments and highly qualified personnel for the expansion and modernisation of its infrastructure. These two conditions, however, could not be met in a state-controlled system. This was the reasoning behind the high-level advisory Committees Swarttouw and Steenbergen. Finally, during the parliamentary discussions over the recommendations of the two Committees (1981–7), it became clear that a market-based structure for Dutch telecommunications and a revision of the legal status of the PTT was needed.

The new governance regime, as put in place by the adoption of the 1989 Telecommunications Act, was based on the combination of a regulated privatised monopoly, market co-ordination and industrial self-regulation. A concession regime was established whereby the infrastructure and the reserved voice telephony services were exclusively provided by PTT. To cope with the distinction between concession-based and commercial services, PTT was forced to keep separate accounts for the operation of its dedicated public services and its competitive activities. After a transitional period of four years, PTT Telecom's concessionary and commercial activities would eventually be separated into two autonomous entities. The markets for value-added services and terminal equipment became liberalised, thereby meeting the demands from consumers, big business users and the services-oriented industries. The 1989 regime was still premised on the notion of a natural monopoly, characterised by economies of scale, universal service, unity of control and a privileged position for the state-controlled PTT. The notions of an alternative network, made up of interconnected cable network operators and utility companies, and the introduction of competition in the core voice telephony market, were not yet viable. Instead, the idea of bringing the local networks of PTT and cable operators together into one integrated broadband infrastructure to be controlled by PTT in the mid-term (thereby following the recommendations of the Zegveld Committee), was actively promoted by the government.

In the 1989 framework PTT's operational tasks were separated from its former regulatory functions and consequently a new institutional structure of separate bodies for consultation, advice and rule making has been established. The newly created administrative structure included the HDTP unit of the Ministry of Transport & Public Works as regulator, the RAPT/CAPT as independent policy advisers, and two consultative bodies. This institutional framework succeeded the neo-corporatist PTT Raad, that had functioned as an intermediary body for regular consultations between PTT and its stakeholders, and also gave recommendations to government on postal and telecommunications issues. The administrative bodies were originally designed for the exchange of information between PTT and its stakeholders, consensus seeking and policy formation. The combination of administrative functions in HDTP of the Ministry of T&PW, being the sector's watchdog, KPN/PTT's shareholder and policy-making body, and the half-hearted implementation of competition policy across the national economy ('the Netherlands as cartel-paradise') further constrained the actual enforcement of the new institutional framework. In 1995, the Supervision Networks and Services Unit (TND) was established next to the Policy Unit and the Operational Affairs Unit of the Ministry. In the summer of 1997 TND was spun off from the government administration and set up as an independent regulator. In the post-1998 setting, safeguarding fair competition will be shared between the future Independent Telecommunications Authority and the future semi-autonomous anti-trust authority (NMA, related to the Ministry of Economic Affairs). A precise division of labour between the two institutions still needs to be elaborated.

The prolongation of exclusive rights in telecommunications, together with the political decision to keep the mail and telecommunications integrated within a larger holding, would allow KPN/PTT to start from a solid financial and commercially viable base. As a newly established private company, KPN/PTT was no longer constrained by a civil servants' statute or obliged to meet political requirements. The Dutch government has floated KPN in tranches, with the intention of disposing of most of its shares by 1998. In June 1994 the state sold 30 per cent of its shares in KPN. A second tranche of KPN shares were sold in October 1995, reducing the state's interest to slightly less than 45 per cent. In response to shareholders' demand for better corporate transparency and the accumulated strength of the postal and telecommunications subsidiaries, KPN is at present preparing a demerger into separately quoted telecommunications and postal services businesses.

In addition to being the exclusive provider of the public infrastructure and the basic services, PTT has diversified successfully into value-added networks, teleservices, cable and satellite television and audio-visual industries and internationalised itself through foreign acquisitions and strategic alliances. In 1992 PTT Telecom BV and Telia of Sweden established Unisource, a joint venture to provide the largest of multinational corporate customers with pan-European services in competition with the largest global operators. Swiss PTT joined Unisource as an equal partner in 1993, and Telefonica of Spain joined under the same conditions in 1995. In December 1994, Unisource, together with AT&T, set up a joint company called Uniworld to provide pan-European services, and WorldPartners (a partnership between Unisource, AT&T, KDD of Japan and

Singapore Telecom) will provide worldwide coverage. Unisource aims to compete for revenues from international traffic to offset expected losses in the relatively small home markets of its shareholders, especially after the 1998 EU deadline for the liberalisation of voice telephony.

The institutional structure of the Dutch telecommunications sector has been transformed from a regulated public monopoly to a regulated competitive market. KPN/PTT Telecom is privatised and all the market segments except voice telephony have been liberalised. An independent telecommunications authority and a semi-independent anti-trust authority are now being put into place, and in July 1997 the key voice telephony market was opened up to competition. This regulatory framework will have to prove itself in implementing and ensuring fair competition in the future. As experience elsewhere has shown, where there has been a shift from administered to open competition, this new framework will be put to the test when it faces (potential) disputes over interconnection, number portability, pricing schemes, universal service obligations etc.

6 The liberalisation, privatisation and regulatory reform of French telecommunications

Still in care of the state?

6.1 Introduction

In the framework of this comparative research on the changes in telecommunications policy in the 1980s and 1990s, the French government has pursued a go-it-alone strategy, reflecting a high degree of state intervention and pursuing objectives aimed at promoting technological autonomy and catching up with international market leaders. The French political economy has been characterised by an active central government taking a leading role in industrial development. In France, competition policy is not an important element of industrial policy; as Darmon (1985:129) has put it: 'In France everything is initiated by the state.' Hall (1986) mentions four elements that make up the unique position of the French government: the state as a cohesive and centralised unit, relatively insulated from the demands of other social actors, effectively representing the public interest and equipped with the capacity to impose its policies on society. France has pursued a state-led growth strategy, in which the central government imposed its will on the market through central administrative planning, nationalisation, (selective) protectionism and the active promotion of strategic industries. The aim has been to ensure that domestic firms would be successful in catching up with industrial/technological leaders and achieving a position of industrial leadership (Zysman 1975, 1977, 1983; Katzenstein 1985).

In the 1980s, when most West European countries were already considering and implementing cuts in public expenditure and deregulation policies, the French Socialist government (1981–6) pursued an opposite policy, expanding the powers of the state in the national economy. The public sector grew between 1981–6 through the nationalisation of the country's key industries (manufacturing, banks, electronics and so on); in the telecommunications industry, the domestic equipment suppliers CGE/Alcatel, CGCT (an ITT subsidiary) and Thomson were nationalised. While other European countries were seriously considering privatising and liberalising their domestic telecommunications industry, France sought to expand and strengthen the position of its national administration, i.e. Direction Générale des Télécommunications (DGT), later renamed as France Télécom, by allowing it, in addition to its legitimate role of operating telecommunications services, to take the lead in restructuring and modernising the electronics industry, telecommunications services and the information industry. As a consequence, DGT/France Télécom's public monopoly expanded beyond the infrastructure and voice telephony to include large parts of the markets for value-added services, datacommunications, cable systems and mobile services.

From the mid-1980s onwards, however, the French government had to accept the increasing constraints of the international political economy, that limited the manœuvrability of the national government in its home market. Albeit moderately and gradually, the French also had to implement more market-oriented and deregulatory measures, as well as responding to the EU directive aimed at liberalising markets and ensuring open network provision. Changes in government, the cumulative opening up of the French economy by European integration, and the globalisation of the equipment and services markets made it clear that the role of the state and the size of the public sector had to be diminished in favour of free trade, competition and privatisation.

In section 6.2 of this chapter the French national framework of government–industry relations and economic policy making is analysed: an overview of its institutional setting will be given on the basis of its production profile, the organisation of interest mediation, the organisation of the state, the economic approach and cultural ideology, and the international context. Section 6.3 discusses the development of economic policy in the 1980s and 1990s. Section 6.4 introduces the established regime in French telecommunications, and discusses the process of demonopolisation, that began in the mid-1980s and proceeded into the late 1990s. Section 6.5 continues with a more detailed discussion on the liberalisation of the French telecommunications market. The privatisation of the public operator is discussed in section 6.6, followed by an account of the implementation of regulatory reform of telecommunications in France in section 6.7. The chapter closes with an organisational and institutional analysis of the restructuring process in French telecommunications with subsequent concluding remarks.

6.2 The wider political-economic setting of French telecommunications

Before discussing and assessing the established organisation and structural changes in French telecommunications polity, we will examine the wider political and economic setting of the national economy. This institutional environment will be outlined in a brief discussion of the following national variables, characteristic of government–business relationships in France: the production profile, the organisation of government, the system of interest mediation, approaches to economic policy, and the international dependence of the French economy.

Production profile

France is the fourth largest economy in the world, after the USA, Japan and Germany. Compared with the relatively open economies of the UK and the

Netherlands, the French economy is relatively protected. Generally speaking, however, in terms of its inward and outward openness (or, more concretely, export rates and foreign direct investment in France), it seems to hold an intermediate position: the level of 'multinationalisation' in France was significantly less than in Germany, but more than in Japan and the United States (Savary 1984; Sally 1995). Although there are obviously exceptions to the rule (e.g. major investments by Matsushita and Toyota), France has a well-established tradition of keeping foreign investment to a minimum and utilising state enterprises to accomplish national objectives. In 1998 national industries included firms Elf-Aquitaine (oil), Renault (car manufacturer), Thomson (electronics), UAP (insurance), some of the major banks (e.g. Crédit Lyonnais), and Air France, Electricité de France, Gaz de France and the other utility monopolies. Traditionally French companies have been insulated from both national and international market forces through protectionist measures (i.e. tariffs, quota), government subsidies and cartel-like business arrangements. The French have obtained good results with their state-led strategies in those markets that depend on the central state for R&D support, public procurement and export diplomacy: nuclear power, aircraft manufacturing and aerospace, military electronics, communications technology and transport equipment. Like the USA and UK, France has a large military-industrial complex, in which the Defence Ministry, the space and telecommunications research centres CNES and CNET, and the leading electronics companies collaborate on military R&D (35 per cent of public expenditure) and the production of defence equipment (Chesnais 1993). In the 1980s and 1990s, the leading French firms in the high-technology and capital-intensive industries (e.g. Cap Gemini (computing services), Matra-Hachette (communications), Thomson (electronics), and Alcatel-Alsthom (telecommunications manufacturing, nuclear power and railway equipment)) have left this shelter strategy to become increasingly internationally oriented, developing joint ventures and alliances with European, Japanese and American partners (Sally 1995). The French strategy was relatively successful in the key industries controlled by the state through large public procurement programmes. In less monopsonistic markets, characterised by competition and mass consumption, the French strategy to design and rely upon an independent domestic manufacturing base has been disappointing. For example, the Plan Calcul in the 1960s to develop a French base in computing to rival IBM and the Plan des Composants to build an internationally competitive semiconductors industry in France in the 1970s, resulted in clear failures (Adams & Stoffaes 1986; Cohen 1992).

Organisation of the state

France has been defined as a 'strong state' society, in which the central government plays a crucial and all-pervasive role in determining the direction of the major industrial activities. The various roles of its public bureaucracy in the economic sphere have been described as 'tutor, patron and main risk taker' (Wright 1987:98). The French political economy exhibits a leading role for the state both as an idea, referring to a 'public service state' and as an institution of political rule shaped through a technocratic and centralised administrative system (Dyson 1980). This notion of strong public power is epitomised by the concept of state interventionism, in French: *dirigisme* and *étatisme*. The executive branch of government, i.e. the presidency, is entrusted with considerable powers to mobilise private interests to realise national and/or public ambitions, staying relatively immune from the detailed interference of Parliament and political negotiation. Every political shift in government (presidency and/or cabinet), from left to right and back again, has been accompanied by a reshuffle in the top echelons of the major state enterprises and certain ministries, and a setting up/rearranging of administrative entities (e.g. in the highly politicised field of broadcasting). Beneath a changing politicised upper layer and its concomitant spoils system, the central administration has significant discretion with regards to the implementation of public policies. Sometimes, additional powers are provided by the party or coalition in power, but the civil service has a tradition of relying upon élite recruitment and searching for technocratic engineering-like solutions; as nicely put into words by Cohen (1992:74): 'The engineer outshines the minister, the civil servant prescribes politics'. The country's senior staff of ministerial cabinets and the public administration are selected and socialised through the Grandes Ecoles (i.e. the Ecole Nationale d'Administration (ENA), Ecole Polytechnique and Ecole National Supérieure des Postes et Télécommunications) and professional membership of the Grands Corps with either an administrative or an engineering background (e.g. Corps des Mines (energy), Corps des Ponts et Chaussées (transport) and Corps des Ingénieurs des Télécommunications).

Traditionally, industrial policy making in France has relied upon a system of indicative economic planning, in which priorities are set for the long-term development of the national economy, particular industries and leading individual enterprises. Rather than delegate responsibility to business and rely on market competition, the Planning Commission, in collaboration with the Ministries of Finance and Industry and organised business, draws up a framework that sets the social and economic priorities for the mid-term. A typical element in French planning is the definition and implementation of *grands programmes*, designed to re-equip the national economic structure, develop new markets and prepare the domestic industry for international competition (Bauer 1988). Additional instruments used by the state to govern industrial activities and (re)organisation are large subsidies, an exclusive public procurement policy, large R&D funds, long-term credit and tax cuts. The Planning Commission (CGP) played a major role in the administration of Marshall Plan aid and the modernisation of the country in the 1950s and 1960s. After France had effectively caught up with the leading international economies in the late 1960s, the need to set priorities and indicate the directions in which the national economy should go was less strong. As a consequence the interests for indicative planning started to diminish in the 1970s and 1980s.

The Ministry of Finance has played a very significant role in the development of French economic policy by controlling the allocation of public credit to particular sectors and leading companies. The Ministry of Industry plays the important role of monitoring events in specific industries and examining their performance and potential in order to 'catch up' with the world's market leaders: it seeks to identify the most valuable sectors and economies, the most successful technologies and the most promising target market niches. Notwithstanding its reputation for being a strong and efficient state with a powerful executive, French government has another side, one that presents a sometimes highly politicised and fragmented picture of decision making. The Ministry of Finance (Trésor), the Planning Commission, other government departments and state-controlled companies are at times more engaged in waging turf wars ('*la guerre des administrations et les batailles des compétences*') than in effectively collaborating in the formation and implementation of long-term economic strategies. In particular the relationship between the omnipresent Treasury and the highly fragmented Ministry of Industry shows the darker side of the efficient and effective French bureaucracy (e.g. Bauer & Cohen 1981; Cohen & Bauer 1985).

Compared with Britain, where a far-reaching privatisation programme was implemented with relative ease, France has had intensive legislative debates about the nationalisation and denationalisation programmes of the alternating Socialist and Liberal-Conservative governments. Besides the recurring stalemate caused by the cohabitation between a left-wing President and a right-wing Prime Minister in the late 1980s, or in the late 1990s between a right-wing President and a left-wing Prime Minister, the manœuvrability of the national government in restructuring state monopolies and selling public property was constrained by constitutional requirements and administrative institutions acting upon the policy process. In the French constitution there are some provisions and requirements that have to be met in the case of (de)nationalisation. The constitution stipulates that undertakings with a public service or natural monopoly character should be state property (Graham & Prosser 1991). Unlike the UK, large public utilities (such as telecommunications, energy, airlines) were not privatised in France. The French privatisation programme was restricted to enterprises operating in competitive markets. Another constitutional provision is that the sale of public assets to private parties must meet criteria of equality and guarantee the protection of property rights: the former refers to set minimum levels of the valuation of enterprises being privatised and the latter to the preservation of national independence and the protection of national property. In order to meet legislative and administrative requirements and to prevent the sale of public shares below their true market value, reviews are carried out by the Conseil Constitutionnel, the Conseil d'Etat and the Commission de Privatisation. After receiving their positive verdict, the Minister has the power to dispose of state interests in enterprises. The Privatisation Commission has used its power to accept or refuse buyers: for instance, in 1996 the privatisation of Radio Monte Carlo was cancelled because the bids were unsatisfactory. In the same year, the Commission issued an influential negative verdict on the plans for a traded sale of the Thomson assets (including the Thomson CSF and Thomson Multimedia subsidiaries) involving the French industrial group Lagardère and the Korean conglomerate Daewoo. Eventually the sale of Thomson was vetoed by the government following widespread concerns about foreign participation in facilities (i.e. defence electronics) that were viewed as critical to national security. The sale of public property can be arranged through a mutually agreed private sale (sale by *gré à gré*) or by setting up *noyaux durs* of privileged investors willing to participate actively in the future development of the privatised company and to secure continuity of ownership (i.e. with temporary no-sell agreements). Important clusters of these 'hard core' shareholders' groups are Alcatel-Alsthom/Société Générale, Paribas, and Saint-Gobain/Elf Aquitaine.

System of interest intermediation

Government–business relations at the top levels of the French economy could be characterised as an all-dominant state forcing its will upon the private sector, and the existence of strong and often informal relationships between representatives of the public administration and private interests. The government's influence upon management of the nationalised and state-owned firms has traditionally been high. The potential powers of the state, however, are, paradoxically, limited, serving the private interests of the major industrial groups. Bauer and Cohen (1981) have coined the term 'subcontractor state' to refer to the state being dependent upon the expertise, information and compliance from industry, resulting in the state more or less underwriting the investments carried out by French businesses. For example, the position of EDF in the domain of electro-nuclear policy could be qualified as that of a state within a state. The personal contacts between business and administration are created and reinforced by way of an intricate élite structure of high-ranking government officials and business leaders, who tend to have a shared educational background at one of the aforementioned Grandes Ecoles. The convergence of public and private sector interests is further strengthened by the practice of *pantouflage*, i.e. the 'institutionalised' rotation of senior managers between state agencies, businesses and financial institutions which has contributed to a strong loyalty to the French nation. These 'old boy networks', bridging the state bureaucracy and industry, play an important role in the formulation and implementation of public policy in France; they encourage – in addition to the official interorganisational ties – the essential interpersonal communication structures required for an effective policy initiation, adjustment and execution.

The framework of industrial relations in France comprises a semi-institutionalised bargaining relationship between a relatively strong employers' association and a divided labour movement, and an all-pervasive state (Bunel & Saglio 1984). The large majority of French capital is organised in the Conseil National du Patronat Français, in which both the industry/trade associations and the employers' associations are represented. Labour in France is organisationally weak with low membership rates, a high inter-union rivalry and a lack of internal union discipline (Lane 1989; Segrestin 1990; Goetschy & Rozenblatt 1992). The French labour movement is fragmented, lacking an encompassing federation to which the various trade unions jointly belong. Organised labour fell from about 20 per cent in 1981–2 to less than 10 per cent in the late 1980s (Goetschy & Rozenblatt 1992). The Conféderation Général du Travail (CGT), with strong links to the Communist party (PCF), exhibits a strong anti-capitalist culture based on the ideology of persistent class struggle. As a consequence CGT has traditionally been inclined to militant industrial action and has always vigorously opposed collective bargaining. The CGT is relatively well organised in manufacturing, the chemical industry, engineering, public transport and public utilities. Since 1981 the power of CGT has diminished slightly, partly as a result of its militant attitude in industrial disputes and orthodox Communist views, partly because of an increasing rift between the union and its rank and file. The Force Ouvrière (FO) and the Confédération Française Démocratique du Travail (CFDT) are two smaller and less radical unions. The FO favours worker participation and collective bargaining, but rejects state-enforced wage restraints. CFDT supports the idea of self-management or self-control by workers. This organisational weakness does not necessarily imply that the unions are without influence.

In some sectors and on some aspects of socio-economic issues labour's institutional powers are significant. For instance, in the 1980s the unions had a stronghold in the 'older' industries (steel and coal), construction, the chemical industry, state-owned companies and the public service industries. The unions were also strongly involved in overall work-related electoral consultations and company-level bargaining within both public and private enterprises. Furthermore the (Socialist) government compensated for labour's organisational weakness by legal intervention in the 1980s. It assured the power and rights of labour unions through fostering social legislation and employee representation and granted them access to some public bodies and corporate decision making. The state-nourished involvement of unions in industrial relations matters was inspired by the 'public service' role of labour, serving as a vehicle for overall workers' representation and the introduction of collective bargaining at both plant and industry level (Goetschy & Rozenblatt 1992).

French capitalism has a tradition of small-scale family-owned entrepreneurship, characterised by conservatism and self-sufficiency, and largely preoccupied with the domestic market and protected from foreign inroads by government-imposed duties and prohibitions. Landes (1949) has referred to the relative retardation of French industrial development until World War II. These small and medium-sized firms, however, in the post-war period were rapidly developed into large and hierarchically integrated enterprises that favoured the mercantilist combination of domestic protection and selective international competition. The unitary (U) organisational form, characterised by high levels of centralisation, formalisation and functional specialisation, has been identified as the dominant configuration of the larger businesses in France (Lane 1989). However, besides being bureaucratic and hierarchical, the French-style U-form was also paternalistic and authoritarian, with extensive powers for the company's single-tier Board (Conseil d'Administration), chaired by the all-powerful Président Directeur Général (PDG). Given the relatively underdeveloped character of the French stock exchange (compared to the USA and the UK), impeding the sale of a company's securities, the larger French firms had to rely for funding on self-financing and borrowing from financial institutions. Large institutional investors (pension funds) and the stock market have merely served as additional sources of external corporate financing. Another interesting feature of French capitalism is its cartel-like and network-like character: there is a dense web between the major industrial groups and financial institutions through a multitude of cross-shareholdings, strategic participations, companies owned jointly by several groups, and interlocking directorates (Sally 1995).

The shares of the major companies were often held by large investment banks and/or industrial holding companies. Zysman (1983) characterised the French financial system as state-dominated, credit-based and relying strongly upon administrative pricing. The task of manipulating the economy's flow of funds in order to forge changes in the organisation of French industry fell under the jurisdiction of central government. Control over the channels of borrowing and lending, credit allocation and price levels in crucial markets were ultimately determined by the Treasury and the Planning Commission. In the French system of indicative planning the government allocated credit to sponsor national priority projects and key industries through a network of state-controlled, state-related and para-public financial-economic institutions. The Planning Commission drew up the details for the reorganisation and rationalisation of industrial sectors, forging strategic alliances and promoting new high-technology ventures. The Treasury decided which financial institutions would have access to the private capital markets to provide French big business with the funds required: semi-public long-term-credit institutions like the Crédit National or Caisse des Dépots et Consignations, state-controlled investment banks (Paribas and Suez) and/or deposit banks (BNP, Crédit Lyonnais, Société Générale). Recently, however, government control over the domestic capital market has been eroded by the globalisation of capital flows, the privatisation of nationalised banks and other state-owned companies, and the concomitant growth of the Paris stock exchange.

Economic culture and ideology

A principal political goal vehemently pursued by the French government in its economic policy over the years has been to safeguard and/or promote the national sovereignty and autonomy of its 'high-technology' and key industries. The development of a French national industrial strategy was inspired by the idea that without active state intervention the French economy would be highly dependent upon foreign capital and technology and would eventually fall behind its major competitors (USA, Germany and Japan). The poor conditions of French industry and management expertise in the post-war period (compared to the superior organisation and management of US firms) meant that the leading position of France in the world could not be maintained by French business, and was therefore the exclusive task of the state. The central government controlled the structural development of the national economy and initiated the restructuring of domestic industries or the promotion of emerging industries in order to prepare them for global competition. France does not have a strong tradition of using competition policy as an instrument for industrial restructuring. So far, safeguarding fair competition has not been given a high political priority by either leftist or right-wing governments; instead preference has been given to furthering concentration and consolidation and establishing extensive national monopolies (Demarigny 1996). The powers to undertake an investigation into anti-competitive practices are not located in an independent national competition authority, but final decision-making power is in the hands of the central administration (i.e Directorate General for Competition, Consumption and Prohibition of Frauds and, ultimately, the Minister of Economic Affairs).

Concerning its infrastructural sectors, there is a strong commitment to safeguarding the provision of public services. As became clear in the dispute caused by the government's announcement to corporatise and privatise France Télécom, this notion of the state looking after the basic interests of its citizens and organising the provision of basic services, *l'état providence*, was strongly supported by senior management and the workforce of state monopolies, as well as the public at large (Mescheriakoff 1991). These requirements to ensure the geographical availability of a package of basic services, that are imposed on the exclusively concessioned public operator, include rather general and symbolic notions about the right of being served and the equal treatment of users, as well as rules to ensure the permanence and continuity of services (cf., the UK, where public and universal provisions are often more detailed, for instance including affordability and minimum quality levels; see Garnham & Mansell 1991). In practice, of course, the public service concept also incorporates the particular interests of private stakeholders, such as protection from competition, preservation of civil servants' privileges, and jobs for life etc.

French industrial policy has generally been aimed at restructuring declining sectors, restricting imports (i.e. reconquering the domestic market) and selecting the more promising industries and technologies. Shonfield (1965:87) has referred to the French approach to economic organisation and industrial policy as 'ersatz capitalism or synthetic capitalism'. Starting from an inherent mistrust of the market economy, the dominant belief is that it is better to develop and produce these high-technology products synthetically, as if in a laboratory (i.e. insulated from competition). Also the marketing and the commercial development rely on a '(re)constructed' version of the economy, in which market signals are suppressed or poorly understood. The projection of market demand is deduced from the theoretical conditions of a competitive market and modelled predictions about the diffusion of these products, rather than seen as the result of unpredictable market forces, and the opportunities and hazards generated by real competition.

Two central elements in French industrial policy are *créneau* and *filière* (Adams & Stoffaes 1986; Groenewegen 1989). The first refers to a specific market niche dominated by one domestic industrial conglomerate, the second to the effective establishment of domestic control over techno-industrial complexes and successive production stages within one particular sector. French industrial policy has been oriented towards the creation of national champions in distinctive sectors by actively encouraging mergers and take-overs that provide the required economies of scale. The reorganisation of particular sectors and companies was facilitated by a massive infusion of capital for investments and the concentration of expertise and capabilities in order to promote new technologies. An alliance of the government and the state-controlled financial sector engineered the amalgamation of particular industries through tax incentives, credit allocation and a swap of stocks. After more than a decade of restructuring, Thomson-Houston emerged as the French electronics champion, incorporating CII, Hotchkiss-Brandt and CSF. France steadily broadened its initially narrow range of industries. For instance, it initiated its own nuclear energy programme to gain energy independence and explicitly stimulated micro-electronics and telecommunications as a key to the future information society. The national government sought to catch up with its international competitors by pursuing international expansion while at the same time protecting its national economy and culture.

Another element in the French industrial strategy are the *grands programmes*, in which central government gives clear priority to the development of a particular technology or an industrial sector in an effort to accomplish particular goals. These ambitious programmes are built upon a policy package, in which the idea of national independence (*francisation* of technology), the involvement of public research institutes, the concentration of industrial production in national champions, public procurement of equipment and state-supported exports are integrated (Groenewegen & Beije 1989). Examples of these '*nouvelles cathédrales*' are the Concorde project, high-speed trains (TGV), nuclear energy, and aerospace (Airbus, ESA/Ariane). In its industrial mission, the state plays a leading role as entrepreneur, protector and facilitator in ensuring technological autonomy and promoting foreign exports. The government actively encourages the development of French-controlled and advanced knowledge and manufacturing bases in these important sectors through reorganisation programmes, financial assistance schemes, the erection of trade barriers and the creation of export markets in developing nations (*la diplomatie industrielle* in French-speaking countries, former colonies, third-world countries and the Communist bloc). An illustration of such a strategy can be found in the colour television industry, in which the French government employed its national SECAM standard as a non-tariff barrier, rather than accepting foreign-set standards or attracting foreign investments. While a large majority of European states had already opted for the German PAL system in the late 1960s, the French government continued to protect and promote its SECAM colour television standard and the domestic electronics sector from foreign-produced TV sets (Crane 1979). From the early 1980s, with ever-increasing free capital movements and integrating capital markets, the viability of French state-directed capitalism has come under scrutiny (e.g. Adams & Stoffaes 1986). This subject matter re-emerged on the political and public agendas in the early 1990s when socio-economic stability was lost due to long-lasting industrial disputes, mismanagement, abuse of power and corruption (e.g. Renault, Alcatel, Crédit Lyonnais and Air France).

Positioning in the international political-economic system

With regard to its international relations, France's strategy has been based on preserving national sovereignty with respect to foreign/security policy and strategic industries and intergovernmental co-operation in Western Europe through the European Union. Traditionally, French governments have strongly rejected the notion of supranationality in the domains of foreign, security and economic policy and instead supported the objective of national sovereignty and techno-industrial autonomy (Dreyfus *et al.* 1993). For instance, from De Gaulle's rule in the 1960s until the Mitterand and Chirac presidencies in the 1990s, France successfully developed an independent nuclear striking force (*force de frappe*) and created a strong military equipment and aerospace industry. Under Socialist rule in the 1980s, France sought to establish a '*force de frappe industrielle*', a striking force in high-technology sectors, services industries and future-oriented manufacturing. The ambitions of the Mitterand-Mauroy government to modernise the French economy through huge public investments and nationalisation, however, had to be scaled back, because the French Keynesian approach went against the global grain of budgetary rigour, deregulation and privatisation (MRI 1982).

France was relatively slow to recognise the supremacy of European law over national legislation. The French government supported an intergovernmental approach, often based on informal Franco-German agreements, to speed up the process of European integration. The increasing internationalisation of the world economy and the coming into force of the Common Market from the 1960s onwards implied that France, a leading member of the European Economic Community, had to reform its tradition of 'industrial patriotism', exemplified by mercantilist and protectionist trade policies, in which the state functioned as intermediary between national industries and the world market (Hayward 1986). The constraints of the Common Market, the European Economic Monetary System and the subsequent Economic and Monetary Union project were essential in the retreat of the French state. The international deregulation of financial transactions and the obligation to reduce the budgetary deficit to meet the requirements for economic and monetary union, strongly restricted the possibilities for the French government to intervene in the domestic economy and sponsor ailing industries (Bull, Air France, Crédit Lyonnais, Renault) and to use credits and subsidies for large-scale investments. The strongest pressures for economic change in France appears to have been generated by external factors, such as the Commission (scrutinising state aid), the impact of free-flowing capital, and the requirements of multinational firms to set up a transnational division of labour and promote global standards and service provision. The domestic financial markets became more and more deregulated and the interests of the French stock exchange were furthered by large privatisation programmes between 1986–8 and after 1993. To finance their expansion, French businesses could rely more easily on issuing shares through the capital market and reducing loans from their home banks. Furthermore, the French government could no longer automatically use import restrictions or impose trade barriers designed to help the balance of payments or support its strategic industries through large public contracts and state aid schemes.

In the early 1980s, the French government acknowledged that innovation policies implemented at the national level were no longer feasible in the world technology race, as they were too expensive and complex. It therefore attempted to extend the tradition of nationally oriented research and industrial policies to the European level. To reduce the risk of duplicating efforts and to give future *grands projets* the required economies of scale and scope, France proposed at EC level co-ordinating scientific and industrial efforts in electronics and the related industries. As a consequence a 'Europeanisation of national innovation policies' was needed (Humphreys 1990a:208). Challenged by the American Strategic Defense Initiative (Star Wars), President Mitterand launched the Eureka programme in 1984, suggesting that European states should collaborate in research and development in the domain of high-technology. The French proposal for European collaboration was an explicit acknowledgement that the Community level was a more appropriate platform for the elaboration of large technological projects than the nation state.

6.3 The politics of economic adjustment in France

Immediately after World War II, when the French economy found itself short of capital and materials and its economic infrastructure was destroyed, the state bureaucracy exerted its leadership to rebuild the country. The Conseil National de Résistance, led by General De Gaulle, launched an ambitious programme to nationalise the country's key industries (railways, electricity and gas), the large banks and insurance companies (such as Banque National de Paris, Crédit Lyonnais and Société Générale) and those firms that had actively collaborated with the German oppressor (e.g. Renault). This anti-capitalist programme was accepted virtually unanimously. After the first wave of nationalisation, two more followed: in the 1960s under De Gaulle, and in the early 1980s under Mitterand. In the 1960s new state enterprises, like Elf-Aquitaine (oil), Société Nationale Industrielle Aérospatiale (aerospace) and Compagnie Internationale pour l'Informatique (CII: information technology), were created with the purpose of rationalising industries and used as instruments in ambitious technological projects. In the 1980s, the Socialist-Communist Mitterand-Mauroy government completed the two previous nationalisation projects by bringing the remainder of the industries and companies of strategic importance, like banks, the telecommunications equipment industry, under direct state tutelage.

In the post-war period the French economy was still dominated by a large agricultural sector and a stagnant industry, sheltered from foreign competition and fragmented among many small and medium-sized producers, most of them family-owned firms. At that time French industry was handicapped by a shortage of capital and materials and a risk-averse and inward-looking business culture. Organised business at that time was weak and relatively ill-equipped to reconstruct and restructure the country. The task of modernisation and industrial leadership was accomplished by high-ranking government officials such as Jean Monnet and Robert Schuman who began reforming the post-war bureaucracy and the national economy and embarked upon a system of indicative economic planning and interventionist industrial policies. Government officials successfully expanded and modernised the post-war economy by guiding business and labour in their collective national search for economic growth. In thirty years of constant economic growth with annual growth rates of more than 5 per cent (*les Trentes Années Glorieuses* 1945–75), France transformed itself from a traditional capitalist economy into an industrial nation with competitive firms that depended on foreign sales and were able to compete with imports in their home markets. In 1958 the Fifth Republic was established in which the political-institutional framework of the Fourth Republic was reformed by extending the powers of the President, reorganising the state bureaucracy and trimming the powers of Parliament.

Industrial activities were guided through five-year Plans to accomplish certain economic objectives. In the first two Plans, between 1946–56, the emphasis was on the reconstruction and modernisation of the economy. The purview of the Third and Fourth Plans, between 1957–65, was extended to include the whole economy, including the preparation of French industries for international competition by the rationalisation and concentration of industries and the reform of social security and education. In the Fifth Plan, from 1967 onwards, the French government initiated a new industrial policy based on nurturing national champions in the key industries, and launching large state-funded programmes in the advanced sectors. After the first oil crisis, the French government was reaffirmed in its opinion that, in order to adapt the domestic economy to the new international economic conditions, more public control over certain key technologies was needed. In the telecommunications domain this view led to the (gradual) nationalisation of the foreign subsidiaries of ITT and Ericsson by Thomson; in the computing industry it resulted in the merger of CII with Honeywell Bull. As a consequence of the gradual liberalisation of world trade and the coming into being of the European Internal Market from the mid-1960s onwards the overall importance of indicative planning decreased. When the French economy was confronted with the technology gap and the successful entry of leading American firms in the 1960s (Servan-Schreiber 1967), the Sixth Plan, for instance, gave priority of investment over consumption and drew attention to the need to strengthen the international competitiveness of French domestic industry and promote external trade. The Seventh Plan (1975–80) proposed a less encompassing programme in which twenty-five high priority projects were identified.

With the Socialists and Communists coming into power after the presidential victory of Mitterand in May 1981 and the ensuing parliamentary elections, a radical strategy was adopted by the Socialist-Communist government, aimed at expanding the public sector and extending state participation in the national economy (MRI 1982). French public policy changed drastically with the implementation of a policy mix consisting of an ambitious nationalisation programme, the resurrection of planification, an interventionist strategy to rescue France's industrial base, an expansion of the public administration, and redistributive measures (a legal minimum wage, an increase of social benefits and progressive tax reform). Also a more active industrial relations policy was carried out, aimed at supporting the position of labour through workers' and union rights in both the public and private sector. While a majority of the European countries were considering or already implementing supply-side economic remedies to the economic recession of the early 1980s, such as deregulation, privatisation and cuts in public spending, the French leftist government followed a 'go-it-alone' strategy, targeted at strengthening public consumption and the public sector. The following industrial holdings, private firms and banks were fully nationalised: Thomson Brandt (electronics), Rhône-Poulenc (chemicals), CGE (manufacturing), Pechiney-Ugine-Kuhlman PUK (aluminium), Saint-Gobain (glassworks), Paribas (banking) and Suez (banking). Besides acquiring full ownership in these companies, the Socialist government took majority shares in Matra (defence), Dassault (aerospace), Usinor Sacilor (steel), CII-Honeywell Bull (computing) and CGCT/LCT (ITT subsidiaries). After the nationalisation operation was eventually executed, the state holdings accounted for 24 per cent of the employees, 32 per cent of the sales, 30 per cent of the exports and 60 per cent of the annual investment in industry and energy in France (Hall 1986).

Within two years, however, the leftist Mitterand-Mauroy government had to reverse its policy because it was found to be based on erroneous presuppositions. The Socialist policies anticipated a prompt upswing of the world economy and regarded the French economy as relatively autonomous from international economic developments. The radical leftist policies lacked all feasibility and turned out to be counterproductive: the nationalisation and tax proposals gave way to a flight of capital, booming inflation, increasing production costs, a run on the franc, and a record trade balance deficit. The consequent loss of competitiveness and serious deterioration in the balance of payments made the French government realise that it had to bring its economic policies more in line with international market conditions. Furthermore the state had to rely on the active involvement of business in the creation of economic growth: President Mitterand publicly stated: '*C'est l'enterprise qui creé la richesse*'. In 1982–3, the government made a spectacular U-turn with three successive devaluations of the franc, a freeze in wages and prices, cuts in public expenditure, and a tightening of exchange controls. These austerity plans (*la politique de rigeur*), together with an emphasis on financial accountability in the state-owned industries, paved the way for a successful comeback of the Frénch economy. After the Communists left government in 1983, the Socialists even pioneered a modest form of privatisation: some nationalised firms were allowed to sell off subsidiaries either directly or through a public flotation. Non-voting shares were issued on the Paris stock exchange to enable public enterprises to raise funds through the capital market. Furthermore, the nationalised firms in the key industries were allowed to invest abroad and form strategic alliances with leading foreign producers.

After the right-wing victory in the 1986 parliamentary elections, a Socialist presidency cohabited with a Liberal-Conservative government. The Chirac government (1986–8) gave clear priority to the promotion of the private sector and competition (*reculer l'état*) through a privatisation plan, deregulation measures (labour markets), cuts in public spending, and the liberalisation of prices and interest rates. An ambitious privatisation programme would furthermore increase the value of the Paris stock exchange and enlarge the number of small shareholders. Chirac followed a high-risk strategy: the sale of state assets took place on a larger scale than in Great Britain, its time frame was much shorter and the stock market four times smaller (Dumez & Jeunemaitre 1994). On the list of denationalisations were companies nationalised by the Socialists in 1981 but also some that had been nationalised just after the war by De Gaulle. Two categories of firms were excluded from the denationalisation programme, because they lacked political feasibility and/or economic viability: the public utility monopolies (SNCF, DGT/France Télécom) and those firms with poor economic performance (e.g. Bull). The minimum selling price of each state enterprise was to be fixed by the independent Privatisation Commission (on the basis of valuation and performance audits and according to generally accepted principles). Later, the task of fixing the floor transfer price was extended, when the Privatisation Commission became involved in deciding the composition of controlling interests in the companies in question.

The original privatisation list contained twenty-eight state-owned enterprises: with the relatively short time span of two years before the next elections, less than half of the companies on the shortlist could actually be privatised. Another major constraint on the number of flotations to be carried out was the small scale and hence the inability of the Paris stock exchange to absorb too many simultaneous privatisations. The industrial conglomerates Alcatel/CGE, Matra, Saint-Gobain, the banking institutions Société Générale, Paribas and Indosuez, and Havas (advertising) were sold off completely. To prevent a hostile take-over, the government imposed restrictions upon ownership. Foreign ownership of the privatised companies was restricted to 20 per cent and a substantial part of the stock of these state-owned companies was sold to a selected 'hard core' group of domestic institutional investors and businesses (*les noyaux durs*). These oligarchic shareholders would commit themselves to providing the privatised firms with stability of control and relative autonomy for strategy formation. The French broadcasting sector was also partially privatised in the mid-1980s through the granting of a special licence for pay TV to Canal Plus (jointly owned by Havas and Compagnie Générale des Eaux) and the sale of the first national TV channel (TF1) to a consortium, led by Bouygues (the construction company). After the collapse of the stock market in October 1987, and the troublesome flotation of the Suez Bank (November 1987) and Matra (January 1988), further privatisations were postponed until after the 1988 elections.

In 1988, the cohabitation ended because the Socialists won the elections both for the presidency and Parliament, and hence returned to power. The leftist government declared that there would be 'no more nationalisations, nor more privatisations' (the *ni-ni* policy). Bauer (1988) has qualified the privatisation operation in France from 1986–8 as 'capitalism without a financial market and without sanctions', in which non-market elements such as mercantilist objectives and political appointments of senior management, prevailed over economic incentives and sanctions by the market. The large-scale privatisation which took place between 1986–8 and in 1993 challenged the close-knit complex of interlocking personal networks, cross-shareholdings, controlling blocks and political allies. Induced by a fear of hostile take-overs and the perceived small size of French groups compared to German and American companies, the flotation process was administered. The denationalisation process was constrained by strong restrictions on foreign ownership and the 'friendly' controls, executed by an interlocking network of reliable domestic holding companies in the privatised business groups. The French government regularly ran into trouble with the European Commission about accusations regarding state aid to the three ailing state-owned companies, Bull, Thomson and Air France.

In 1993, the Liberal-Conservative Balladur government, succeeding the Socialist Cresson government, announced an ambitious privatisation programme, expected to raise more than FFr 350bn. While the goal of the 1986–8 privatisations had been to reduce the national debt, the post-1993 privatisation was intended to balance the budget (i.e. prepare for European economic and monetary integration) and finance the government's employment programme (Dumez & Jeunemaitre 1994). The government relaunched the 1986 privatisation programme and planned to sell off the remainder of more than twenty shortlisted state-owned enterprises, whose state equity would be sold on the stock market or to private investors. The final privatisation list included the following state-owned enterprises: Air France, Renault, Elf-Aquitaine (oil), Rhône-Poulenc (chemicals), UAP (insurance), Crédit Lyonnais, and Banque Nationale de Paris. It was then believed that these revenues would not only help to balance the budget, but also boost the Paris stock exchange and thus stimulate the competitiveness of French business in general. As a consequence of the general political and economic uncertainty caused by the cohabitation and the presidential elections of 1995, only a few of these companies' assets were actually sold (and often only minority shareholdings) between 1993 and 1995: Seita (tobacco manufacturer), Rhône-Poulenc, Elf-Aquitaine, BNP, Renault and UAP. Originally the ailing companies Thomson and Bull featured on the privatisation list, but the sale of state assets was postponed until after the 1995 presidential elections. Just as in 1986, the public monopolies (France Télécom, railways, energy, etc.) were excluded from the whole operation: the combined shift in ownership (from public to private), from a state monopoly to a more open market environment, was not feasible. The support for changes in these 'natural' monopoly sectors was not great: the ideological resistance to 'Anglo-Saxon' free trade and an open European market place was widespread, the performance and quality levels of the utility industries were above average, a large majority of staff were satisfied with being civil servants (Dumez & Jeunemaitre 1994).

After their victory in the 1995–6 elections, the Liberal-Conservative Juppé-Chirac government resolutely set out to ease the country's budget deficit by reducing social security spending, a move that created widespread unrest and paralysing strikes, and by continuing the privatisation project. Majority stakes in Usinor-Sacilor (the steelmaker), Pechiney (the aluminium manufacturer), and AGF (insurance) and minority shares in Renault, Elf Aquitaine and Total were sold to the public and institutional investors. Furthermore, a private sale of a 39 per cent share in Bull was negotiated between the French government and the company's (new) shareholders Motorola, NEC and IBM. After the Liberal-Conservative government's 1997 election gamble backfired, France saw a new cohabitation between a Socialist-Communist cabinet and a right-wing President. Like its predecessors, the Jospin cabinet also sought to stimulate economic growth and combat unemployment, and to meet the strict Maastricht public deficit criteria for economic and monetary union, albeit by a different path. The left-wing government used policy measures aimed at reducing the legal maximum working week from thirty-nine to thirty-five hours in order to create jobs while lifting corporate taxes and tapping the reserves of utilities (e.g. EDF and Caisse des Dépots et Consignations) to balance the budget and qualify for monetary union. As such, the Jospin cabinet went ahead with the plan of the previous right-wing government: to use the proceeds of a public offer of the flagship of the domestic public services, France Télécom, to recapitalise ailing state-owned companies. Although postponed for almost half a year, because of opposition from the labour unions and divisions in the cabinet, 20 per cent of France Télécom's assets were privatised. More privatisations have been planned to follow: the left-wing government has scheduled the sale of shares in Renault, Usinor, Air France and the CIC bank, starting in 1998.

6.4 Demonopolisation of telecommunications in France

The established telecommunications regime

In France, telecommunications was a statutory monopoly of the Ministère des Postes, des Télécommunications et de la Télédiffusion; in other words telecommunications was the exclusive responsibility of a state administration, the Direction Générale des Télécommunications (DGT), later renamed France Télécom. In addition, the Ministry included a Postal Services Directorate (DGP). Before the 1990 reforms the annual telecommunications budget, as an annexed budget, was part of the overall state budget and therefore had to be approved by the National Assembly. Traditionally the budgets of the French PTT depended on year-by-year choices made by government and Parliament in their macro-economic and fiscal policies. The workforce of DGT/France Télécom belonged to the civil service, which implied occupational safety and uniformity of rules. France Télécom has never produced telecommunications equipment; instead all terminal is purchased from private industry, using general procedures for state procurement. The PTT was partly a public administration and a public company, including two vertical directorates: Telecommunications, with a workforce of more than 150,000 employees, and La Poste (including the money services) with more than 350,000. The PTT furthermore included three horizontal directorates, notably Budget and Accounting, Personnel, and Estate/Transport. The DGT strategy was to use a vast expansion of the telephone system as a procurement base on which to rebuild the French information technology industry and open up new types of consumer markets for innovation-oriented firms. France Télécom is also involved in broadcasting and cable transmission through one of its subsidiaries, called Télédiffusion de France (TDF).

The bulk of R&D in French telecommunications is carried out through the Centre National d'Etudes des Télécommunications (CNET). Officially, CNET, modelled after the famous Bell Labs of AT&T, was merely the PTT/DGT's autonomous R&D centre, but it gradually expanded its powers to include public procurement decisions and technology/industrial policy formation. CNET was required to forge a consensual outlook and close collaboration between the various key players in French telecommunications (Sally 1993). Through conducting and/or contracting out fundamental research to public and private partners, CNET was able to modernise the domestic telecommunications industry by nourishing new technologies (like the E-10, one of the first digital switches, and ATM high-performance switching) and promote the interest of the French telecommunications equipment industry (i.e. CIT-Alcatel). Another relevant institution in French telecommunications was the Caisse Nationale des Télécommunications, established in 1974, to enable DGT external borrowing. Other institutions involved in the financing of France Télécom's operations were the Sociétés de Financement (e.g. Francetel, Finextel, Codetel).

The DGT's workforce was divided between senior managers organised in *corps des ingénieurs des télécommunications* and lower-ranking civil servants, including civil engineers and operating staff. The first category, including the company's administrative-technocratic élite (*le corps)*, was represented through the Association des Ingénieurs des Télécommunications (AIT). Especially during the large-scale modernisation of the French network in the 1970s, the reputation of the *corps* engineers grew significantly, ending up ranking behind only the traditional corps of *mines* (energy and nuclear power) and *ponts et chaussées* (transport and public works) (Vedel 1984). The second group was organised in the Force Ouvrière and the CGT. Although precise data about union membership of France Télécom are not available, there are some indications as to the presence and power of the unions in corporate and public policy making and their individual strength at the professional elections. Between 1970 and 1990 the distribution of votes swung between 35–40 per cent for the Communist CGT, and 17–27 per cent for both the reformist CFDT and the Socialist FO (Rugès 1970; Lyon-Caen 1993; Barreau 1995). Telecommunications users were organised in the Association des Utilisateurs du Téléphone et des Télécommunications (AFUTT), representing both residential and business customers. The interests of large business users in information-intensive industries were organised in the Club Informatique des Grandes Enterprises Français (CIGREF). This small but effective interest group was the intermediary between individual multi-site companies on the one hand and France Télécom and central government on the other, promoting large user's demands for special treatment and policy privileges (e.g. definition of closed user group).

Traditionally, the manufacturing market in France has been dominated by subsidiaries of foreign firms: LMT and CGCT/Thomson-Houston (US-based ITT) and SFTE (Swedish-based Ericsson). The domestic industry included the CIT subsidiary of CGE and AOIP as only small suppliers to the DGT. Together these companies were organised in cartel-like agreements, like Sotelec for transmission (set up in 1947) and Socotel for switching (set up in 1958), through which they pooled patents and agreed on market distribution and quota arrangements. From the 1960s onwards, however, the French government, directly and through its agents the CNET and the DGT, embarked on a concentration and rationalisation path that was meant to bring the supply of electronic equipment into domestic hands. This strategy was inspired by a desire to improve fragmented market structures and poor business practices and to promote and develop information technologies at a national level (Zysman 1977). In 1969, the electronic giants Thomson and CGE agreed on an 'electronic Yalta' to divide their activities into spheres of influence, respectively electronics (components, consumer and defence electronics) and telecommunications (Bauer & Cohen 1981). The next step was Thomson's take-over of the French subsidiary of ITT, LMT, and Ericsson's SFTE in 1976–7; CGCT, ITT's second subsidiary, was not affected. In 1982, the French state acquired the two telecommunications equipment subsidiaries of ITT, CGCT and LCT. In 1985 a second Yalta was announced when the Thomson Telecommunications division decided it would merge with CIT-Alcatel; in return, CGE-Alcatel would restore Thomson's troubled balance sheet. After its withdrawal from telecommunications Thomson retained its core activities in defence electronics (Thomson-CSF) and consumer electronics (Thomson-TCE).

Historical legacy: Le Plan de Rattrapage Téléphonique of 1970

During World War II, under the Vichy government, a new institutional structure for telecommunications was laid down. This proved to be very effective in the longer run, providing the basis for the goals of economic modernisation and industrial catching up in the 1970s and 1980s. First, a national public research centre, named Centre Nationale des Etudes de Télécommunications (CNET), was set up to strengthen the technological base of French telecommunications; this institution would promote and carry out R&D jointly with equipment manufacturers and the public operator. Second, a specialised institute for higher education in telecommunications engineering and management training was established, called École Nationale Supérieure des Télécommunications (ENST). In addition to its 'normal' task of educating and producing technical experts for the state bureaucracy and the state-controlled industries, ENST contributed to the creation of convergent views about industrial management and state leadership, and the development of professional ties among the future ruling élite (*esprit de corps*). Third, immediately after the war the telephone and telegraph administrations were amalgamated in the Direction Générale des Télécommunications, and granted more corporate autonomy from central government.

Although telecommunications modernisation was regularly mentioned and discussed in the various Plan consultations and negotiations, it only received sufficient support to be targeted as a priority sector at the end of the 1960s (Libois 1983). From the First to the Fifth Plan priority was given to the reconstruction and modernisation of the key industries of steel, energy, electronics, computing and aerospace and to the promotion of trade and exports in these sectors. Through the Plan Calcul, originally launched in the early 1960s, the French government pursued an ambitious strategy to catch up with American leadership in information technology. The need for this Plan became manifest when the US government prohibited the export of two supercomputers, that would be required for the development of a nuclear force (Gilpin 1975). By combining large public procurement programmes and a related concentration programme to rationalise the industry, the French government expected to build a solid domestic base in computing. To that end, the computing activities of CGE, Thomson and Schneider were combined in CII. When the newly created CII decided to link up with Philips and Siemens and create the pan-European Unidata venture in 1973–4, the French government initially supported the project. A year later, however, the government, decided to pursue a go-it-alone strategy, blocking the cross-national venture and giving priority to fostering a national computing champion by consolidating CII with the Franco-American company Honeywell-Bull.

The development and commercial expansion of telephony in France was a long way behind the rest of Europe between the two world wars. While countries like Germany and the United Kingdom had density rates of about 6 per cent in 1938, France had less than 4 per cent (Pinaud 1985:57). Compared to countries of about the same size and/or level of economic development, the French statistics were striking. Productivity and quality of service levels were extremely low, tariffs were among the highest in the world and the network density figures showed an uneven pattern (cf., extremely low outside the Paris/Ile de France region). The official explanation for the state of affairs referred to heavy network damages caused by three wars within a century and the absence of real demand before 1960. At that time the French telecommunications system was highly dispersed, lacking interregional connections and unequally distributed across the country. There were huge waiting lists and the technical performance of the telephone service was disappointing. The goal of universal service provision was given little priority: until the end of the 1960s telephony was still regarded as a luxury service or a 'gadget' the French did not really need. Attali and Stourdze (1978) have qualified this phenomenon as 'the slow death of monologue in French society'. It has been argued that the diffusion of all new information and communication technologies (the telephone, colour television, and video recorders) has been significantly slower in France than in other countries (Mattelart and Stourdze 1985).

Jannès (1966, 1970) and Rugès (1970) have referred to more structural causes, like the obstruction of investments by the Treasury, the cross-subsidisation of the loss-making postal services by the profitable telephone services, bad co-ordination between the central and local authorities, political scandals, the capture of the DGT by dominant foreign equipment manufacturers and the *corps* of engineers (*polytechniciens*). For the expansion of its network the PTT administration could not be certain of adequate investment funding, given the constraints on domestic tariffs set by the government, so instead it had to finance its own investments. Furthermore, in the post-war period telecommunications suffered from the poor supervision of the postal branch and lack of internal co-ordination and integration within the telecommunications branch, both severely handicapping its functioning. Traditionally, the French PTT had been dominated by the postal branch, in terms of workforce, appointed senior officials, standardised job qualifications and better access to ministerial cabinets and local/regional governments. Furthermore, there was a strong rivalry between the corps of engineers on the one hand and the corps of administrators on the other: the former had a technical-engineering background, were educated at the École Nationale Supérieure des PTT and appointed at the central R&D departments or other staff units, or at the regional/local management levels of the organisation; the latter were as a rule educated at the École Nationale d'Administration and were selected to work at the Paris headquarters. Finally, the performance of the French telecommunications system was disappointing because of the highly dispersed and collusive character of the domestic equipment industry (to a large extent foreign-owned), not allowing for any substantial economies of scale or any real competition among producers and leading to overpriced charges for equipment.

Although most of the symptoms of the French telephone crisis were already identified in an official report as early as 1947, the modernisation of the telecommunications infrastructure only received governmental priority in national economic planning in the 1960s (Libois 1983). After the reconstruction and rationalisation of the key industries (e.g. coal, steel, chemicals, etc.) had effectively been dealt with in the first four Plans, the rationalisation and modernisation of the domestic telecommunications industry now moved to the forefront. The underperforming French telephone system ran into structural problems, when demand suddenly increased in the early 1960s, making the waiting list go up from 100,000 to 370,000 by 1966 (Jannès 1966). The public and political debate was dominated by '*la crise du téléphone*'; people joked that 'half of France was waiting for a telephone, while the other half was waiting for a dialling tone' (Ergas 1992). Even NATO became concerned over the safety and military implications of the structural weaknesses of the French telecommunications network. As an indication of its concerns, NATO contributed to network modernisation by providing more than 40 per cent of central government funding of telephone investments between 1951–5 (Ergas 1983). In the Fifth and Sixth Plans the deficiencies of the French telecommunications sector were acknowledged by the Pompidou and Giscard governments and the political target for the longer term became to catch up with European competitors. Between 1967 and 1975 a coherent framework of credit institutions was created to aid DGT in raising capital for large-scale network investments and hence reduce its self-financing ratio. This consisted of the Caisse Nationale des Télécommunications (CNT) and its related agencies Finextel, Codetel, Agritel, Créditel and Francetel.

Between 1967–78, the French PTT Ministry was reorganised: the Direction Générale des Télécommunications was officially separated from the Directorate Générale de la Poste, each having separate accounts. In 1969, the Minister of Finance, Giscard d'Estaing, suggested carrying the reform measures even further by proposing a Bill to establish an independent public telecommunications entity with legal status and financial autonomy from the state, to be called Compagnie Nationale du Téléphone. These radical suggestions were channelled into more moderate proposals by a high-level interdepartmental committee that had been established in 1971 to make recommendations concerning the reorganisation of the PTT administration. This Committee, including representatives of the Prime

Minister's Office, the Treasury, the PTT Ministry, the Home Office and the Council for State Enterprises, proposed keeping the postal and telecommunications services integrated within the PTT administration, but separating the two functions in terms of budgeting and labour contracting. In 1973, the Commission Parlementaire de Controle de Gestion de Service Publique de Téléphone was installed to examine any reforms that would improve the management of telephony (Bertho 1981). It suggested transforming the DGT into a government corporation like EDF. The opposition to the core elements of the restructuring plan, to separate the telecommunications function from the postal services and allow it the structure of a public enterprise, was fierce, manifesting itself in the paralysing strike of 1974. The government considered the recommendations of the consultative bodies too radical, opting instead for a massive infusion of investment.

In its famous Plan de Rattrapage Téléphonique of 1970, the French government put forth an impressive proposal to expand and modernise the domestic telecommunications system to reach the capacity and automation levels of its neighbours Germany and the United Kingdom within ten years. It stipulated clear objectives for network growth and the modernisation of the public infrastructure. This programme also aimed for market leadership and large exports for the French telecommunications equipment industry. In the Sixth Plan, that started in 1971, the government reserved FFr 45bn to give the entire population access to the telephone service and transform the equipment manufacturing sector into a high-technology industry that would compete successfully in export markets. The CNET was in charge of developing new technologies and making French telecommunications manufacturing independent of large American firms.

In the Seventh Plan the CNET was stripped of all industrial policy functions apart from R&D; the DGT became the central co-ordinator of the ambitious programme to upgrade and expand the network. In order to give the DGT sufficient leeway to make these plans possible, the network operator could borrow on favourable terms and no longer had to suffer the constraints imposed by the Ministry of Finance, reporting directly to the President. Within the government administration, post and telecommunications were functionally separated. The main instruments used by the state and the DGT to modernise and expand the telephone network were direct subsidies (a massive R&D programme in the field of public switching) and public procurement. For its financing PTT relied upon internal resources through debits and profits, and loans through the public Caisse Nationale des Télécommunications (CNT) and additional funds from private investors. The national infrastructure was transformed from one of the most backward in the Western world to one of the most advanced: between 1973 and 1983 the number of telephone lines increased from 5.6 million to 20.9 million at a growth rate of 12.2 per cent (Germany 7.2 per cent). In the same period the PTT Ministry itself was also modernised to give DGT more corporate autonomy and flexibility to realise its objectives: for example DGT was allowed to set up subsidiaries under private law in commercial fields such as terminal equipment, value-added services and datacommunications.

The large-scale modernisation programme of the French telecommunications infrastructure was not only directed at expanding the telephone system throughout the country, but also served the industrial objective of catching up in telecommunications technology. The public programme envisaged a large procurement base for the French equipment and electronics industry for newly emerging markets. In 1975, the government made it clear that it wanted to retain control over the nation's key industries and subsidy flows by creating strong national champions in the electronics industry and minimising foreign competition. The government accused CGE/Alcatel, ITT and Ericsson of cartel-like behaviour, identifying this as one of the causes of the relative backwardness of French telecommunications (Nouvion 1982). In 1976, the government actively intervened in the purchase of the subsidiaries of the foreign equipment manufacturers ITT and Ericsson (LMT and SFT respectively) by the French firm Thomson-CSF.

Capitalising on the telematics revolution in the 1970s and 1980s: le Plan Télématique

In the Seventh Plan (1976–80) telecommunications was defined as a national priority and a strong emphasis was laid on network expansion to the level of other European countries, improvement of the quality of service, the development of new applications (e-mail and datacommunications) and tariff controls linked to RPI. The budget for upgrading French telecommunications tripled to approximately FFr 140bn (from an investment level of FFr 5.2bn in 1971 to FFr 15bn in 1976, and FFr 26bn in 1977). The DGT was freed from the tight financial controls set by the Treasury to restructure the French telecommunications sector: the annually controlled budgets were replaced by a five-year period of budgetary freedom. After the *grand projet* of the 'telephoning of France' a new target was set for the mid-term with the Plan Télématique.

In December 1976, President Giscard d'Estaing appointed a small committee, headed by Simon Nora and Alain Minc, to investigate the computerisation of society. The Nora-Minc Report (1978), submitted to the President in January 1978, presented a lucid analysis of the economic crisis in France, the impact of new information and communications technologies upon society (unemployment, job creation), and the opportunities it provided for increasing productivity in particular sectors (banking, insurance, education, social services, public administration) and overall economic growth. The Nora-Minc Report referred to the dominance of foreign multinational corporations in the world electronics market: the American multinationals IBM and AT&T were market leaders in computing, telecommunications and semiconductors and the Japanese companies in consumer audio-visual products. The Nora-Minc Report was very much concerned about IBM's move into the telecommunications domain through the creation of its satellite ventures SBS and Comsat. Assuming a techno-economic convergence of these technologies in the near future, the Report warned about the danger of a loss of national sovereignty in these critical industries to foreign multinationals and the serious challenge to the public monopoly of the European PTTs. The Report, identifying telecommunications as a key industry, called for a comprehensive national and European strategy to meet the challenge of American domination in both telecommunications equipment and information services. The European authorities, the central governments and the PTT administration, could counter the ambitions of IBM and AT&T by working together in setting norms and standards (e.g. in the CEPT framework). Given the divergent interests of the computing and the telecommunications industries, the European PTTs could catch up with the two American giants by joining forces with AT&T and competing against IBM.

The Nora-Minc Report argued that given France's weak position in the computing industry and consumer electronics, the future of communications or telematics (the cross-fertilisation of telecommunications and informatics) lay in state control of the telecommunications infrastructure. Therefore, France should maximise its advantages via the control and modernisation of the network. For example, the usage of advanced tele-information services would be encouraged by way of promoting the development of new high-capacity channels, like satellites and datatransmission networks. Nora and Minc suggested that the convergence of telecommunications, informatics and office automation had far-reaching policy implications and therefore proposed a reorganisation of the ministries and public sector organisations concerned. In order to respond more quickly to the techno-economic and international challenges, an organisational restructuring of the French PTT was required. First, the DGT needed to adopt a more commercial and business-like approach and decentralise its responsibilities. Second, a separation of the postal and telecommunications business of the French PTT was also regarded as necessary. The traditional consolidation of the telecommunications and postal functions in the PTT Ministry was challenged by the aforementioned structural developments, revolutionising the capital-intensive telecommunications services and leaving the labour-intensive postal services relatively untouched. The identification of telecommunications as a strategic industry called for the DGT to take a leading position in the promotion of a national telematics strategy. For that purpose a state enterprise could be established with special privileges regarding its budget and investment policy and a flexible personnel policy to recruit and select highly qualified personnel. Third, the relationships between DGT, TDF (the broadcasting transmission authority) and also CNES (space), then unclear and not well defined, needed to be redefined. Formerly the three belonged to the PTT Ministry, the Prime Minister's Office and the Ministry of Industry respectively. Nora and Minc suggested that a large Ministry of Communications needed to be formed that would integrate these crucial agencies in the domain of telematics. The government endorsed the broad outlines of the Nora-Minc plan, in which the state actively promoted the supply of and demand for computers and data services.

The idea underlying the Nora-Minc Report, namely the development from 'telephony' to 'telematics' was further elaborated in the Eighth Plan (Pigeat & Virol 1980). After the targeting of network expansion and modernisation, as defined in the Seventh Plan, the new objective became to adapt the French telecommunications industry to the new techno-economic conditions. A *filière* approach was suggested, aimed at strengthening the various areas and linkages in the emerging telematics sector (Lorenzi & Le Boucher 1979). A collaborative effort was needed that included the co-ordination of initiatives in the domain of integrated circuits, opto-electronics, fibre optics, software, terminal equipment, network upgrading, service development, and the education and training of systems engineers and personnel (Pigeat & Virol 1980). Besides supporting the ongoing expansion of the French telephone network (from 12 million subscribers in 1980 to 24 million in 1985), the Planning Commission gave priority to the following issues: the further upgrading and digitisation of the infrastructure, the nationwide provision of basic telematic terminals (i.e. Minitels), the active promotion of telematic services for business customers (e.g. dataprocessing, video conferencing) and the active pursuit of new high-quality telematic technologies (videocommunications, satellite and radio technologies, pay TV etc.). The Nora-Minc Report, and the consequent planning proposals, inspired the Minister of Industry to put forward a national action programme for the French electronics industry. This Programme d'Action par la Filière Electronique (PAFE) sought to restructure the industry through nationalisation, increase R&D spending by both government and industry, promote the diffusion of technologies through large national projects, oblige French companies to buy French, and reform the education system to produce more qualified personnel (MRI 1982; Groenewegen 1989). The DGT, controlling the vital telecommunications infrastructure, was instructed to mobilise and revitalise the French electronics industry, leading it towards greater international competitiveness.

The successful society-wide penetration of the telephone in the early 1970s had made the DGT one of the most powerful administrations in the French civil service. In the new plan DGT received large subsidy schemes to upgrade and digitise the network and to prepare an effective industrial shift towards telematics. The DGT acted both as an administration, given the responsibility for the state's industrial policy towards *la filière électronique* and as a flexible public sector company, actively promoting and developing new systems and services. The DGT was also allowed to establish a number of separate subsidiaries under private law with private sector participation. These mixed affiliates (or '*sofres*') were considered an effective instrument to stimulate diversification of the public enterprise on the one hand and utilise public resources for the stimulation of the private sector, on the other. Under the Socialist government the leading firms of the French electronics industry were nationalised and restructured into the following national champions: CII-Honeywell-Bull (computing), CGE-Alcatel (telecommunications equipment and office automation), Thomson and Matra (semiconductors and electronics).

As early as 1978, the DGT had launched one of the first packet-switched datanetworks in the world, called Transpac, on which the various enhanced data services could be transmitted. One year before, Transpac was created as a subsidiary of DGT, set up as a hybrid public–private organisation of large users and the DGT: 67 per cent of the shares were in the hands of the state, the workforce of the company owned 5 per cent and the large users (Crédit Agricole, Havas, UAP, Renault, etc.) possessed the remainder of the shares. In 1984–5 DGT acquired the private shares and Transpac was placed under control of the state holding Cogecom. Transpac turned out to be a commercial success: its high quality level and distance-independent tariffing attracted a large number of subscribers.

In the late 1970s and early 1980s the government pursued a sequential targeting strategy regarding the French telematics and electronics industry: the first aim was to establish mass consumer markets for relatively simple IT applications; the second aim was to export these applications successfully. The ultimate aim was to produce more sophisticated and high-tech applications. In the *Grand Projet Plan Télématique* the national government (through the DGT) played a powerful role: it determined the technology, selected or created the firm(s) suited to the development desired, created guaranteed demand for the product and/or assumed a significant share in the financial burden of development in an initially protected market. After the military, DGT became the second biggest investor in the country. Besides an integrated network that could provide a variety of telematics services, four main service groups were identified: videotex, the electronic directory, mass fax and tele-writing. In this Priority Action Programme DGT was given a five-year concession with budgetary freedom. DGT effectively managed to pursue its own organisational-political goals of consolidating and expanding its public monopoly by providing additional services to the mass public. The Telematics Programme had the objective of protecting and supporting the interests of the domestic industry by developing an effective market at home and abroad. The state-directed commercialisation of information technology worked at home, giving the French industry a captive market, but failed to make major inroads in international markets. The exception to these export failures are Transpac (France Télécom's datanetwork), ISDN and, to a certain extent, Télétel/Minitel (DGT's videotex system), three spin-offs of the Telematics Programme.

When the DGT had successfully modernised the telecommunications infrastructure in 1978, two new challenges emerged: the state of the directory assistance services was rather poor and the POTS market (plain old telephone services) was nearly saturated. The DGT saw the development of its videotex system Télétel/Minitel as adequate remedy to both problems. The electronic provision of inquiry services would be more efficient than printing and distributing telephone directories and running a telephone information service that was regularly overbooked and out of order. A videotex system would not only generate more traffic, it would facilitate new information and communication services for the general public and business users. The DGT followed an ambitious implementation strategy, aimed at creating a mass market by purchasing videotex terminals in large quantities and giving them away for free to residential and professional users.

The Télétel/Minitel system was connected to Transpac, the packet-switching datanetwork of DGT, enabling subscribers throughout the country to use the various information and communication services regardless of distance. The French videotex system was one of the first offsprings of the Nora-Minc Report, aimed at identifying the potential of telematics and its impact upon society, to be put into practice. After its experimental try-out in 1978, Minitel was officially introduced by the DGT in 1983. DGT's expectations for the society-wide diffusion of Minitel terminals was optimistic: it expected to have 30 million terminals installed in 15 years

(Cohen 1992). One application that could justify the development and introduction of videotex, as identified by the DGT, was the provision of a national electronic directory service through Télétel/Minitel, that would replace printed telephone directories and the DGT's traditional inquiry services. Being at the same time entrusted with the responsibility to stimulate the interests of the French telematics industry, the DGT asked the domestic computer hardware manufacturers and the computer services sector to develop an appropriate low-key terminal that could be easily mass-produced. After the prototype of the Minitel terminal had been developed, DGT decided to combine the aforementioned objectives of establishing an integrated electronic inquiry service and the large-scale diffusion of terminals. The DGT followed a supply-driven strategy that was aimed at enhancing the availability of electronic directory services through cheap and easy-to-use terminals that would be supplied free to subscribers (Charon 1987).

In its experimental stages the DGT's experiment with videotex was opposed by the publishers, who saw the freedom of expression and the written press culture threatened (Charon & Cherki 1984). The introduction of telematics would imply the availability of alternative distribution channels and the draining away of advertising revenues to new players in the information market place, notably the DGT, computing companies and new information providers. After an agreement was reached between the DGT and the press about a clear-cut division of labour, the Télétel/Minitel turned out to be a commercial success, with 4.2 million subscribers in 1988 and more than 7 million in 1996 (France Télécom *Annual Reports* (1988–96)). Besides acting as a common carrier providing the appropriate infrastructure and user terminals, the DGT restricted itself to the provision of gateways, common billing and menu functions, and the 'basic' electronic directory services. The publishers, computer service providers and information providers were involved in the rendering of a variety of information, communication and transaction services (e.g. teleshopping, message services, tele-reservation, etc.) and supplying software applications and the non-standard terminals and peripherals. After the failure to develop billing systems based on smart cards and the unfeasibility of introducing subscription-based charges, the DGT and its business partners developed a transparent billing system. This so-called 'kiosk' system is based simply on the consumption of the telematic service to be collected by the DGT, who passes a certain proportion of the usage charges on to the service providers. This proportion is set through bilateral agreements between the DGT and the various information and service providers. The kiosk system allows for flexible price setting and transfer pricing between the DGT and service providers through the creation of different charge bands. This saves users from having to subscribe for individual services, while preventing the DGT from intervening in the content of services (Charon 1987). Later the same 'kiosk' billing system was implemented by France Télécom/DGT for the provision of audiotex services.

Under the Socialist Mitterand-Mauroy government the DGT lost its financial autonomy, granted by the previous government, and as before was used to subsidise the general budget. The DGT also became involved in the new objectives set by the Socialist government to promote the French electronics sector and cable policy.

In 1982, the DGT was given full control over the ambitious five-year Plan Filière Electronique, scheduled to promote the French electronics sector and to rationalise the loss-making firms Bull and Thomson. However, it also implied that the DGT's revenues were used by the French government as a 'cash cow' to restructure and finance other troubled segments of the French electronics industry (notably computing (Bull) and semiconductors (Thomson)). Confronted with the international deregulation trend, the Socialist government was supported in its desire to preserve the *status quo* by the workforce of the French PTT. The labour unions defended the necessity of the DGT's monopoly by referring to public service requirements and the French nation's independence in high-technology matters. The Corps des Ingénieurs, organised in the AIT, including a high proportion of the DGT's senior management, largely agreed on these matters, but demanded at the same time more corporate autonomy for the DGT in shaping technological and market conditions. The AIT opposed the practice of the DGT's revenues being used to subsidise the unprofitable post and nationalised sectors. It demanded that the postal and telecommunications branch be separated and that the DGT be given more corporate autonomy from the government. Supported by the Director General of Telecommunications, the administrative-technocratic élite of the DGT argued for a public enterprise status (Cohen 1992; Barreau 1995).

The first steps to introduce competition

In the years of the Liberal-Conservative government (1986–8) the status of the powerful DGT – as public operator, commercial service provider, executor of industrial policy and regulator – was questioned. Value-added services stayed in the hands of the all-powerful DGT/France Télécom, concerned both with safeguarding its common carrier monopoly status and, even more ambitiously, with ensuring its future control of the promising new value-added services. The Liberal-Conservative Minister for PTT was in favour of telecommunications reform to bring France in line with international developments. One of his first acts was to remove the competence the Socialists had given to the DGT for the direction and execution of industrial policy in the *filière électronique*.

The Liberal-Conservative government proposed legislation that would establish modest competition in the telecommunications field by the end of 1987. The Minister for PTT Affairs, Gérard Longuet (1988) identified three structural weaknesses of DGT/France Télécom: its large debt, social policy constraining its flexibility and international presence, and lack of competitiveness. To bring France more in line with the changing economic reality of the times, the markets of value-added network services, cable TV networks and mobile communications were to be liberalised. The proposed measures did not really touch upon DGT/France Télécom's monopoly on the installation and operation of the public infrastructure and the basic voice and databearer services, however. According to a legislative proposal, presented to Parliament in 1987, the DGT should be transformed from a government department into a state-owned public limited company (Société Nationale), operating with a much greater degree of commercial autonomy from the state, like the French railways (SNCF) or the electricity utility (EDF). This telecommunications company would be separated from the Post Office branch. France Télécom would be supervised by an Administrative Council in which government, staff and independent experts were represented. Furthermore, Longuet suggested that two existing institutions should become involved in French telecommunications policy making: CNCL, the semi-autonomous broadcasting transmission agency, would regulate access rights to information and the Competition Authority (Conseil de la Concurrence) would regulate the provision of public infrastructure and basic services.

Notwithstanding complaints about an overcentralised organisation, lack of financial autonomy, enormous debts caused by France Télécom cross-subsidising the loss-making electronics industry, the majority of the workforce was against any major changes. France Télécom's staff was strongly attached to its civil servant status and public service tradition and opposed such a prospective corporatisation (or pseudo-privatisation) and any neo-liberal solutions (e.g De Guers 1987). The influential Corps des Ingénieurs des Télécommunications and the powerful public sector unions made their opposition to the government's radical plans very evident: more than 50 per cent of the entire workforce went on strike. After two years of Liberal-Conservative rule the policy outcomes were mixed. The new administrative body CNCL proved unable to perform its task adequately or to act autonomously from the state. Nevertheless, some impressive results were achieved by the right-wing government: the creation of cellular duopoly (opening the network monopoly of the DGT), the liberalisation of the VANS market, and private sector participation in cable network operations. The government also sought to increase market transparency by separating DGT's operational from its regulatory activities and creating separate subsidiaries for DGT's competitive activities (i.e. preventing anti-competitive cross-subsidies). The relatively organised workforce strongly opposed the proposal to separate the postal and telecommunications services, and objected to any attempt to turn the organisation into a state-owned company through massive industrial action campaigns. The small but influential trade unions, the CAT and FO, vehemently defended the unity of the postal and telecommunications administration, the civil servants' statute, and its attractive 'fringe' benefits (pension schemes, sick benefit and job security).

Under the Socialist government the proposal to change France Télécom into a commercial company with public capital was dropped and replaced by the idea of a powerful public service. When the Socialists returned to power the Minister in charge, Paul Quilès, emphasised the DGT's public service responsibilities and the proposals to prepare France Télécom to become an independent public corporation, along the lines of EDF, were shelved. At the end of 1988 Hubert Prévot, a top civil servant, was asked to organise a public debate on the present and future role of the PTT services. Prévot drew up a draft report that formed the basis for a long series of national and regional consultations with management and employees of the two PTT administrations, industry, consumers and other stakeholders between December 1988 and June 1989. The Prévot Report recommended the formal separation of the postal and telecommunications services, replacing the state administration statute by that of an industrial and commercial public enterprise (Prévot 1989). The policy debate focused on the definition of public service missions, regulation and competition, human resource management issues in the postal and telecommunications administrations, and corporate autonomy. After six months of intensive discussions, a consensus was reached on the aforementioned issues (Prévot 1989):

- a further separation between the postal administration and the DGT and greater autonomy was suggested, but the two companies would remain under the state's sponsorship and budgetary supervision (i.e. subject to public law);
- the role and mission of public service was to serve customers, modernise the economy, and increase the competitiveness of domestic enterprises;
- a balance between public monopoly, controlled competition and full competition: for the basic infrastructure and voice telephony an exclusive regime (justified by cost scale and scope economies, network integrity, public service, and strategic resources); for cellular and databearer services and independent 'private' networks, managed competition; and for value-added services and terminal equipment, a free and open market regime;
- more corporate flexibility for the two companies in their internal management ('*autonomie de gestion*' in hiring staff, and in financial matters and business operations) and in their external management (e.g. setting up new businesses and subsidiaries, allowing for strategic alliances at home and abroad).

Although a few minor amendments were made to do with job security for employees, the proposed plan passed the National Assembly and the Senate in May 1990. The new telecommunications legislation, as laid down in two Acts that replaced the 1923 Act, was in line with the Community Rules on the effective opening of distinct markets and the separation of operational and regulatory functions. The first Act (No. 90–578), which was passed in July 1990, focused on the organisation of the provision of the public postal and telecommunications services, the second (No. 90–1170), which was passed in December 1990, referred to the regulation of the postal and telecommunications sectors. The July Act stated that France Télécom would be separated from La Poste and its sponsoring Ministry. From early 1991 onwards France Télécom would be established as a relatively autonomous public corporation ('*exploitant autonomes de droit public*'), entrusted with corporate legal status. Although remaining under the guardianship of the Ministry of Post and Telecommunications, France Télécom finally had a large degree of autonomy to possess assets, draft budgets and take responsibility for its own management and business policy. The composition of the Board of Directors (Conseil d'Administration) would be tripartite, made up of state officials, staff representatives, experts and representatives of user groups. The government would nominate the President of the twenty-one-member board for a three-year term.

France Télécom's significant business decisions were still subject to state scrutiny and its longer-term plans needed political approval. France Télécom management had to draft a mission statement that needed government acknowledgement. The statement referred to the company's provisions to achieve the goals set out in its mission, the contribution of France Télécom to the objectives set by the national government and its relation with consumers and local authorities. The company's workforce would retain its civil servants' statute and labour unions representatives would have a say in salaries and staffing levels through their membership in the Commission Supérieure du Personnel et des Affaires Sociales. Two new consultative bodies were established as an interface between France Télécom and its institutional environment: the Commission Supérieure du Service des Postes et Télécommunications and Conseil National des PTT. The former represented independent experts and the various socio-political and business stakeholders in the French telecommunications sector, the latter the Ministry, management and labour of La Poste and France Télécom discussing R&D and training issues.

The influence of the state and the public sector unions in the newly established public enterprise was still substantial. France Télécom would still be answerable to the Ministry of Post, Telecommunications and Space. France Télécom and the supervisory Ministry also had to agree upon a three-year planning contract (*contrat du plan*), dealing with corporate strategy, the quality and tariffs of its services, wage policy, investment levels and other substantial financial decisions. Within the PTE Ministry, the Direction des Services Publiques was responsible for the reserved services and the newly created Direction de la Réglémentation Générale was in charge of the competitive services. Furthermore, France Télécom had to negotiate its budget with the government and contribute towards the huge financial losses of the state companies Thomson and Groupe Bull. In addition, the Cogecom subsidiaries, including TDF, Transpac and FCR were all removed from the control of the Finance Ministry. The position of the PTE Minister was downgraded to that of Under-secretary of Finance, Economy and Budget.

After the landslide election victory of the Liberals and Conservatives in 1993, the issue of hiving-off France Télécom to the private sector was once more put on the political agenda. To reduce its heavy debt burden, the government considered partial and gradual privatisation as a possibility: it envisaged a flotation of 10 per cent within a three-year period and a maximum of 30 per cent sold to private investors within seven years. Anticipating opposition from the powerful public sector unions, the government considered allowing France Télécom's workforce to retain its civil servant status. In 1993, Longuet, the Minister in charge of trade, industry and telecommunications, argued that in order to respond effectively to the liberalisation of the European telecommunications market by 1998 and to allow France Télécom to link up with other telecommunications companies, a change in its legal status was necessary. The government proposed transforming France Télécom from an 'autonomous operator under public law' into a joint stock company. This would allow France Télécom to swap equity stakes with industrial partners and enable central government to privatise the company. To integrate the strategic alliance with the Deutsche Bundespost even further, ownership in France Télécom could be made available to industrial partners.

The government, however, would keep majority control over France Télécom and only a minority of its shares would be sold. The plans of the Balladur government (1993–5) to privatise France Télécom produced an outcry, after which the proposals to reform the national telecommunications company were shelved yet again. More than 80 per cent of the workforce went on strike (including a large number of senior officials) in October 1993, fearing layoffs and the end to the civil servants' status. France Télécom's staff argued there was no need for structural change: the country already had outstanding levels of quality of service provision and one of the best equipped networks in the world (e.g. Leray 1994). A potential privatisation would go against the public interest and challenge the industry's public service tradition and would only benefit private investors like Alcatel, national banks and foreign investors who considered taking a minority stake in France Télécom. The French labour unions, together with their German counterparts, also acted against the proposed strategic alliance of the Franco-German Atlas joint venture with the US-based company Sprint, because of its anti-union activities. The French and German unions expressed fears about the intimidation of workers by Sprint and demanded from their Board of Directors that it hold up the negotiations until Sprint would recognise workers' rights to organise.

In 1994, the French government asked Gérard Théry, the driving force behind the successful Minitel diffusion and former head of the DGT, to formulate an appropriate strategy concerning the viability of the integrative broadband networks in an increasingly deregulated telecommunications market. Although the functioning of competition and the utilisation of some market-led mechanisms were widely acknowledged abroad, Théry (1994) suggested a further continuation of the French approach by emphasising state guidance towards the development of a nationwide fibre-optics infrastructure by 2015, that would be accessible to and link every citizen and corporate user. The costs of upgrading and creating an advanced broadband network, estimated between FFr 150bn and FFr 200bn, should be borne by France Télécom. The Report exhibited a clear technology push strategy, in which infrastructure provision would act as a catalyst for new uses and services. Little attention was paid to existing and future demand for these so-called 'broadband' services, and the alternative of competitive network provision was totally ignored (e.g. coax cables, satellite, mobile networks, the Internet).

In his analysis, Théry made it clear that the traditional low-capacity services, such as telephony and Télétel/Minitel were in the saturation stage of their product life cycle and that an integrated approach was needed towards the entire telecommunications and multimedia *filière*. Priority needed to be given to the upgrading of the public infrastructure by the large-scale utilisation of fibre optics, the targeting of cable development, the promotion of new switching technologies (ATM) and the commercial application of a variety of multimedia services. While stimulating the commercial development of high-capacity networks and services such as digital television and video services, Théry suggested that universal service and open network provision requirements were safeguarded in the next decade. To promote the development of new technologies, infrastructures and services effectively, the Report asked for clear political and financial support from the government. For instance, additional resources were needed to fund applied R&D and the upgrading of network engineering expertise and the promotion of pilots in which public and private actors could develop new information and communication services in the domains of health, education, culture and public administration.

Théry's inquiry into the public infrastructure of the future was complemented by a sectoral analysis of the major technological, economic and international developments in France's tele-services market in the short and longer term. This prospective study investigated the commercial possibilities of a multitude of service applications in sectors like health care, education, commerce and public administration (Breton 1994). The Report also warned against the potential risks of an emerging global tele-services market. For instance the outsourcing of information-processing activities to neighbouring and/or developing countries and the delocalisation of business activities could lead to a brain drain of technological expertise and have a negative effect on domestic employment. The overall aim of these two reports was to establish a technologically advanced and competitive telecommunications industry that would fully utilise the high-capacity infrastructure and trigger the large-scale provision of a multitude of tele-services for both business users and households. Although less explicit than before, the public administration was expected to lead the diffusion of new multimedia services throughout the country.

Between 1994 and 1995, a law was prepared to promote experiments with the provision of telephony services over cable networks, building new metropolitan optical fibre networks, and permitting the licensing of experimental multimedia services. Also the national railway operator received permission to resell network capacity to the second mobile operator SFR.

The lead the French were thought to have in the provision of on-line services through the widespread usage of the home-grown Minitel technology turned into a major setback with the diffusion of the Internet in the late 1990s. At present, the level of personal computers and Internet hosts in France is among the lowest in the OECD area (OECD 1996; Spectrum 1996). Furthermore, the adoption of Internet services in France suffers from relatively high access charges, a lack of local points of presence (making alternative on-line services expensive in rural areas), and, obviously, from the large installed base of obsolete terminals. It was only recently that the French government became aware of the Internet challenge and decided to boost Internet usage across the nation: for instance in January 1998, the Jospin government announced that it would promote the integration of Internet technologies in the educational system, governments and business through a massive subsidy of FFr 1.5bn.

In the mid-1990s, the French government organised several consultation rounds to develop a consensus on a fully competitive domestic market by 1998 and to prepare a new bill, establishing new rules (Lasserre 1994; MTIP 1995). In the discussions leading up to the new legislation five objectives were put forward: the protection of public services; the introduction of new technologies and services; the fostering of the sector's competitiveness; the development of France Télécom's future competitiveness; and the creation of an effective regulatory framework.

The new legislation, which replaced the 1988 Telecommunications Act, was passed in the summer of 1996. It provided new arrangements for the industry and reformed the corporate structure of France Télécom. Furthermore, the 1996 Act sought to open up the market for alternative network provision immediately and for voice telephony by 1998.

Next to the Ministry for Post, Telecommunications and Space, new institutions were planned to regulate and police the newly emerging market place: early in 1997, an independent regulator, the Autorité de Régulation des Télécommunications (ART), and a semi-autonomous body for frequency management, Agence des Fréquences, were established. After a number of major strikes over a three-year period, France Télécom was corporatised in January 1997 and granted the Societé Anonyme status with 100 per cent state ownership. After the government had made several concessions to the company's strike-prone workforce (e.g. keeping pension rights, employee share ownership, a gradual phasing out of civil servants' status), approximately 25 per cent of France Télécom's stock was sold on the stock market in the autumn of the same year.

6.5 The liberalisation of the French telecommunications market

The Cable Plan of 1981 envisaged an interactive fibre-optics network that could transmit sound, voice and data. According to central government the Cable Plan should accomplish the following goals (Vedel 1987; Vedel & Dutton 1990): develop innovations as a solution to the economic crisis facing France at that time (e.g. the promotion of new markets and employment); support cultural aims (e.g. domestic production and distribution of broadcasting programmes, and hence the preservation of French culture); and implement decentralisation policies (e.g. by giving more powers to the local and regional authorities). Initially, the development and the construction of cable systems received little priority from the key players in the domain (the PTT/DGT, the TDF and the publishers). The PTT/DGT was fully occupied with the upgrading of the telecommunications infrastructure and the development of telematics, the TDF showed a bigger interest in direct broadcasting by satellite as alternative transmission channels, while the local press was concerned about the potential impact of cable distribution upon advertising revenues.

In the developing institutional setting, the granting of licences for building and operating cable networks would be given by the Haute Authorité de la Communication Audiovisuelle (HACA) and the actual instalment and diffusion of broadband cable networks was carried out by France Télécom, in collaboration with the local authorities. Originally, cable systems were the responsibility of the national broadcasting transmission authority (TDF), but after the acquisition of the TDF in 1980–1 by the DGT, PTT was given full control over the construction and technical operation of these cable systems. For the development of cable networks the government promoted a partnership between the DGT and local municipalities in the financing of the investments. The actual operation of these cable systems was in the hands of so-called Societé Local d'Exploitation du Cable (SLECs). These companies, in which both municipalities and private investors could participate financially, would lease the cable facilities required from the DGT. Especially the Caisse des Dépots et Consignations (CDC) and two water utilities, namely Compagnie Générale des Eaux (CGE), and Lyonnaise des Eaux have become active in the French cable market. The Cable Plan, due to disappointing penetration, failed to be a success. The cabling project suffered from an institutional separation between the construction and exploitation of the network and a reluctance on the part of local communities to engage in sponsoring the various cabling projects. Between 1987–94 the market for cable networks was modestly liberalised: the government encouraged the involvement of private investors in installing and operating these cable networks through enlarging private sector participation in these public–private cable ventures. Early in 1994, CDC decided to leave the cable market altogether and sold its cable networks to France Télécom and CGE.

Although a beginning was made with the liberalisation of value-added services in 1987, it was not until 1993 that full liberalisation was implemented. This 'delay' was due to the government imposing significant restrictions on firms entering the sector, arguing that VANS would be a serious threat to network integrity and open the door for 'bypassing' the public infrastructure. In the mobile market, effective competition was introduced from the early 1990s onwards. In 1989, a second mobile communications licence was awarded, in addition to France Télécom's Radiocom 2000, to the SFR group Société Française du Radiotéléphone. In the SFR group the following major partners are represented: Compagnie Générale des Eaux, Vodafone, Bell South and some smaller participants. The rationale behind this duopoly on cellular communications was to create a level playing field, in which the two providers would operate under strictly identical conditions with equal rights and possibilities. Initially, competition between the two mobile operators was restricted: the expansion of SFR's network was constrained by limited spectrum availability and a lack of collaboration from France Télécom, both contender and supplier of fixed network links, to ensure interconnectivity. After complaints from SFR about France Télécom's high interconnect and leased lines fees, the Minister of Communications gave permission to use the cable networks and microwave links as alternatives to France Télécom's network. In the cellular market, SFR within five years acquired a market share of approximately 25 per cent. In the mid-1990s radio-based service competition had not posed a major threat to the fixed infrastructure and the local loop to the customer in France. In October 1994, the licence to operate a third mobile network was awarded to the Bouygues consortium (including besides the constructor Bouygues, VEBA, Cable & Wireless and, later, Alcatel).

Although competitive network provision was prohibited, limited resale of capacity was allowed for some data services in 1987. To prevent arbitrage, tariffs were set by the government. A few years later the rulings on the sharing of leased circuits and traffic capacity resale were loosened for the supply of value-added services and datacommunications services to closed user groups (commercial voice resale was prohibited). The limited liberalisation of terminal equipment and value-added services was legalised in 1990 with the enactment of the new telecommunications legislation. When the European Commission in 1989 proposed the liberalisation of data service provision, France protested, arguing that network operators should be given authority to issue licences and levy access charges on private VAN operators. France also led the opposition to the EC Commission's Open Network Provision. All value-added networks required a licence and tariffs were set to prevent arbitrage. After BT had already been licensed, the Swiss/Dutch/Swedish operator Unisource was granted a licence in the second half of 1994 to operate value-added network services in France. The decision to introduce network competition was, however, postponed, following a nationwide strike by France Télécom employees, protesting against the government's privatisation and liberalisation initiatives. After AT&T's 'Project Atlantic' came to nothing, involving France Télécom and Deutsche Bundespost merging their global activities, the American operator showed a clear interest in the French telecommunications market. Later in 1994, AT&T wanted to buy shares of the loss-making computer manufacturer Bull on the condition that it would be granted a telecommunications operating licence. The French and German operators cemented their strategic partnership by aligning themselves with Sprint, the third largest American long-distance/international carrier, in the Atlas/Phoenix joint venture, later renamed as Global One.

In the 1996 Act the government had agreed to open up the provision of alternative networks at once and liberalise voice telephony two years later. Three contenders have put themselves forward to take on the incumbent operator France Télécom. The first new carrier providing alternative fixed and mobile services to the public is Cégétel: it includes the telecommunications activities of CGE (SFR), British Telecom, the national railway operator SNCF's optic network, and the German industrial conglomerate Mannesmann. Early in 1998, the company unveiled a price-cutting strategy focusing on business users in the larger urban areas of France, notably Paris, Lille and Lyons. The second alternative operator is '9 Telecom', a collaboration of Bouygues (41 per cent), VEBA (20 per cent), and Telecom Italia (39 per cent), which is preparing to provide fixed-line services and hopes to integrate this with its existing mobile offerings by the end of 1998. Despite having established its French subsidiary 'Siris', AT&T-Unisource has suffered setbacks (i.e. CGE/SFR left for Cégétel/BT) and has yet to find a solid domestic partner.

While retaining complete monopoly on the public telecommunications network, the DGT did not have a monopoly on terminal equipment sold at market conditions (the first telephone set was excluded from the DGT's monopoly in 1986). From 1974, the French government loosened the terminal market and concentrated its efforts and investments on switching, transmission and network installation, while allowing modems, PABX and customer premises equipment to be sold at market conditions. The government allowed competition among French manufacturers (e.g. Alcatel, SAT, Jeumont Schneider and Thomson) and made it difficult for foreign manufacturers to enter the market through certification procedures from which foreign firms were virtually excluded. Barriers to competition were only slowly and selectively removed, with new entry by foreign suppliers and competition administered by government authorities. Through its private subsidiary EGT, the

DGT competed with the equipment suppliers for the delivery of terminals. In the mid-1980s the two most important equipment manufacturers were Compagnie Générale d'Electricité (CGE) (CIT-Alcatel), with 84 per cent of the market, and CGCT (an ITT subsidiary) with the remaining 16 per cent. As part of its overall nationalisation programme of 1982, the French government acquired the remainder of ITT's telecommunications activities (mainly CGCT). In 1982–3, CGE/Alcatel and Thomson swapped their civil telecommunications and computing and defence activities, in the so-called second 'Yalta of Electronics'. It was believed that this would reduce existing competition and generate economies of scale. The Yalta agreement between Thomson and CGE made CIT-Alcatel the new telecommunications champion and Thomson the national champion in consumer electronics and semiconductors. The DGT was against the consolidation because such a bilateral monopoly would diminish its bargaining power. The DGT did not wish to be dependent on single sourcing. In its view rivalry between competitive suppliers would benefit technological innovation and cut prices.

After purchasing 56 per cent of ITT's telecommunications business in 1986, CGE/Alcatel gained access to ITT's advanced digital switching technology and was allowed entry into a number of previously closed markets in Western Europe: Belgium, Spain, Italy, the Netherlands and Germany. When the strongly diversified ITT conglomerate ran into financial trouble in the mid-1980s, threatened by corporate raiders and the disappointing commercial and technological results of its telecommunications switch (System 12), it decided to withdraw from the telecommunications race and begin to look for a partner. ITT had also suffered from the aftermath of the nationalisation of several of its subsidiaries; in France, for instance, ITT had received only a tenth of its subsidiary's market value when it was deprivatised in 1981–2 ($32m instead of about $325m) (Araskog 1990:127). CGE was clearly interested in acquiring a substantial stake in ITT, which would significantly increase its share in the domestic equipment market, and started secret and bilateral negotiations. Alcatel (CGE/ITT) was originally designed to be a pan-European project, with ITT owning 37 per cent of the shares and a European holding company named Eurotel owning the other 63 per cent. CGE had intended to include partners from Belgium (Société Générale), Spain (Telefonica), Italy (Italtel) and Germany (Bosch) in the Eurotel holding. After negotiations with the Italian and German partners collapsed, the Belgian and Spanish partners also decided to abandon the pan-European project. The French government, at that time the sole shareholder of CGE-Alcatel, only became involved in the ultimate settlements of the CGE-ITT venture. Basically, the French government was confronted with a *fait accompli* by its relatively autonomous subsidiary and had no choice but to agree to the alliance being finalised (Araskog 1990; Sally 1993).

From that moment on, CGE-Alcatel has successfully developed from a French company highly dependent upon its protected home market and former French colonial markets into a strong and decentralised Euro-national with world market leadership in the equipment market. This will have consequences for its R&D activities, currently carried out in France: the company may be forced to focus its R&D strategy on the needs of non-PTOs and non-French customers. Nominally,

Alcatel is based in the Netherlands, partly to illustrate the European character of the company and because of the attractive Dutch fiscal climate. In 1990, Alcatel acquired a 65 per cent stake in the second Italian equipment manufacturer Telettra, a subsidiary of Fiat. Besides getting a market share of 33 per cent in the Italian equipment market Alcatel gained a minority stake in the Fiat holding company. In 1992, Alcatel agreed to purchase the remaining 30 per cent of the ITT shares. In return ITT acquired a 7 per cent stake in the CGE holding company, renamed Alcatel-Alsthom. In this diversified conglomerate, active in transport, energy and communications equipment, the latter domain contained almost 70 per cent of the holding company's sales (Sally 1993, 1995). In order to reduce its dependence upon the PTTs, Alcatel diversified into new market segments directly or indirectly related to public switching, like datacommunications, mobile service operations, software and multimedia applications. In 1993 Alcatel made it clear that it would be prepared to consider taking equity stake in France Télécom (or swap stakes) if the operator were to be privatised and if vertical integration in telecommunications was allowed by the European Commission.

In the 1980s, the future of France Télécom's second preferred supplier CGCT was highly uncertain: it relied upon a minor market share of 16 per cent without having a central office switch of its own and was confronted with operating losses of approximately FFr 1bn between 1976 and 1983 (HBS 1988). Between 1982–6 the government considered two options to ensure a continuation of CGCT: the acquisition by a major French company or prolongation with state support. The attempts to integrate CGCT into CGE, Thomson or Bull proved fruitless and consequently the troubled company relied upon large state subsidies for its survival. The sale of CGCT by the government in 1987–8 caused a serious battle between three powerful foreign bidders: AT&T, Ericsson and Siemens. The DGT saw its strong support for the AT&T/Philips lobby, based on superior technology and solid market experience, rejected. The government avoided severe trade confrontations with the Germans and the Americans by awarding CGCT to a Franco-Swedish consortium, in which Matra, Bouygues and Ericsson participated. The government's decision was based on financial and socio-economic arguments (switch price, acquisition price and minimum number of layoffs) and the selection of foreign partners, that would include collaboration with domestic telecommunications businesses and allow for a substantial transfer of technology (HBS 1989).

6.6 The privatisation of France Télécom

Until 1988, the Ministry of PTT and Space controlled French postal and telecommunications services through the Direction Générale de la Poste and the Direction Générale de Télécommunications. The DGT, established as an autonomous directorate in 1946, employed about 170,000 people (of whom about 12,000 were working in its subsidiaries) and was composed of twenty-two regional directorates, which were in turn divided into operational sub-regions. France Télécom/DGT saw the workforce shifting between 155,000 and 165,000 between 1981 and 1996 (see Table 6.1).

Table 6.1 France Télécom: key data

*Key data France Télécom***	*'81*	*'82*	*'83*	*'84*	*'85*	*'86*	*'87*	*'88*	*'89*	*'90*	*'91*	*'92*	*'93*	*'94*	*'95*	*'96*
Turnover before tax (in billion FFr)	na	na	na	na	na	na	na	na	na	108.4	115.8	122.6	127.0	142.6	147.8	151.3
Profit before tax (in billion FFr)	0.3	0.3	0.7	0.9	1.1	1.1	1.2	1.2	2.4	2.6	2.0	3.3	4.8 (9.1)*	14.3 (9.9)*	14.0 (9.2)*	14.5* (2.1)
Workforce (x 1,000)	156.5	161.3	162.5	164.6	166.2	165.0	161.4	158.6	156.5	156.6	156.1	156.9	154.5	167.9+	164.7	165.2

na: not available

* New fiscal and accounting regime from 1 January 1994.

\+ This figure includes the average number of employees in the France Télécom Group (i.e. the core company and its holding and semi-independent subsidiaries).

** The financial results of France Télécom, as offered in its Annual Reports, are difficult to interpret, given the use of different bookkeeping procedures, the various changes in the fiscal regime France Télécom was subject to over the years, and the far from transparent ownership and control structure of France Télécom and the Cogecom holding.

Source: DGT/France Télécom *Annual Reports* (1981–96)

One could say that the DGT and the French government brought about a kind of hidden privatisation, by creating various separate subsidiaries and public/private *filiales*, subject to company law, and therefore not controlled by the Cour des Comptes. Through the pursuit of such a filialisation strategy, DGT/France Télécom was able to cope with a more demanding and complex market environment and acquire the flexibility in terms of better access to capital and labour for effective diversification and foreign investment, without being constrained by government prohibitions and parliamentary scrutiny. Constrained by the civil servants' statute, budgetary requirements set by the Treasury and operations confined by the French nation state, France Télécom has established a number of subsidiaries or *filiales* that have a certain managerial autonomy and flexibility in their marketing, investment and personnel policy. In addition to its core functions as network supplier and basic service provider, France Télécom diversified by establishing semi-independent subsidiaries to penetrate emerging product markets, to promote innovation and entrepreneurship, and internationalise its business activities by establishing international joint ventures. This filialisation policy had clear advantages, especially with respect to ownership, authority structures and market strategies (Vedel 1991). Compared with France Télécom, with no legal status or assets of its own, these special subsidiaries did not belong to the state and were not controlled by the Treasury and the National Audit Office. They had a large degree of corporate autonomy in personnel, financial and market policies, and in foreign investments.

Over the years, the following *filiales* were created:

- Transpac, established in 1977, the public-switched datanetwork;
- Sofrecom, established in 1967, international consultancy and project development;
- Entreprise Générale de Télécommunications (EGT), established in 1965, the provision of advanced terminal equipment;
- Télésystèmes, established in 1969, the supply of telematic systems and services;
- France Câbles et Radio (FCR): new services to large users and international telecommunications;
- Compagnie Auxiliaire de Télécommunications (CAT), created in 1973 to provide credits to small and medium-sized businesses and venture capital for France Télécom.

In order to achieve synergy effects and increase co-ordination, these subsidiaries were regrouped in 1985 and placed under the control of the holding company Compagnie Générale de Communication (Cogecom) (99.9 per cent owned by France Télécom). In 1987 a new subsidiary was created, called Télécom Système Mobiles, later renamed France Télécom Mobile Services, in charge of providing computer services and software and developing the mobile communications market. In 1988–9 the TDF (Télédiffusion de France) was incorporated as a new subsidiary into the Cogecom holding. After 1988 the DGT was given greater autonomy from the civil service. Its status as government department was replaced by that of a national public enterprise, owned entirely by the state. Renamed France Télécom in 1991, it expanded internationally by entering new geographical and product markets. The aim for the mid-term became to establish itself as an *administration entreprenante*, complementing its responsibility as a public administration with private enterprise characteristics. To facilitate and provide the financial means for its overseas investments, France Télécom established the subsidiary FTFI. Other important subsidiaries are France Télécom Cable, CNET, France Télécom Mobile International, and France Télécom Mobiles Radiomessagerie.

The precise legal status of DGT/France Télécom has been unclear since the early 1980s (Bonnetblanc 1985). The (semi-)corporatisation of France Télécom limited the state's abilities to interfere in day-to-day management and to stipulate grand projects. After the 1990 modification of its statute, France Télécom has followed a collaborative strategy and established several links with domestic businesses and partners from abroad (e.g. Deutsche Telekom and the US-based company Sprint). In 1991, France Télécom was semi-corporatised: it was not a fully government-owned company under company law, but it had some commercial autonomy. One of the political constraints imposed upon its business was its contribution to the Treasury and its payments to a multitude of industrial policy commitments. In accordance with the Telecommunications Act, France Télécom had to pay a levy to the government, that would be used to finance the state's commitments to national security, innovation and industrial policies. This could vary from subsidising the public R&D laboratories for telecommunications (CNET) and space (CNES), to financing a large percentage of the manufacturer Alcatel's R&D activities (through CNES and CNET), to propping up troubled French companies in strategic industries, like Bull, Thomson and SGS-Thomson. These funds were channelled through Cogecom for non-telecommunications related investments in subsidiaries and via the acquisition of equity in Bull, Thomson, SGS-Thomson (see Table 6.2), and the recently privatised insurance companies AGF and UAP (2.2 per cent and 0.6 per cent respectively), the advertising and communication company Havas (5.5 per cent), Banque Hervet (8 per cent) and the COFIRA holding company (8 per cent). Since early 1994, however, the levy contribution to the state's budget was replaced by a normal fiscal regime, making the company subject to the French corporate tax regime and requiring it to pay corporate and other taxes.

In April 1994, France Télécom (*Annual Report* 1994) renewed its contract with the state (*cahier des charges*). For the period 1995–8, the following objectives were agreed: tariff rebalancing, accelerated debt reduction efforts, and productivity improvements (efficiency). As a consequence of the large investment in the diffusion of Minitel, the subsidisation of ailing firms and high-cost ventures, France Télécom was heavily in debt. To free resources for expansion and to relax financial charges, France Télécom started to pursue a policy of debt reduction. The effort to cut back debts paid off when the debt/equity ratio was reduced from 102.9 per cent in 1992 to 70 per cent in 1994 (France Télécom *Annual Report* 1994). The recent policy of France Télécom to cut back its debts substantially has had a negative impact on its

Table 6.2 France Télécom: strategic linkages

	Participation (%)	*Major activity*
Bull (France)	18	computer manufacturer
Thomson (France)	20	consumer electronics supplier
SGS Thomson (France)	11	semiconductors manufacturer
Casema (Netherlands)	100	cable operator
Mobistar (Belgium)	57	mobile service provider
Sema Group (UK/France)	21	IT services company
ITJ (Japan)	3	network operator/service provider
Sprint (USA)	10	US-based operator
Panafon (Greece)	39	second cellular operator
Nortel Inversora (Argentina)	16	holding
TelMex (Mexico)	5	national public operator
SKD (Sweden)	75	specialist in network intelligence
Eutelsat (Europe)	14.5	European satellite services organisation
Intelsat (global)	4	international satellite services organisation
Inmarsat (global)	5.3	international maritime services organisation
Community Link Associates (USA)	na	tele-information services provider
Intelmatique (Europe)	na	promoter of Minitel/videotex abroad
Centertel (Poland)	24.5	cellular service provider
Atlas/Global One (Europe)	50	pan-European service provider
General Magic (global)	na	multimedia products and applications
Telecom de Argentina (Argentina)	10	national public operator

na: not available
Source: France Télécom *Annual Reports* (1990–6)

relationship with Alcatel, effectively leading to tougher negotiations, more competitive tendering, and a postponement or cancellation of investments. This short-term reorientation and 'more value for money' approach by France Télécom caused a sharp fall in profits in 1994. In order to reduce its critical dependence on equipment manufacturing, Alcatel decided to spread its risks by gradually moving into service provision and network operation.

After a period of industrial unrest with a number of large strikes between December and May 1995, a bill was passed to convert France Télécom into a joint stock company and pave the way for the sale of a minority of its shares. The government insisted that it would keep 51 per cent with another 10 per cent reserved for employees. France Télécom was fully corporatised on 31 December 1996 and was privatised almost a year later, in October 1997. Although the workforce and senior management of France Télécom, actively supported by labour union representatives, the corps of engineers, and the social democratic factions in Parliament, were unable to block corporatisation (and later flotation), they successfully delayed or minimised some of the more radical measures concerning job security, pension liabilities, and reduction of public and universal service requirements. The right-wing Juppé government made several concessions to ensure support from France Télécom's top managers and employees: for instance, only three out of twenty-one directors would be appointed by outside shareholders and many employees would remain civil servants until 2002. France Télécom's flotation was originally due for early May 1997, but it was postponed until after the election. The privatisation was further delayed after the election, when the leftist Jospin government stressed the importance of public services remaining in state hands. After it became clear that the unions were divided about striking against privatisation in September 1997 (i.e. CGT was pro-strike; CFTC and FO were against), and final evidence was offered about the flotation of France Télécom (Delabarre 1997), the Jospin government sold 21 per cent of France Télécom's assets (raising FFr 42bn) and another 6 per cent for staff; another 7.5 per cent of France Télécom may be part of an equity swap with Deutsche Telekom, its foreign strategic partner.

The current strategy of France Télécom is to become a strong overseas competitor, while remaining integrated at home in order to provide high-quality universal services. The company is free to distribute video signals and use radio in the local loop (unlike BT in the UK). To prepare for international and domestic competition, two burning issues needed to be addressed in the early 1990s: the still centralised and functional structure of the organisation (hampering quick and effective decision making and its strategic and operational activities), and the vague legal status of the state handicapping its international acquisitions, alliances and access to financial markets.

The first problem was addressed in 1994 when France Télécom announced an action plan to transform the group's business by focusing on three priority areas: customers, innovation and performance. The second problem is still pending. From 1990 onwards, France Télécom has targeted business customers as a group and established a special division to serve its largest 300 customers (Services Grandes Comptes), like Société Générale, Rhone-Poulenc, Eurodisney and Elf-Acquitaine. France Télécom is taking a leading position in the promotion and diffusion of ISDN: it was the first to launch ISDN commercially (in 1988) and had established a full national coverage by 1992. France Télécom pursued a collaborative strategy by offering ISDN services in close collaboration with large user groups: for instance, business customers could enter into 'partnership arrangements' whereby France Télécom would cover up to 50 per cent of the costs of applications and up to 20 per cent of user-specific development costs (Ergas 1992:15). The company has started to increase customer responsiveness and moved closer to the market and its different customer groups. To that end, it has decentralised its activities through specialised account managers for large corporate accounts and multi-site companies, smaller businesses and professionals, and residential customers. In 1994, the company created a separate subsidiary for multimedia matters, called France Télécom Multimedia, to respond quickly and effectively to the techno-economic convergence between broadcasting, telecommunications, electronic publishing and computing. In the 1990s, France Télécom started to use price and quality of service as weapons in a more competitive environment; for instance, in October 1997, anticipating full competition by 1998, the company introduced more cost-based charging schemes for basic services and Internet access.

As a consequence of its former leadership in the modernisation of the domestic telematics and electronics industry, France Télécom is still heavily involved in the cross-subsidisation of the loss-making national champions Bull and Thomson, receiving equity shares in these companies in return. In 1994, France Télécom had a 17 per cent stake in Bull, and other minority stakes in Thomson Consumer Electronics (about 20 per cent) and in the semiconductor manufacturer Thomson SA (about 11 per cent). In 1993, France Télécom entered into an alliance with the SEMA Group, a leading French system integrator. France Télécom attempted to capitalise on its Télétel/Minitel videotex service by seeking to export it to various European countries and the USA. The designated subsidiary of France Télécom, Intelmatique, was established to commercialise French videotex abroad. Intelmatique has acquired minority participations in the leading Dutch system provider Videotex NL and the US Community Link Minitel Associates (together with US West). However, the successes at home could not be repeated and equalled abroad. Furthermore, the ever-increasing popularity of Internet access services was a real threat to the domestic and international position of Minitel. France Télécom realised that an upgrade of the Minitel system was needed: new multi-functional terminals with built-in smart cards were developed, a much faster Minitel service was launched (Télétel Vitesse Rapide) and links with Internet-based computer systems and PC-based on-line services were established. France Télécom has also sought to build up a presence in newly emerging products markets: it acquired participations in new fast-growing cellular businesses, like Martin Dawes (mobile service provider) and RAM Mobile Data (mobile datanetwork operator).

France Télécom has become very active in setting up branch offices all around the world (either on its own or as part of the Global One venture with Deutsche Telekom). In its activities in developing countries, emphasis was laid on rolling out and operating fixed and cellular networks abroad. For instance, France Télécom has acquired a 10 per cent share in US-based Sprint, 2.5 per cent in ITJ Japan, 10 per cent in Telecom de Argentina, 5 per cent in TelMex (Mexico). In 1994, France Télécom set up a partnership with the national operator to expand Vietnam's telephone network and to develop a cellular network. In 1997, France Télécom acquired a majority stake in CI-Telecom, Ivory Coast's national operator. The company was also successful in its bid for the largest cable operator in the Netherlands, Casema, herewith opening the way for the acquisition of the second operator Enertel, a federation of utilities and cable operators, in which Casema participates. Together with Deutsche Telekom and ENEL (oil company), France Télécom set up WIND, which is expected to develop into one of Italy's leading alternative carriers. The company participates in cellular operations in Poland (Centretel), Belgium (Mobistar), Greece (Panafon), South Africa (SIM), Lebanon (Liban Cell), Romania (MobilRom), Slovenia (Slovtel), and India (BPL Mobile). Between 1997 and 1998, France Télécom was awarded cellular licences in Moldovia, Denmark and the Netherlands. Furthermore, France Télécom participates in the internal and European satellite consortia Intelsat and Eutelsat, and was involved in Inmarsat (international maritime services).

From the early 1990s, France Télécom embarked upon a path of foreign expansion by setting up partnerships with Deutsche Telekom (Eunetcom and Atlas) and, in early 1996, with Sprint, the third US operator, forming an international company called Global One. After it was ratified by the European Commission and the American Federal Commission, the Global One venture became one of the leading worldwide telecommunications carriers, competing with Concert (BT and MCI). France Télécom and Deutsche (Bundepost) Telekom – and to a lesser extent Sprint – expecting shrinking market shares in the mid-term as a result of further European deregulation and worldwide competition, decided to join forces to become a key player in the global telecommunications market. Besides forming a joint venture that would offer international companies a full range of voice and data services around the world, the alliance would also allow the two companies to negotiate for more corporate and financial flexibility in their home base and allow for a safe privatisation path of cross-sharing equity.

As early as 1989, France Télécom and Deutsche Telekom started to collaborate by setting up Eucom, a venture that would provide value-added networks and one-stop-shopping services to business customers; Eucom was later followed by Eunetcom, offering corporate network products. In March 1992, France Télécom and Deutsche Bundespost Telekom announced that they would provide joint telecommunications services for international customers through the globally managed network service company Eunetcom. The possibility of cross-over equity stakes in the near future, following the privatisation of Deutsche Telekom and France Télécom, was not excluded. The 50/50 Franco-German venture began looking for a potential American partner to facilitate a worldwide provision of telecommunications services for international corporate businesses. The operator MCI, with whom the French and German PTTs already collaborated in Infonet Services Corporation, was intended to join the Eunetcom venture. When MCI signed a far-reaching agreement to collaborate more closely with major European rival British Telecom, DBP Telekom and France Télécom had to consider alternatives.

In November 1993, DBP Telekom, France Télécom and AT&T drafted a broad memorandum of understanding on merging their European and transatlantic businesses. The so-called Project Atlantic would add a European branch to the AT&T Worldsource venture to provide global business services (also including KDD/Japan and Singapore Telecom). This agreement was opposed by a group of actors for various reasons. The European Commission made it clear that it would advise against the alliance because it would clearly result in a dominant market position in international telecommunications, thus nipping competition in the bud. Alcatel and Siemens felt that their privileged position as, respectively, France Télécom's and Deutsche Bundespost Telekom's preferred equipment suppliers could be threatened by their biggest rival in that market segment, namely the equipment branch of AT&T. The two European operators decided to tighten their partnerships and one month later in December 1993, France Télécom and DBP announced a strategic alliance in the common provision of data services, managed network services, satellite services, virtual private networks and multimedia. This joint venture, provisionally named Atlas, later renamed as Global One (including US Sprint) would extend the already established Franco-German partnerships of Eucom and Eunetcom by integrating France Télécom's Transpac and DBP's Datex-P datanetworks to provide a trans-European datanetwork. Switched voice telephony would be excluded from this Euro-alliance. Depending on the liberalisation of the European telecommunications market and the privatisation process in France and Germany, the alliance might even lead to an equity swap between the two operators and long-term network unification.

In the summer of 1994, France Télécom and Deutsche Telekom announced the $ 4.2bn acquisition of a 20 per cent stake in Sprint (the third largest US long-distance operator); the new company was provisionally called Phoenix, later renamed Global One. This would substantially expand the geographical coverage of their networks for the provision of global one-stop-shopping telecommunications services. The plans of the French and German operators to collaborate in the Atlas project and their joint venture with US Sprint were not fully supported by the European and American regulatory and anti-trust authorities. These authorities have made their approval contingent on the degree to which the still strongly protected French and German markets would be opened up to competition. Given the closed character of the French and German markets, the two bodies hesitated in giving their authorisation and the parties involved started to negotiate with the European Commission to take away the aura of a non-aggression pact between two of Europe's largest carriers, operating from a sheltered home base. By persuading their home governments to meet the EU infrastructure and voice liberalisation requirements before 1998, and by excluding some units from the joint venture (e.g. considering the exclusion of each other's datanetwork subsidiaries from the Atlas venture), France Télécom and Deutsche Telekom managed to break the regulatory deadlock and the Global One venture was approved.

6.7 Regulatory reform of French telecommunications policy

In 1982, a regulatory body, the Haute Autorité de la Communication Audiovisuelle (HACA) was set up to monitor public broadcasting, license local and regional radio and TV stations, and protect broadcasters from external pressure from the government, politicians and industry. HACA sought to establish itself as the guarantor of independent public broadcasting without direct state interference. With the shift in government from left-wing to right-wing (cohabiting with the Mitterand presidency) in 1986, HACA, closely associated with the outgoing Socialist government, was replaced by an independent regulatory authority, Commission Nationale de la Communication et des Libertés (CNCL), intended to supervise both broadcasting and telecommunications. Initially, CNCL had only regulatory powers over the media sector, but then it took over the administrative responsibilities of the broadcast transmitter TDF, when that company was incorporated in the Cogecom/France Télécom holding. In addition to its licensing and supervisory tasks in the broadcasting domain, CNCL was put in charge of licensing the so-called first category of telecommunications networks (i.e. dedicated wired and radio infrastructures). With the abolition of the monopoly on local cable infrastructures, as a consequence of the adoption of the 1986 Communication Act, the authorisation to operate cable networks was delegated to the CNCL; however, the use of local cable systems was strictly limited to one-way audio-visual services. With the return to power of the Socialists in 1988, the 'politicised' CNCL was replaced by the Consul Supérieur de l'Audiovisuel (CSA). This new administrative body was put in charge of authorising licences to non-public telecommunications systems, like private networks, cellular telephony, on-the-air diffusion and satellite communications. Notwithstanding another change in government to the right, the remaining political sensitivity of broadcasting policy, and ongoing techno-economic developments in the field, this independent broadcasting authority has managed to survive until today (Chamoux 1993).

In 1986, a task force was set up within the Ministry for Post and Telecommunications, named the Mission à la Réglementation, chaired by Jean-Pierre Chamoux, to draft postal and telecommunications legislation and develop an administrative structure in which the regulation and operation of postal and telecommunications services would be separated. In the draft Bill of 1987 the Mission à la Réglementation (1988) defined basic services as the transmission of signals that were not processed, including telephony, telex, switched datatransmission and leased circuits. Concerning network provision two separate categories were proposed: a public network that would meet socio-political obligations and alternative private networks that were supposed to compete with and complement the core network (e.g. in the cellular domain). The control of the public network and the provision of basic services would remain the exclusive responsibility of the DGT, which was also allowed to offer alternative complementary networks. The issues of interconnection and the definition of network interfaces would be defined by the Ministry to ensure the public network's integrity. The task force suggested relocating the DGT's administrative activities to the Regulatory Task Force and to the independent newly established CNCL. The former would be responsible for the regulating of the public network, while the latter would regulate the private networks and be responsible for frequency allocation. The task force furthermore drew attention to the fact that French telecommunications policy was relatively in line with the outlines of the European Commission's Green Paper: the terminal equipment market was already open to competition and the separation of the regulatory and operational functions would be implemented shortly. The task force also raised doubts about whether the 'unity within the PTT House' should be maintained: the presumed synergies between the postal and telecommunications functions could be questioned, the sole reason to keep PTT integrated was to be found in the administration's civil servants' statute. In the same period the government also reduced the direct involvement of the DGT in industrial policy by transferring the ultimate responsibility to the Ministry of Industry. The institutional landscape changed once more when the Regulatory Task Force became established as a separate department within the Ministry of Post, Telecommunications and Space.

After the legislative reform of 1990–1, the administrative responsibilities of the CSA in the telecommunications domain were transferred to the P&T Ministry. In 1988, the regulatory and operational functions of the DGT were officially split: France Télécom, the new name for the DGT, was freed from its administrative tasks and the Directorate General of Regulation (DRG) would handle the regulation of modest competition between the DGT and private companies. The DRG, drawing some of its staff from the DGT, would have authority over type approval, value-added services and the electromagnetic spectrum; it would also represent France in international negotiations. The prices concerning services for which France Télécom had exclusive rights or for which no competitor had received authorisation, were submitted for approval by the Minister in charge of telecommunications and the Minister responsible for the economy. A price cap system was set for France Télécom within the scope of the contract of the plan between the state and the operator. The system operating in France consists of the retail price index (RPI) –3 per cent for three years, a formula that is slightly different from the one existing in the UK. A beginning was made with the separation of regulatory and operational activities and the bringing in line of telecommunications legislation with the EC Green Paper. Marcel Roulet, the Director of the DGT, recommended ending the dual role of network operations and administrative activities, and establishing the DGT as a state enterprise rather than a government administration, able to participate in joint ventures with private companies and to purchase its equipment more freely. The Direction de Service Publique (DSP) is responsible for overseeing France Télécom's investments, debts, personnel policy and tariffs. Under the Balladur government in 1993–4, the administrative structure of the DRG, the DSP and the larger Ministry of Industry and Post and Telecommunications was transferred to a new Direction Générale des Postes et Télécommunications (DGPT), which belonged to a larger Ministry of Information Technologies and Post.

In the new regulatory framework of 1990, France Télécom became a public body under the guidance of the PTT Ministry entrusted with commercial, industrial-economic and social goals. The company carried out its activities according to a planning contract (*cahier des charges*), stipulated by the Minister in charge, in which the licence obligations of France Télécom were stipulated. For instance,

France Télécom was pressed by the government to balance a number of policy objectives:

- adopting a more decentralised structure and becoming more customer oriented;
- rebalancing its tariffs;
- guaranteeing job stability for its employees;
- investing in international ventures and new technologies (France Télécom *Annual Report* 1990–1).

France Télécom's financial autonomy is still severely constrained by the Treasury: it still has an *ad hoc* and unique status, and is not yet fully subject to private business laws. In order to prepare itself for both domestic and international competition France Télécom made a beginning with tariff reform in 1992. Given the already strong competition at the international level and the emerging challenge from cellular operators as well as possible cable operators and other fixed operators in the future, France Télécom, like any other public telecommunications operator expecting or confronted by new entry and rivalry, decided to rebalance the charges for local, long-distance and international services. The company began by cutting international telephone rates by more than 20 per cent between 1992–4. Under pressure from large users, service providers and the regulator DRG, it also cut its prices for long-distance leased lines. Early in 1994, although residential customers were compensated for the fact that the local zones were enlarged and benefited from cheaper long-distance call services, local calling tariffs went up by 100 per cent and monthly subscription rose by more than 30 per cent.

After the strikes and political turmoil over the 'hot potatoes' of 'privatising France Télécom' and 'liberalising the public telecommunications service' in 1993–4, the Minister responsible for PTT matters, Gérard Longuet, came up with the idea to set up discussion platforms for the exchange of information and views, in a bid to find support for structural change. Also confronted with the requirement of bringing French legislation in line with the liberalisation directives of the Commission by 1998, the Minister asked Bruno Lasserre, head of the Regulatory Unit (DRG) of the Post and Telecommunications Department, to organise a 'public consultation' of all the public and private stakeholders about the future organisation of the French telecommunications industry (Lasserre 1994; MTIP 1995). When the direct way to propose and implement legislation was not available due to strong opposition from the major operator's workforce, and to some extent supported by factions in the government and Parliament, the idea was to forge a consensus on particular controversial issues before any crucial political decisions were taken. By the end of 1994, the DGPT (1994a, b) had collected evidence from all the parties concerned and had started to draw up (more modest) legislative proposals to prepare the incumbent operator and others for competition at home and abroad, and to set up an appropriate regulatory structure.

The notions of infrastructure competition and opening up voice telephony by 1998, as required by the Commission, were supported wholeheartedly by the incumbent, the user community and potential new fixed infrastructure providers. Unlike BT's ban on entry into the cable market, interactive entertainment services and hybrid cellular/fixed systems, France Télécom was allowed to set up and participate in various multimedia testbeds, cable telephone, interactive television, pay-per-view broadcasting, and wireless in the local loop (all related to the government's information superhighways initiative). The corporate large users were pleased with the overall effects of introducing competition, spurring innovation, choice, quality and price cuts, but also with the permission given to large companies to set up closed user groups and provisions for information sharing (Cohen 1992). New operators, such as the SNCF (the railways company), highway companies, utilities and new foreign entrants (e.g. BT, AT&T/Unisource) were offered the possibility to sell capacity to third parties, while CATV operators and Alcatel were now allowed to become (multi-)service providers. In setting up an appropriate institutional framework to facilitate fair competition in French telecommunications beyond 1998, policy makers and regulators also had to accommodate the interests of other stakeholders. A large majority of France Télécom's workforce insisted on retaining the civil servants' statute, and demanded that some of the public operator's privileges (e.g. universal service provision, and its commitment to national security and industrial policy) be maintained in the future organisation of the industry.

The French regulatory framework, based on the separation between the public operator and the government, showed its built-in bias towards the incumbent position and lack of experience in dealing with abuse of market dominance. In October 1994, the French regulator DGPT had to solve a dispute between SFR, the new entrant providing competitive cellular services, and France Télécom, concerning interconnection charges and the abuse of market power. Competition between the two mobile operators was restricted and the expansion of SFR's network was constrained by limited spectrum availability and lack of collaboration from France Télécom, both contender and supplier of fixed network links, to ensure interconnectivity. SFR complained about the high prices and interconnect fees it had to pay to France Télécom. The company argued that because of these high charges, it was unable to compete effectively with France Télécom. Eventually, the Minister gave permission to SFR for using cable networks and microwave links as alternatives to the incumbent's public infrastructure and put France Télécom under pressure to cut its network interconnect fees by 50 per cent.

The policy of infrastructure and voice telephony competition by 1998, as required by the European Commission, was supported wholeheartedly by the incumbent, the user community and potential new fixed infrastructure providers. In setting up an appropriate institutional framework to facilitate fair competition in French telecommunications beyond 1998, policy makers and regulators also had to accommodate the interests of other stakeholders. A large majority of France Télécom's workforce insisted on retaining the civil servants' statute, and demanded that some of the public operator's privileges (e.g. universal service provision, and its commitment to national security and industrial policy) somehow be maintained in the future organisation of the industry.

After the rules for the post-1998 French telecommunications markets were defined and the new Act was passed, a new institutional framework was introduced. Early in 1997, an independent regulator, Autorité de Régulation des Télécommunications (ART) , and a body for frequency management, Agence Nationale des Fréquences (ANF) were set up. The ART, headed by three independent experts who are appointed by the government for a period of six years, had already taken more than 100 staff from the Ministry when it started its operation. The French independent telecommunications regulator is responsible for detailed regulation (e.g. numbering and interconnection issues), monitoring and policing the domestic market place, and providing specialised support for policy making for the Ministry. ANF, the separate national radio frequency agency, has a spectrum planning and management function and prepares any future move to realise the commercial value of spectrum resources. Together, these two bodies to some extent replace the involvement of the Direction Générale des Postes et Télécommunications of the Ministry of Industry, Post and Telecommunications. The latter remains in charge of formal granting of licences for public networks and voice service provision, developing policy and overseeing all aspects of universal service, including tariffing and funding. In August 1997, the French Competition Tribunal manifested itself in regulating the domestic market, when France Télécom was fined for anti-competitive practices, going back to 1993. Transpac, France Télécom's subsidiary, won a contract to provide data services to the insurer AXA by renegotiating the contract and securing the deal by scaling back the prices of its existing contract.

6.8 Summary and concluding remarks

France has a strong interventionist tradition, in which the national government, together with state-owned and state-controlled institutions (nationalised enterprises and investment banks), have often imposed policy changes in the face of opposition from organised private interests. The government would often take a mercantilist stance by actively supporting its domestic industries, protecting them from fierce international competition by large state subsidies, and various tariff and quota measures. The overall aim of government policy in France in the period investigated in this study was to avoid the duplication of efforts and funds and to produce economies of scale that would improve the competitive prospects of leading French companies. Active state involvement in the strategic industries was justified because rationalisation and modernisation were required in order to make these companies (or industries) more efficient and commercially oriented. Large horizontally and vertically integrated industrial groups were created, that encompassed more than one industry: for example Alcatel-Alsthom (transport, power and communications) and France Télécom (telecommunications, information industry, cable and entertainment). In the 1980s, French industry effectively expanded its base through mergers and acquisitions abroad. Clear priority was given to the emerging Internal European Market, shunning head-on competition in the two other big regional markets in the Global Triad, the USA and Japan. So far it looks as though the European Community has been used as a springboard for global markets. The French government has convincingly stuck to the traditional state monopoly model by maintaining public control over the operator, the infrastructure and basic voice and data services. For instance, the high-performance ISDN was strongly promoted by DGT/France Télécom and was scheduled to meet public service goals and be introduced on a nationwide basis. The fringe markets of value-added services and peripheral equipment, however, were liberalised.

In their industrial strategies the French have given clear priority to strengthening their manufacturing base and core sectors (high-tech industries, like aerospace, information technology) and to ensuring that domestic firms are able to achieve world leadership in those key markets. The policy instruments used to accomplish these objectives have been state planning in the formulation of targets, credit provision and an active and dirigiste role of the government in interorganisational and sectoral restructuring. The major reason for the state's involvement in the telecommunications domain in the early 1970s was the poor performance of the domestic telephone network with low density rates and low quality levels. The government delegated this priority programme for the modernisation of the domestic telecommunications industry to the DGT/PTT and to the corps of telecommunications engineers. The DGT was given a high level of autonomy and political responsibility to develop and expand new services and equipment. The modernisation of the public network provided the French national manufacturing champions with massive R&D subsidies and guaranteed market shares. An illustration of the French government's pursuit of national autonomy in critical sectors can be found in the telecommunications manufacturing industry. The state managed to restructure an industry once dominated by foreign firms (ITT and Ericsson), and create the CGE subsidiary Alcatel as a clear national champion. In this process of nationalising the French telecommunications industry, or *francisation* (Aurelle 1986:59) the position of the DGT/PTT was ambiguous: on the one hand the DGT, as part of the state apparatus, followed the overall industrial strategy of the French government of controlling all the components of the domestic telecommunications industry, while on the other hand it favoured competition between the incumbent manufacturer and new entrants to reduce its techno-economic dependence on one supplier.

The French emphasis on the manufacturing aspects of telecommunications has led in some respect to a neglect of the service element. The DGT has emphasised the introduction of new services and facilities to the low-growth consumer market, giving less priority to the more sophisticated demands of corporate businesses and foreign multinationals, which would have generated high growth margins for the DGT. The Télétel/Minitel project was part of a well-co-ordinated industrial strategy, that integrated typical French elements such as a large-scale and high-tech infrastructure provision by the state, the promotion of the interests of the DGT and the French telecommunications equipment industry by facilitating a mass market at home and exports abroad and overall informatisation policy of the French government. The French state also continued to promote the international competitiveness of the domestic equipment industry by supporting the ITT/LCT and CGCT take-over, by CGE-Alcatel and Matra/Ericsson respectively. The relative success of the large-scale introduction of Minitel aside, the state-led strategy was not always effective. For example the Télécopieur Grand Public (i.e. facsimile) was unsuccessfully aimed at the consumer market, and it was only later that the service was introduced to the professional business market (Cohen 1992). The Cable Plan did not really take off before 1990, when a number of administrative hindrances had been removed. Furthermore, as a result of the ambitious Minitel distribution policy and the cross-subsidisation arrangements between the profitable DGT/France Télécom and the ailing firms Bull and Thomson, the public operator is still heavily in debt.

The results of the French state-led approach of industrial adjustment are mixed. State intervention was successful in the development of one of the most modern infrastructures in the world (in terms of 100 per cent digitisation of transmission and switching) and the society-wide expansion of new services (e.g. approximately 7 million videotex terminals installed, and 1.3 million ISDN channels) (France Télécom *Annual Report* 1996; OECD 1997). Another success was the *francisation* and consolidation of the fragmented domestic telecommunications equipment industry into the national champion Alcatel. The French interventionist approach, however, has been less successful in the domains of cable television, mobile telecommunications and global service provision. In the 1970s and 1980s state intervention in France increased not only in the telecommunications industry, but also in the areas of computing and information. France Télécom's public monopoly has expanded beyond the infrastructure and voice telephony to include large parts of the markets for value-added services and datacommunications. In terms of organisational manœuvrability and flexible service provision, France Télécom has been, however, constrained by the civil servants' statute and restrictions upon financial borrowing. To overcome these limitations set by the public administration, the central government in 1978 established Cogecom as a holding company, allowed to establish subsidiaries under private law. A network of affiliated commercial companies was created in the fields of audio-visual communications, business communications, telematic services and IT industries. Coordination between the core company and its affiliated subsidiaries can be poor, as the rivalry between France Télécom and the TDF has shown.

The problems normally associated with state monopolies, such as poor innovation record, low productivity and inflexible bureaucracy, do not apply to the French situation. Although France Télécom may be hierarchical and lacking internal transparency, this case study has shown that the vertically and horizontally integrated structure of the French telecommunications system, including network operations, manufacturing and R&D, was effective in catching up technically and economically with world market leaders. Looking at its successes and failures, the French system of state entrepreneurship might not be very efficient, but it was certainly effective in modernising the domestic industry from the 1970s onwards. Whether or not to corporatise/privatise France Télécom and when and how to open up the provision of basic communications services to competition, has remained controversial until well into the late 1990s.

The relative success in the upgrading of their backward telecommunications infrastructure in the 1970s and the society-wide diffusion of telematic applications clearly exhibited the capability of the French to implement society-wide technology programmes. These programmes, aimed at developing an administrative market, namely public telecommunications, are carried out by a technocratic élite, mobilising the required resources, and supported and sponsored by leading politicians. In the telecommunications domain the modernisation programme was first carried out by the public research body CNET, later succeeded by the French PTT, the DGT. French industrial policies have given a clear priority to top–down approaches, producer-oriented and technology push programmes over bottom–up, demand-led and commercial diffusion-oriented. In cases where markets cannot be controlled at the national level and where innovative entrepreneurship and international competitiveness are required, such as computing, semiconductors and consumer electronics, the French results have been less impressive.

State intervention was also aimed at co-ordinating and encouraging rationalisation and collaboration between national producers. This strategy was based on the premise that domestic firms would be able to pool their resources constructively and benefit from newly achieved economies of scale and scope. The state might further seek to sustain these companies and help to establish a firm base for an assault on international markets by developing large-scale ambitious and centralised projects for development and diffusion of their products in the internal market. The strongest pressures for change in France, however, appear to have been generated by external factors, such as the European Commission, the availability of new technologies (e.g. the Internet), and the requirements of multinational firms.

7 A comparative institutional analysis of British, Dutch and French telecommunications policy

7.1 Introduction

Traditionally, telecommunications was organised at the national level whereby government, the telecommunications administration and the equipment industry were sovereign in the provision of domestic equipment and services. When confronted with the current far-reaching technological, economic and international challenges, however, policy makers have become aware that telecommunications can no longer be seen as a domestic affair with policies mainly oriented towards home markets and national constituents. The techno-economic rationale for a state-controlled monopoly has eroded and the iron triangle of PTT, government and equipment manufacturer(s) has begun to lose political control. European states have started to reorganise their established telecommunications administrations and transform monopolistic markets into more open and regulated forms of competition. In this chapter we will compare the strategic responses of France, the Netherlands and the United Kingdom to the new techno-economic and international conditions of the 1980s and 1990s and see whether there is any evidence of a typically Dutch, French or British approach to telecommunications restructuring.

The usual British strategy to industrial adjustment is characterised by a market-oriented and company-led trajectory, with clear-cut liberalisation, deregulation and privatisation policies, arm's-length relations between government and industry, and the pursuit of free trade strategies in the international domain. The characteristic French response to far-reaching structural changes in markets and technologies is a state-controlled adjustment strategy, in which the national government acts both as an entrepreneur and administrative guide in modernising the economy. To achieve the goals of enhanced efficiency and innovation, the French government relies upon making large-scale infrastructural investments and manipulating money flows and ownership rights to rationalise its key sectors and establish national champions, while protecting the home market from foreign entry. Furthermore, it strongly supports its national champions at the global level through strategic trade policy and active export diplomacy. The typically Dutch response to ruptures in the world economy and technological change is to emphasise the search for a trade-off between political and economic stability at home and international trade flexibility. The inclusive deals elaborately negotiated between public and private actors in the domestic realm and the collaborative practices in the home market, function as a springboard to the successful export of goods and services and thereby facilitate the international expansion of Dutch firms.

In this comparative organisational and institutional study, these rather general notions about the different national adjustment strategies have been investigated in the telecommunications domain. The study has revealed interesting material about the actions and reactions of France, the Netherlands and the UK to the same set of stimuli, posed by techno-economic and international changes in telecommunications in the 1980s and 1990s. Some of the responses of these countries to the new threats and opportunities were identical and converging. For instance, they all abandoned the traditional monopoly and replaced it with a more competitive model. On the other hand, there were differences in the degree and timing of the implementation of structural change.

7.2 Liberalisation, privatisation and regulatory reform of Dutch, French and UK telecommunications

There has been a growing tension between the continuing globalisation and Europeanisation of business on the one hand and a decrease in national sovereignty on the other hand. Techno-global pressures, combined with deregulation and European integration, threaten the relative autonomy of the European nation state in developing its domestic policies and they require a redefinition of the national sovereignty and an adjustment of the national practices in the light of these international pressures. The traditional manœuvrability of European governments in designing their national industrial policies and the efficacy of their interventions has declined considerably: technologies of freedom, globalisation and differentiation of markets, international deregulation and European integration have made evaluation and reform of the traditional PTT-dominated complex necessary. The options for pursuing traditional and nation-specific policies have been seriously reduced for many countries, not only for small European countries that are strongly dependent upon international markets (like the Netherlands), but for large European countries like Britain and France as well. Notwithstanding these structural forces constraining national stakeholders in their manœuvrability, the degree and pace of implementing these institutional changes in European telecommunications was different in each of the countries investigated (see Figure 7.1 and Table 7.1).

The ongoing process of European integration brought forward another constraint for the member states' governments in the development of telecommunications policy. From the mid-1980s onwards, national policies needed to be more in line with the policy directions spelled out by the European Commission. With the pre-eminence of Community law over national law, more and more legislative competencies have been transferred to the EC level. As assumed in this study, European integration is one of the four structural forces that have significantly enlarged the impact of Community regulations on national

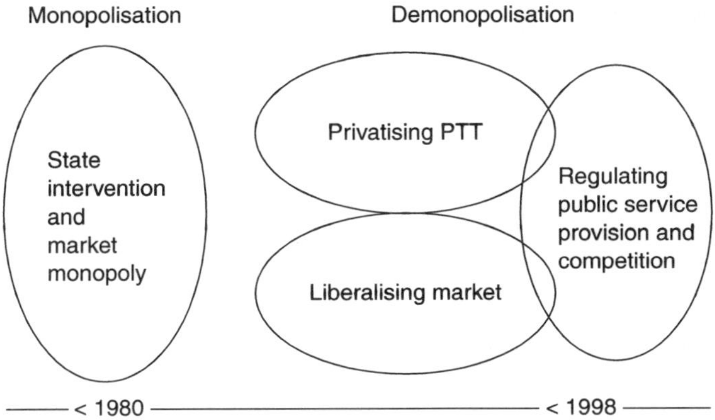

Figure 7.1 Governance transformation in European telecommunications, 1980–98

policies and reduced the decision-making manœuvrability of the member states in designing their own policies. In its aims of integrating the national markets into one internal market the Commission, next to the continuous harmonisation of member states' telecommunications policies, increasingly relies on economic regulation instead of governmental intervention. The liberalisation of the markets for telecommunications peripheral equipment and services was accomplished by 1993. National monopolies are now restricted, with the core public network subject to open network provisions, the operational and regulatory function of the PTT clearly separated and the heavily protected government procurement programmes (switching equipment) opened up. Given the complexity of regulating telecommunications and the restrictions on the Commission in terms of budget and work load, this might even lead to the hiving-off of the Commission's jurisdiction to an independent regulatory agency (a kind of FCC at European level), analogous to the London-based European agency for the testing of new medical drugs.

In terms of liberalising the telecommunications markets and developing a new institutional structure based on regulation and competition the UK is far ahead of both France and the Netherlands, who at the most have adjusted only incrementally to the structural changes. As part of an all-encompassing, neo-conservative strategy of rolling back the frontiers of the state, the Thatcher government pursued a radical policy shift as early as 1979, starting with the separation of the telecommunications branch from the Post Office into British Telecom (BT), quickly followed by a privatisation of BT, and the overall liberalisation of the markets for telecommunications equipment and services. BT was established as an independent but regulated entity, and confronted with competition from two competitive carriers: Mercury in public domestic and international telecommunications and Vodafone/Racal in the cellular radio market. The British telecommunications market, was placed under the supervision of a new regulatory agency, which was created to promote competition and efficiency. This agency, called Oftel, is an

Table 7.1 Market, ownership and regulatory changes in British, Dutch and French telecommunications, 1980–98

Degree of restructuring		*United Kingdom*	*The Netherlands*	*France*
Liberalisation	public monopoly	1980	1980	1980
	terminal equipment	1981	1989	1974–87 (first telephone set excluded)
	value-added networks and services	1982–7	1989–93	1987–93
	cellular services	1984	1994	1987
	voice telephony	1984	1997	1998
	alternative public network and resale capacity	1984	1995	1996
Privatisation	state administration: civil service	before 1980	before 1980	before 1980
	organisational separation between the postal and telecommunication administration	1981 (British Telecom)	1989–98 (demerger of KPN holding into separate post and telecom entities	1990
	corporatisation: managerial autonomy but state owned	1969 (Post Office)	1989	1990–6 (from semi-corporate to full corporatisation
	privatisation	1984–93	1994–8 (?)	1997–??
Regulatory reform	government intervention and political interference	before 1980	before 1980	before 1980
	separation between operational and regulatory functions	1981	1989	1989
	independent regulatory agency to secure the public interest and promote competition	1984	1997	1997

example of innovative institution building through a single-industry regulatory body, backed by a sector-independent regulatory commission in charge of fair competition (the Monopolies and Mergers Commission).

The Dutch policy aimed at gradually following international trends of deregulation, liberalisation and privatisation. In the Netherlands the process of establishing a regulatory commission for the telecommunications industry is still in its embryonic stage, a new institutional framework is slowly evolving in which direct political involvement by the cabinet is reduced. The responsibility of the ministry in charge of telecommunications has been divided into separate bodies for independent policy advice, consultation with stakeholders in the field and regulation. These semi-governmental bodies slowly move away from the sphere of influence of the government apparatus into the direction of an independent administrative agency. A policy alternative has been suggested for the mid-term to integrate the three para-governmental telecommunications bodies and an already established independent administrative agency for media policy to arrive at a kind of Dutch version of FCC or Oftel.

France is a special case in terms of European telecommunications policy, because in many ways it still follows a protectionist/mercantilist strategy with strong state intervention imposing its will on the market place and basically securing its established institutional structures. The French political strategy in telecommunications still aims at supporting the interests of its leading domestic stakeholders in the sector: the ministerial bureaucracy (DGT/PTT) and its *grands corps* of telecommunications engineers, together with the national champion of the equipment industry (Alcatel). While nearly every western industrialised democracy regarded deregulation and privatisation as effective policy measures to combat the economic crisis at the beginning of the 1980s, the French Socialist-Communist government distinguished itself by implementing an ambitious interventionist programme aimed at extending the role of the state in the economy through nationalisations, huge subsidies for strategic industries and the launching of ambitious *grands projets*. These measures implied that the privileged role of the PTT bureaucracy and public control over telecommunications was strengthened even further. The state-controlled PTT (DGT/France Télécom) followed an 'enlightened approach' in the large-scale development of new telematic services like videotex, while securing industrial interests of promoting them vigorously and at the same time responding to customers' needs with a cunning marketing strategy. Thanks to the large investments in the digitisation of the telecommunications network in the late 1970s and the development of new teleservices in the early 1980s, DGT/France Télécom claims to have the highest network digitisation rate in Europe, giving France a major competitive advantage over other countries in the commercialisation of ISDN services in the near future (France Télécom *Annual Report* 1994–7). This has been confirmed in several comparative studies into the performance of seven industrialised countries, including the UK, France and the Netherlands: in terms of the number of subscribers connected to digital exchanges France was number one, Britain number two, while the Netherlands ranked sixth (PA Consulting Group 1994; Spectrum Strategy Consultants 1996; OECD 1997).

Liberalisation

The formerly integrated telecommunications industry has been divided into several submarkets: fixed public networks, the basic voice service, value-added networks, enhanced services (e.g. dataprocessing) and tele-information services, terminal equipment, switching equipment, cellular and satellite services. These segmented markets were in varying degrees responsive to demonopolising forces. In Western Europe, for example, the core public network, consisting of the fixed infrastructure, local telecommunications services and long-distance and international transmission, is still exclusively operated and exploited by the former PTTs (with the exception of the United Kingdom). At the end of the 1980s several fringes of the European telecommunications market have gradually and selectively been opened up to competition. Although the degree of competition varies substantially across Europe and the dominant position of the PTTs is still largely untouched, the markets for value-added networks (datacommunications, videotex, EDI, e-mail), telematic services, the market for peripheral equipment, cellular services and satellite transmission have been liberalised.

In 1984–5, after the majority of the domestic telecommunications market had been opened up to competition and BT had been privatised and listed on the stock market, the United Kingdom had established the most liberal telecommunications regime in the world. The early lead of the UK in the creation of a market-oriented regime was inspired by free market ideology (clearly against statism and corporatism), responsiveness to the needs of the large and internationally oriented business services community of the City of London and by the sheer pragmatic drive to make BT and the domestic telecommunications industry more efficient. The Thatcher government followed a segmented liberalisation strategy, aimed at establishing different market structures in British telecommunications with the crucial market segments regulated by a soft regulatory regime. The markets for equipment and value-added services were clearly liberalised, with hardly any barriers to entry left. For the provision of satellite communications, cable networks and value-added networks, licensing schemes were established with modest thresholds for penetrating those markets. In fixed network provision and cellular communications *de facto* duopolies were established, in which the incumbent BT would compete exclusively with respectively Mercury/C&W and Vodafone for a fixed period of time.

In 1990–1, the newly established managed competition regime of UK telecommunications was evaluated. The conclusion of the Duopoly Review was that workable competition existed in those markets that were fully liberalised (e.g. equipment, value-added services), but that competition in local, long-distance and international communications and cellular services was far from effective. The regime of administered competition that had been established in those markets, clearly favoured players with vested interests, like the first generation incumbent (BT) and, to a lesser extent, second generation incumbents (Mercury and Vodafone). To foster efficiency, innovation and customer responsiveness in those crucial markets even further, a new governance regime based on asymmetric regulation was established. With the exception of international telecommunications, a second round of liberalisation measures was announced based on freedom of entry for new service providers into the exclusive domains of BT and Mercury, furthering detailed regulation and supervision by the Oftel agency and handicapping the incumbents BT and Mercury by restricting or even precluding their entry into the emerging markets of, respectively, PCN, cable systems and home entertainment services.

While the UK government was establishing its market-responsive regime of administered competition, the French and Dutch were still considering and opting for an *enlightened* approach, in which the PTT administration, on behalf of central government, would take the lead in paving the way for both business and residential consumers to the information society. In this interventionist strategy the government targeted the development of an advanced public infrastructure that would be controlled exclusively by the state-controlled PTT. At that time, PTT's exclusive domain included nearly every segment of the telecommunications market with the exception of terminal equipment and computing services (including both the fixed infrastructure, basic services, datacommunications, value-added networks (videotex and cable networks) and some parts of the enhanced services market). The British had already established their market-oriented regime with a privatised BT, equipment and service liberalisation, modest entry barriers and moderate regulatory provisions in the core segments, when the French decided that only corporatisation of the PTT administration and a gradual liberalisation of the equipment and the services market were feasible in the near future. When these plans were finally accepted and adopted in France and the Netherlands, between 1986 and 1991 and in 1989 respectively, the British government had already embarked upon a new liberalisation path that would not only establish effective and workable competition in the entire telecommunications market, but also reinforce British leadership in telecommunications.

The recent initiatives in the Netherlands and France to continue the liberalisation process and include basic services and fixed network provision by 1995–8, show a combination of the former administered competition regime and the free-for-all competition in the UK. The only difference between the UK on the one hand and the Netherlands and France on the other, is that the first has already implemented a working regime that fosters efficiency and innovation, while the two followers still have to establish a market-oriented governance regime that is more responsive to market dynamics in telecommunications. In the late 1980s and early 1990s, when the Dutch and French PTTs were corporatised and granted corporate autonomy, BT had already evolved into a fully fledged and profitable company, that operated multinationally by establishing strategic partnerships and had diversified into new services. The same held true for Cable & Wireless (the former operator in the British colonies, also privatised in the early 1980s), that with its highly profitable subsidiaries Hong Kong Telecommunications and Mercury Communications, has clearly manifested itself in several developing markets and built up a global presence. The Dutch and French PTTs (and their dominant stakeholders, the national governments) were clearly more than five years behind on the road to commercialisation. Unlike a freed BT, that responded more and more to large users at the expense of residential users, the French and Dutch were still restructuring their central and hierarchical administration into a market-responsive corporation, that would be prepared to meet competition at home and abroad. In building up a worldwide coverage in telecommunications the British are also well ahead. While both BT and C&W/Mercury managed to build up a solid market presence outside Europe in the early 1990s (BT in the US with MCI and Syncordia, and C&W with HKT in Asia and Bell Canada), the French and Dutch PTTs are still in the process of transforming themselves from national telecommunications champions into Euro-nationals with loose relationships with American operators (Global One (France and Germany) with Sprint, and Unisource (Sweden, Switzerland, Spain and the Netherlands) with AT&T).

Privatisation

The relationship between the PTT, the government and the telecommunications industry changed from direct involvement in governing the telecommunications industry by a government department, normally a state-controlled PTT, to a governance structure in which the government operated more at arm's length from both the PTT and the telecommunications industry. In a public monopoly that has taken the shape of a nationalised industry, management, regulation and supervision were fully integrated into a single government department, allowing for synergy between the utility, commercial, regulatory and policy functions of the PTT administration, but also causing conflicts of interest between them. Recently, government discretion over PTT has been substantially reduced: government competencies on investment policy, strategic planning, social and recruitment policy and tariffs were transferred and have become the direct responsibility of the PTT organisation. The national governments decided to hive-off the PTT administrations from the civil service by granting them a legal corporate status under private law. This 'corporatisation' (or pseudo-privatisation) implied corporate autonomy for PTT management and allowed for a more market-oriented and entrepreneurial drive within the newly established company.

In all three countries the PTT administration was given a legal corporate status, in which the civil service status was replaced by business autonomy for PTT management. In the UK this happened in 1969, in the Netherlands in 1989 and in France in 1991. Another organisational change within the PTT administration was the separation of the postal and the telecommunications functions. In the UK this occurred in 1981, in the Netherlands in 1989 (although the two functions remain integrated within the larger KPN holding) and in France in 1984. The responsibilities in these new-style PTTs became differentiated between a 'public' branch, in charge of the regulated provision of infrastructure and basic services, and a 'private' branch, which was allowed to enter the market for peripheral equipment and value-added services under the same conditions as any other private sector firm. The separation between the public and private functions of the PTT differed, however, from country to country and varied over time. In France, the UK and the

Netherlands the split was limited to separate accounting between the public utility and the commercial divisions. In the UK a further division of BT into separate subsidiaries for the public utility and commercial functions was created. In the USA the process of devolution went even further: here, an organisational divestiture of the Bell System was accomplished between the local monopoly of the Regional Bell Operating Companies and the commercial functions taken over by a deregulated AT&T. In some other countries government involvement became even more restricted by the privatisation of the PTT organisation into a normal commercial enterprise.

Privatisation, as in the British case, was characterised by the sale of a majority of the shares of the former public enterprise to the private sector, a redefinition of the portfolio of the privatised corporation through separating its regulatory from its operational functions and subjecting its operational activities to a free competition regime. In the UK the public operator BT was privatised in 1984 and its assets were sold on the stock exchange between 1984–93. The Dutch followed the British privatisation example in 1994–5 and the French finally prepared a balanced sale of a minority stake to private investors just before the liberalisation of European telecommunications in 1998. Although they operate more at arm's length from their domestic governments, the French and Dutch PTTs still remain completely in government hands (at least for now). The complex discussions about corporatisation and privatisation in Dutch and French telecommunications have illustrated how and to what extent state and society are intertwined and how hard it is to separate the public and private sectors in the design of appropriate telecommunications policies. In these countries the national government continues to have an interest in the corporatised and the soon to be privatised PTT administration.

After the recognition of corporate autonomy, the corporate transformation of the former PTT administration, with the aim of enhancing market responsiveness and efficiency, was the next step in the process of organisational development. Several reorganisation programmes were implemented to replace the hierarchical bureaucracy of the PTT administration, based on centralisation and functional and geographical divisions, with a more customer-responsive and commercial business culture and a decentralisation of activities and operational responsibilities. PTT management became more responsive to the pressures from the market place as it was confronted with new demands from both traditional stakeholders, such as the state and the user community, and new stakeholders, such as (potential) financial investors and competitors. The increasing differentiation of user demand and intensifying competition meant that the former PTTs had to be equipped with more decentralised organisational structures in order to increase responsiveness, flexibility and innovation capabilities. With regard to balancing the needs of customers and coping effectively with the pressures from potential competitors and volatile markets, the functional and centralised bureaucracy was replaced by an M-form structure based on divisions and business units, that were closer to particular product markets, customer groups and/or regional areas.

The PTTs also became more and more fascinated by, even obsessed with, improving efficiency levels. The ongoing process of denationalisation, from corporatisation to flotation, also required a slimmed-down and market-driven organisation that could realise high performance objectives. A corporatised and privatised PTT would be exposed to the pressures of its financial stakeholders and the capital market: the state would be interested in maximising the sale of its assets on the stock exchange and private and institutional investors would closely examine the performance ratios (sales and profits) and the distribution of dividends. In order to increase efficiency levels, the PTT corporation invested heavily in human capital, automation equipment and total quality management programmes. These measures certainly led to an upgrading of the workforce, to the recruiting of experts in informatics, systems development, marketing and sales, and to a larger involvement and responsibility of personnel in making business decisions. However, there were the negative consequences of the shedding of labour with an ever increasing number of redundancies, making the telecommunications industry even more capital intensive. The next step in the organisational transformation of the West European telecommunications industry was the process of vertical and horizontal integration.

The PTT companies diversified into the emerging markets of cable networks, cellular systems, Internet access and services provision, software/computing, satellites, broadcasting and multimedia applications. In the search for control of these markets, closely related to the telecommunications sector, they were confronted with competition from the incumbents of these domains, such as cable companies, cellular and satellite operators, information service providers, media conglomerates, who not only defended their position on these developing market segments, but were in turn ambitious about penetrating the PTT-dominated telecommunications market. The last step in the process of corporate evolution is the transformation of the PTTs and the telecommunications industry from domestically based production and consumption ('local for local') to international, transnational or even global patterns of economic organisation. Clearly, the former monopolistic telecommunications industry has become a worldwide oligopoly, in which national PTTs have established cross-border alliances (Concert, Unisource and Global One) and are establishing partnerships with equipment manufacturers, software/computing companies and service providers. The new corporate objectives refer to the supply of an integrated package of voice, cellular, data and videocommunications services to large and multinational users, based on one-stop-shopping, full service provision and worldwide coverage.

Regulatory reform

Most Western European countries are in the process of adjusting their national public monopoly to the new technological, economic and institutional conditions of international telecommunications. Although the degree of implementation will differ from country to country, the institutional framework seems to be characterised by a restriction of the public monopoly to the reserved transport services, the liberalisation of the markets for telecommunications products and services, (pseudo-)privatisation of the PTT administration and (some degree of) economic regulation to guarantee universal service provision, reasonable tariffs, interconnectivity and fair competition. The European states and the European Community have left the traditional framework of a hierarchically organised and nationally oriented regime, in which both the operation of the telecommunications service and the regulation of the industry were the exclusive responsibility of the state-controlled PTT administration. After the PTT-dominated governance regime there seems to be a new post-interventionist framework emerging in Western Europe, inspired by American public utility regulation, in which a general anti-trust or competition policy is combined with sector-specific administrative agencies (e.g. the Federal Communications Commission dealing with the special features and conditions of the telecommunications industry). European states have embarked upon a regulatory trajectory, in which the regulatory and the operational functions of PTT are separated and the administrative tasks, dealing with safeguarding fair competition, interconnectivity and quality of service, are executed by the government ministry or by an independent administrative agency. In the Netherlands and France two ministerial departments, HDTP of the Ministry of Transport and Public Works and DRG of the Ministry of Post, Telecommunications and Space respectively, were initially exclusively assigned to supervise and regulate the telecommunications industry. Now the two governmental departments have been established as independent regulatory bodies. The administrative structure implemented in the UK is made up of two autonomous administrative bodies, one in charge of sector-specific regulation and the other in charge of general competition policy, respectively Oftel and MMC. So far the regulatory structures and experiences on the European continent and the UK are different from those in the USA. In France and the Netherlands, the regulatory powers were for a long time located in the ministry (before giving them autonomy from the government), while in the UK they are located in the Director General of Oftel. Public utility regulation in the USA is organised through regulatory commissions operating independently from the relevant ministries. Furthermore, the American and national European regulatory systems differ in terms of their regulatory hierarchies and the price/revenue controls: two-tiered and rate-of-return regulation in the USA, single-tiered and price-cap regulation in Europe.

The introduction of regulation, accommodating the interests of network operators, service providers, consumers and the economy as a whole, that replaced government intervention in the 1980s, is an illustration of the formalisation of government–industry relations in the telecommunications domain. Under the interventionist regime the various tasks (of utility operations, policy and rule setting and commercial exploitation) were integrated within the state bureaucracy of the PTT. The recent differentiation and separation of these tasks, however, has clearly reduced the role of the executive and increased the role of a more autonomous PTT and private sector players. The task of commercial and utility exploitation has been farmed out to the PTO, reducing the role of the government to policy making and rule setting. Although the emulation of the UK regulatory structure of politically independent agencies in the Netherlands and France is controversial, there is a general trend to separate the political task of policy formation from the administrative tasks of rule setting, consultation and supervision. Such a refined structure based on a separation of powers, reduces political meddling and restricts the role of the government to the design of the general outlines, plans and institutional structures of the sector. Administrative regulation attempts to mediate between market forces and state actions by the application and enforcement of public policy and the setting of rules through quasi-governmental agencies that operate outside the political domain.

The shift from an interventionist regime with direct state control and public ownership to a more distant and flexible governance regime with co-ordination through market forces and economic regulation seems to be highly difficult to manage from a government point of view. The governmental or independent administrative bodies were confronted with the structural problem of information asymmetry between the regulator and the regulatee. The latter had more sound information about relevant aspects of the telecommunications industry: the allocation of costs within the network, market conditions (standards, interconnection, prices) and future developments. In order to function effectively these independent regulatory agencies rely on expertise with regard to the various aspects of telecommunications, the stipulation of exclusive concession and licence schemes and franchise contracts and the application of general rules on competition, interconnectivity and public service provision. This process of information gathering, detailed contracting, rule application and the accumulation of jurisprudence contribute to a process of what has been called 'juridification' or 'legalisation' (Volcansek 1992a).

The European nation states were also confronted with the arrival of two new actors on the telecommunications stage. The process of European integration, which was propelled forward in the 1980s, meant the European Commission (through its general competition and telecommunications directorates) and the European Court of Justice were increasingly involved in the harmonisation of national policies and the design of a common European telecommunications market. The efforts of the Commission and the Court of Justice have clearly pulled telecommunications out of the political domain of national governments. They have been essential in redesigning the European telecommunications sector into a more market-oriented sector, based on general anti-trust and mutual recognition rules to be supported by a two-tier regulatory structure at both the national and EC levels. The Commission and the Court of Justice have become players in their own right with substantial jurisdiction and jurisprudence at their disposal: the two institutions have clearly contributed to the prevalence of EC measures over member states' interventions and national practices. The Commission and the Court of Justice have also instigated a process of juridification (or legalisation) at EC level by the creation of an expanding body of jurisprudence on the various aspects of European telecommunications. The actual implementation of these European directives and regulations into national legislation by the member states is a different matter, as the case of adopting the Leased Line Directive (92/44/EEC) shows. At present, the large majority of the member states has not yet implemented the essential requirements (minimum set of leased lines to be provided for each country, regulatory procedures for dispute settlement, application and cost accounting principles), as set forth in the Directive, which is supposed to safeguard open network provision and market transparency (Sauter 1995).

7.3 Explaining the strategic responses of the UK, the Netherlands and France: domestic ruling coalitions and institutional constraints

We have found evidence that national strategies still played a significant part in the formation and implementation of more competitive telecommunications policies between 1980–94. The UK with its governance regime, characterised by market-led adjustment and a relatively open and pluralist political system, has implemented liberalisation, privatisation and regulatory reform policies more fully and more speedily than the Netherlands and France. The neo-corporatist approach of the Netherlands followed international developments haphazardly. It was only after elaborate and time-consuming negotiations between all the parties involved that the Dutch could impose modest reform measures. The French statist regime has so far responded only selectively to the international market and regulatory developments: government intervention has aimed at securing national interests by sheltering the home market, stimulating domestic innovation in key technologies and supporting the international expansion of its national champions. In the research framework we have recognised in Chapter 1 that a country's particular response to international techno-economic restructuring (variable X1) is shaped by the strategies of key actors and constellations of interests at the sectoral level (variable X2), and institutional 'filters' that (pre)select certain issues, arrangements and policy paths, thereby excluding others (variable X3). Institutional conditions at the national level, such as specific national endowments, the positioning of the state in the economy, systems of interest intermediation, dominant economic ideologies and their embeddedness in the world economy, restrict the manœuvrability of actors in the telecommunications policy arena.

The first explanation for the – to some extent – varying responses to global and techno-economic restructuring between France, the Netherlands and the United Kingdom refers to the strengths and weaknesses of the domestic ruling coalition and the overall composition of the sectoral configuration. If we look at the distribution of political and economic power in the three countries, three different pictures emerge of the ruling coalitions that formulated and implemented telecommunications policy. The dominant coalition in the UK included the large telecommunications users, who found political access and support for their demands to introduce competition and improve efficiency and quality of service, and Conservative party officials, who strongly pushed for radical economic changes by 'getting the government off the backs of the people'. The financial services industry, located in the City of London, also benefited from the various business activities generated by privatisation. This alliance of neo-liberal or 'Thatcherite' politicians and the services industry extended to include the floated British Telecom, the newly established industry regulator, Oftel, and new entrants such as Mercury/Cable & Wireless, Vodaphone and cable operators, all of whom have contributed modestly or wholeheartedly to the establishment of the new market situation. The stakeholders not included were the unions, residential consumers, the Labour party and the domestic equipment manufacturers. Their demands for more collaborative and socially responsible labour relations, lower tariffs, renationalisation of the public utilities and protection for the domestic manufacturing base were not met.

The UK was one of the first countries to adjust its telecommunications regime to new techno-global market conditions. The restructuring of UK telecommunications started with the publication under Labour of the 1977 Carter Report which envisaged a separation of the postal and telecommunications services located within the General Post Office, the consequent corporatisation of British Telecom and limited liberalisation of the terminal equipment market. Successive Conservative governments went even further with the adoption of far-reaching policy measures such as the liberalisation of services and equipment, infrastructure competition, privatisation of BT and detailed supervision by the independent regulatory agency, Oftel. By the mid-1980s the UK had implemented a new governance regime, that even in 1994 was too radical for other Western European countries. The drastic reform programme of the Thatcher government was inspired by a neo-liberal ideology, strong demand for high-quality, low-cost telecommunications by large business users in the City, and the poor performance of British Telecom. The early adoption of these stringent reforms gave the UK business community a clear competitive advantage over less innovative followers like the Netherlands and France. As a consequence, the London-based services industry has strengthened its position *vis-à-vis* other world centres, and domestic competition and privatisation have revitalised the two 'sleeping giants' of UK telecommunications, the domestic telecommunications administration Post Office/British Telecom and Cable & Wireless/Mercury, into two leading global carriers. Also, multinational users have benefited from the low charges for international transmission and data services in the UK by rerouting their traffic through the 'hub' of London and/or locating their telecommunications centres in the UK.

In the Netherlands, the dominant élite included the vast majority of stakeholders, all of whom participated in the process of finding compromises between liberalisation and persistent monopoly markets, privatisation and government administration, and (self)regulation and political interference. Consensus was reached after lengthy bargaining and technocratic solutions provided by non-political advisory committees and consultancy companies. The preferred institutional structure included a privileged corporatised PTT and an administered market comprising a core monopoly and liberalised fringe markets. These collaborative arrangements were supported by the majority of the stakeholders in this near-inclusive coalition; however, such package deals were regularly challenged by the European Commission and new market entrants (cable companies, foreign operators and cellular operators). Unlike the UK, which was well ahead of the other European countries and the European Community in establishing an open and competitive telecommunications market, the Dutch and French governments regularly found themselves under pressure from the Commission to liberalise their home markets further, to dispose of the exclusive rights of their domestic operators, and eventually to establish a level playing field characterised by open entry and fair competition.

Initially the Netherlands followed the restructuring process in the UK with great interest and as an early adopter clearly benefited from the British experiments. The Dutch began to examine structural changes in the international telecommunications market in the early 1980s. Several *ad hoc* government committees were appointed to propose policy alternatives to the established public monopoly regime. In these official reports and in ensuing parliamentary discussions, Dutch policy makers often referred to the deregulation and privatisation programmes in the UK and the USA. The British case offered plenty of insights and experiences that were helpful in the design of a new telecommunications polity in the Netherlands. Furthermore, the *mission-oriented* programme of the UK government, glorifying the 'creative destruction' of entrepreneurship, innovation and market dynamics and aiming to give Britain a lead in the information technology and services domain, provided an exemplar to the Dutch, who avoided such a revolutionary model. Compared with the UK 'Big Bang' scenario, the Netherlands followed a 'many little bangs' strategy of gradually transforming its domestic telecommunications regime. Between 1981 and 1986 the Dutch drafted a new telecommunications framework that attempted to strike a balance between the traditional public monopoly model and the radical British scenario. After lengthy negotiations, leading to the new 1989 Telecommunications Act, the Dutch were convinced they had sufficiently addressed the emerging techno-economic and international challenges.

The Dutch path, however, also had its disadvantages. The *diffusion-oriented* strategy was handicapped by the complex accommodation of interests in the Dutch political arena and the time-consuming character of consensus-seeking in the telecommunications policy domain. The focus on balancing internal stability to meet external challenges led, in the Dutch case, to a search for acceptable compromises among domestic stakeholders on the outlines of a common national policy and a consequent meticulous drafting of more adequate legislation. For instance, the Dutch 1989 Telecommunications Act was a kind of 'one law to fit all': it struck a bargain between modest liberalisation and preservation of the PTT privileges of the past, effectively meeting the demands of the PTT, the large users, the union representatives, the consumers and the equipment manufacturers. The one-sided attention in the policy process on domestic pacification and the drafting of once-for-all legislation had the perverse result of overlooking or excluding major ongoing structural developments, such as technological progress, European integration and globalisation of the business environment. This policy lag came to the fore when the newly established telecommunications framework could not cope with the requirements of the European Commission on cellular radio, satellite communications and datatransmission, and the envisaged full liberalisation of the European telecommunications market before 1998. The effects of rapid technological innovation, ongoing market differentiation and European integration upon the newly established framework were underestimated, making the 1989 legislation obsolete within five years.

The dominant coalition in France was clearly state-controlled: the central government was guiding the country's key industries and the close circles around the core ministries including such semi-corporate companies as France Télécom, the national public R&D institute for telecommunications, CNET, and the private national champion, Alcatel. The role of large telecommunications users in France was not as influential as in the UK (or, to some extent, in the Netherlands). The reasons for this were threefold: the lack of centrality of the services sector, the dispersed organisation of large users and the more than adequate performance levels of France Télécom. First, the domestic services industry is less substantial in the French economy and less internationalised than its British counterpart. Second, French business had only minor complaints about the performance and quality of the service of France Télécom (unlike the TMA, speaking on behalf of large British users, who heavily criticised BT's poor quality of service in the late 1970s and early 1980s). Third, compared to the highly professional TMA in Britain, large telecommunications users were less well organised and not fully aware of the costs of telecommunications services. The relationship between central government and France Télécom is close but also contentious: still fully state-owned, France Télécom is regarded by the Treasury as 'the goose that lays the golden eggs' and by the Ministry of Economic Affairs as vital in shaping and carrying out its high-technology policies. The relationship between Alcatel, CNET and France Télécom has been quasi-hierarchical with the companies working together, under state sponsorship, in rationalising the industry, co-ordinating research and development and in the joint efforts of exporting technologies. Another link between these companies is the shared educational background of engineers and civil servants at élite schools, their consequent *esprit de corps* and the flow of staff between central government, CNET, France Télécom, Alcatel and other French state-controlled companies.

In France, the national government followed a clear interventionist strategy, in which France Télécom/DGT was given almost exclusive responsibility for designing a competitive information and telecommunications industry. The French government has given number one priority in its industrial policy to the development of an advanced 'high-tech' electronics industry. Long before the nationalisation programme of the 1981 Socialist-Communist Mitterand-Mauroy government, DGT/France Télécom was actively involved in the modernisation of the French public telecommunications network and the promotion of new telematic services, while the domain of terminal equipment was left to market forces. After nationalising the domestic equipment industry (Thomson, CGE/Alcatel and the subsidiaries of ITT), the French established a kind of equivalent to AT&T, France Téléphone & Télégraph ('FT&T'), based on a horizontally and (quasi-)vertically integrated telecommunications monopoly, including R&D, network operations and equipment manufacturing. Whereas, in the early 1980s, the majority of western countries examined, considered or even implemented telecommunications deregulation, liberalisation and privatisation, France established a state-controlled telecommunications sector, protecting the domestic industry from the forces of international competition. However, this 'shelter' or 'go-it-alone' strategy was partially abandoned with the shift in government in 1986 from the left to the right. The Chirac-Mitterand government carefully adjusted French industrial and telecommunications policies to the demands of the international economic environment: the equipment industry was (partly) reprivatised and the corporatisation of DGT/FT was prepared. It received strong internal opposition throughout the DGT, and the plan was postponed. When the Socialists returned to power in 1988, the more ambitious plans to privatise France Télécom and further liberalise the market were moderated, leading to a semi-corporate France Télécom and a structuring of the domestic market that met the minimum requirements of EU legislation.

Besides the distribution of power in the political economy of telecommunications in the three countries, another important factor in explaining the degree and timing of adjustment was the institutional contingencies at the national level. These variables to some extent constrained the policy-making process by setting limits upon strategic behaviour (restricting the availability of alternatives and selecting those that were feasible), but they also provided incentives to follow established routines, rules and persistent patterns of interaction. These institutional variables, as borrowed from Gourevitch (1986), included: the production profile, the composition of the political system and the role of the state, the system of interest mediation, the established approach in industrial and economic policy, and the embeddedness and the positioning of the national economy within the international system. One could consider the negotiated Dutch approach, the *dirigiste* French policy style and the British market-inspired strategy as path-dependent, reflecting particular traditions, inherited capabilities and persistent ruling coalitions, that together shape a country's economic and political conditions. Normally, institutional development will have a gradual, step-by-step character. For instance, policy development in Dutch telecommunications was path-dependent and critical shifts did not occur. Industrial performance was never really below average, the majority of the stakeholders found their political demands honoured to some extent and the consensualist system was not actually questioned by any of the players. Only in extreme circumstances, triggered by economic shocks and radical political swings, might 'revolutionary' trajectory shifts occur. In such extraordinary cases, new political and economic groups might come to power and the community of policy makers might have to be reshuffled to improve performance levels and to accomplish society-wide infrastructural or ideologically driven projects.

In a market-oriented society like the UK, the creation of a more open telecommunications market, complemented with regulation to protect the public interest, fitted into an inherited pattern of finding solutions to economic problems through the mechanisms of free markets and international competition. As well as relying upon these inherited market-based practices and a traditional hands-off relationship between government and industry, the British could also draw on comparative experiences in the transformation of related infrastructural sectors (such as energy, airways, water, etc.) to install new governance regimes based on market regulation. The availability of routines, competencies and experiences, that had proved effective in practice, is one explanation why the British adapted so smoothly to the new techno-economic conditions. Both the Dutch and the French were constrained by the institutional disadvantages of their regimes in adapting to market-oriented telecommunications. Given their traditions of close government–business exchange patterns and a social market economy, and hierarchical government–industry relationships and state-controlled key sectors of the economy, respectively, the Dutch and the French systems seemed to lack the prescribed ways of thinking and the necessary skills to consider and implement such a new governance regime. While the British gave up the dominant techno-economic paradigm of the natural monopoly and embraced the new model of regulated competition relatively quickly, a large majority of the Dutch and French policy communities ignored the anomalies in the established governance regime and continued to perceive things in a conservative way. It was only in the early 1990s that a growing number of stakeholders in both France and the Netherlands became aware of the revolutionary shift that was taking place in telecommunications. In the politicised and less dynamic equipment market, however, the reputation of the French was, and still is, outstanding, as furthered by inherited engineering competencies, public procurement, acquired scale and scope advantages, and trade diplomatic skills.

The occurrence of path-dependent developments, interrupted by critical junctures challenging the *status quo*, is particularly relevant in explaining some of the breakthroughs in French and British telecommunications policy. At the end of 1960 French policy makers, already convinced of the strategic character of information technology and its positive effects upon national productivity levels and economic growth, were confronted with the need to overhaul and modernise a technologically backward and inefficient telecommunications system. The modernisation of telecommunications in France became a target in the national indicative planning system and subsequent interventionist industrial policies, envisaging *Frenchified* technology and the promotion of a national champion (i.e. curtailing the leading manufacturer ITT), the granting of subsidies and public procurement contracts by government and the co-ordination of R&D and overall guidance by state enterprises and public research institutions (i.e. DGT and CNET). At the end of the 1970s, it became clear that the UK telecommunications system needed to be upgraded and reformed. The large users in the City complained about BT's poor service provision: an underperforming network with an operator lacking a commercial orientation. The coming to power of Thatcher's Conservative party, attacking loss-making and inefficient state enterprises, dealt another blow to BT and the established monopoly in UK telecommunications. In the end this led to the privatisation of BT and the establishment of managed competition, based on a network duopoly, liberalised equipment and services markets and regulatory surveillance to ensure quality of service and fair play. In the early 1980s the French DGT had already caught up with its international contenders, merely as a result of the policies of central government focused on the electronics industries. By that stage DGT was not as vulnerable as BT: it had been strongly involved in the large-scale modernisation of the French network and had strengthened both its direct and indirect control over the domestic telecommunications industry.

The institutional variable of the production profile on constraining or spurring policy development manifested itself especially in Britain and France, where the performance level of the domestic telecommunications industries was far below average and outstanding, respectively. In Britain the poor level of service in the Post Office/British Telecom and the funds needed to upgrade the network acted as major incentives for large business users and the newly appointed Thatcher government to consider the shock therapy of privatising British Telecom rapidly and carrying out a radical market liberalisation programme. High-quality information services at low cost were extremely relevant to the internationally competitive services industry in Britain. In France, the economic position and the political strength of the services sector was not as great. Here emphasis was laid on equipment manufacturing and industrial engineering. The poor condition of the public telecommunications network in the 1960s motivated the French government to give first priority in its planning activities to upgrading the domestic infrastructure. Furthermore, France Télécom was given a special role in carrying out industrial/technological policy. It was responsible for orchestrating the process of catching up with the major international contenders in the electronics domain. After the highly successful modernisation of the French network in the 1970s and 1980s, there was no need for drastic changes to the domestic framework: why embark upon a risky privatisation and liberalisation trajectory if the technological performance levels of France Télécom were already world class? Thus a particular production profile in the domestic economies and in industry can influence the state to leave things as they are (e.g. France) or spur public and private actors to transform the existing situation (e.g. the UK).

Another institutional factor explaining to a degree the path taken by the three countries involved in this study, is the composition of the political arena and the role of the state. While the Dutch system is characterised by proportional representation and coalition governments, searching for society-wide compromises, consensus and continuity of policy development (i.e. controlled liberalisation and privatisation), the British and French polities exhibit more adversarial traits and 'winner takes all' practices. The parliamentary duopoly of the UK and France, relying upon strong cleavages between left and right, and the tradition of majority rule, allows not only for more radical party manifestos and political debates but, more importantly, for direct implementation of politically sensitive proposals, thereby overruling the opponent's point of view. For instance, the electoral victories of the right-wing Thatcher government in 1979, the leftist Mitterand-Mauroy government in 1981 and its replacement by the Chirac-Mitterand government between 1986–8, were sea changes compared to the previous incremental policies in post-war Britain and France. In the UK, Thatcher embarked upon a radical deregulation and privatisation programme that was effectively carried out in the subsequent decade. In France, the Socialist-Communist government launched an ambitious nationalisation programme, bringing the majority of the large companies and the country's key industries under state control. Despite the Socialist presidency of Mitterand, these interventionist measures were undone by the Liberal-Conservative Chirac government (1986–8) and by the Balladur government (1993–4), which strengthened the private sector by denationalising a substantial number of state enterprises. The practice of state leadership and administrative guidance of the critical sectors of the national economy was not abandoned however; the strategic position of France Télécom was not affected.

The role of the state in restructuring telecommunications is different in each of the three countries investigated. The French approach is a clear illustration of a developmental state, where the state combines the role of demiurge and husbandry. The French state was actively involved in the domestic telecommunications industry in a variety of ways, through sponsoring departments, state enterprises and affiliated businesses. First the French government acted as the guide to telecommunications restructuring by nationalising foreign firms and rationalising domestic industries and as the sole producer of strategic goods and services to secure strategic industries and technologies under centralised control. Second, the state acted simultaneously as entrepreneur (developing new markets and taking risks) and financier (providing and facilitating funds and credits). Finally, the state acted as intermediary between the various interests and stakeholders and as the gatekeeper to the French market, attempting to stimulate exports and selectively reduce imports.

The role of the UK government could be characterised as regulatory with the government combining the functions of custodian and midwife. The British state withdrew from active involvement in domestic telecommunications by abandoning its role as producer, entrepreneur and (co-)financier and giving more emphasis to its continuing role as arbitrator or referee, acting as an administrative watchdog overseeing the liberalised and privatised industry. Thus the hierarchical relationship between the state and industry was transformed into a contract-based principal–agent relationship, the state being the principal and the market players the agents. The British government took its role as regulator seriously and set up Oftel as an independent body to stimulate fair competition, investigate anti-competitive practices and ensure that socio-political objectives were met. Through this rule-setting approach the British government stimulated new entry in nearly every segment of the telecommunications market, thereby facilitating the domestic and international expansion of BT, the 'birth' of successful new firms such as Vodafone, Ionica and Cable & Wireless/Mercury. It also allowed foreign entry of large American and Canadian companies that linked up with smaller domestic cable and telecommunications operators.

Compared to the relatively strong state in France and (to some extent) the UK, the position of the Dutch government *vis-à-vis* its stakeholders was less hierarchical. The role of the Dutch state in telecommunications policy was that of a middleman or broker attempting to strike deals and forge a consensus on future policies that included all the major stakeholders. The Dutch government was able, through a prolonged process, to resolve controversial issues and conflicting interests with acceptable comprises, thereby avoiding more radical adjustment strategies that would bring about rapid and disruptive changes which could damage some of the stakeholders in the policy process. In the early and mid-1980s the Dutch strategy of incremental adjustment was relatively effective. However, in the early 1990s, when the full impact of rapid technological change and market developments and the effect of European integration acting upon the carefully reached compromise of 1989 was felt, the handicaps of the Dutch negotiated and consensualist approach became obvious. The Dutch government now faces the problem of planning and developing a new path for the future, taking advantage of new technological opportunities, contending with political and economic conditions, while the state and some stakeholders are still stuck in the traditions of the past.

The third institutional factor which helps to explain the different trajectories followed by the three states is the system of interest intermediation. We have discussed the 'public' element of state–society exchange relationships, now we move to the private side of stakeholders and interest groups also influencing the policy process. To some extent, the (lack of) power of the constellation of private interests provides us with a reflection of the strengths and weaknesses of the state. For instance, the Netherlands is known as a corporatist state, in which centrally organised interest groups and trade associations play major roles in the processes of policy formation and implementation. From the early 1980s, the relatively weak Dutch state was confronted by strong private sector partners in the telecommunications domain, such as the employers' associations, large users, consumers' organisations, labour unions, equipment manufacturers and the PTT administration, all of whom wanted a slice of the 'deregulatory' cake or at least a reward for giving up certain rights or privileges. The Dutch coped with industrial adjustment by compensating the 'losers' through a 'negotiated' redistribution of the 'winners'' benefits. A society-wide settlement was reached on non-radical and evolutionary policies, that reflected a combination of temporarily safeguarding the core monopoly and making a gradual and co-ordinated introduction of competition in the other market segments. The Dutch negotiated an adjustment strategy aimed at finding policy alternatives acceptable to every stakeholder without touching upon vital interests or upsetting the *status quo.*

Traditionally, the strong French state has faced only poorly organised and dispersed private sector stakeholders. Evidence of this can be found in the telecommunications domain. Compared to the omnipresent state-controlled agencies, such as DGT/France Télécom, state-enterprises (Thomson, Bull and financial investors) and the 'private' national champions of the domestic electronics industry (such as Alcatel, Matra, CGS, etc.), the power of organised labour, employers and large users' associations and consumers' organisations, as institutionalised in the policy and decision-making process, was weak and marginal. In the case of France the adjustment strategy in telecommunications was clearly state-led, relying upon political manipulation of the market. Government institutions such as DGT/France Télécom and the national telecommunications research institute, CNET, together with the semi-public electronics industry, stipulated the targets to be achieved. The whole process of goal formation and policy implementation was facilitated by the informal exchange of engineers and senior officials between the public sector and quasi-private national champions. At the shopfloor and grass roots level the interests of organised labour, including both higher and lower ranks within DGT/France Télécom, converged around the notion of an industrial strategy that was led by DGT/France Télécom on behalf of the French state. Therefore support for privatising France Télécom was not strong; the large majority of the workforce wanted to remain in the public sphere and keep the privileges of state officials serving the public interest. Throughout the decision-making process the discussion was dominated by the French PTT and its preferred manufacturer, Alcatel; the interests of large users and consumers were given minor attention or were simply ignored.

In the UK case study we found evidence of the traditional market- or company-led adjustment strategy, where priority was clearly given to the introduction of effective competition, free trade and picking the winners of the domestic market. It was believed that the encouragement of entrepreneurship and rivalry would increase the competitiveness of the national economy as a whole. The British adjustment strategy was to establish a favourable business climate at home that would foster innovation and foreign entry, and facilitate the international expansion of its home-based firms. However, the benefits and costs of telecommunications restructuring were distributed unevenly. The object of the radical privatisation and liberalisation policies in Britain was to meet the demands of large corporate users (such as the City-based financial services industry) and create innovative and internationally successful network operators by transforming former state enterprises, such as BT and Cable & Wireless, nurturing new firms, such as Vodaphone, Mercury and the cable operators, and encouraging foreign direct investment (mainly by US and Canadian companies). The interests of residential users, organised labour and domestic equipment manufacturers were not only largely ignored but seriously damaged (and left uncompensated). The increase in charges for residential services, the erosion of the domestic manufacturing base, the shedding of labour by BT and GEC/Plessey (without sufficient job creation by new entrants) and union busting strategies followed by management of the telecommunications companies, were the down side of the British adjustment strategy.

The fourth institutional factor that had an impact on the particular implementation path chosen by the three countries refers to the economic approach and ideologies prevalent in national telecommunications policy. We have found evidence that the three policy responses in French, Dutch and British telecommunications reflected the established adjustment strategies at national level: the French approach reflected state interventionism and mercantilism, the British approach gave priority to domestic competition and free trade, and the Dutch approach emphasised domestic cartelisation and an export-led orientation towards trade. With the exception of the shift in government policy in Britain with the coming to power of the Thatcher government in 1979 and the interruption of the right-wing Chirac government in France between 1986 and 1988 in a period dominated by the Socialists, the formation and implementation of telecommunications policies in France, the Netherlands, and the United Kingdom did not really deviate from the established economic routines of the three countries. One could argue convincingly that the inherited economic approaches of two of the three countries (the Netherlands and France) and an ideologically inspired modernisation programme (UK) acted as a severe constraint in the formation of policy responses to telecommunications restructuring.

The last relevant institutional factor was the individual country's particular position in the international political economy. Compared to the cartelised federation of national monopolies of the past, the three governments actively contributed to the emergence of a new governance regime regulating cross-national telecommunications. This new model seems to be based on a global oligopoly of diversifying and internationalising telecommunications operators through the acquisition of new business ventures, investments abroad and the formation of international strategic alliances. The role of national governments in this global oligopoly game is to find a trade-off between facilitating and supporting the international expansion of its home-based telecommunications company and attracting foreign operators linking up with new entrants in order to stimulate competition in the domestic market place. In the UK we have seen the rise and corporate expansion of three multinational companies in telecommunications with a myriad of foreign partners, namely BT, C&W/Mercury and Vodafone, and an influx of American firms on the cable market. In France, the state clearly pursued a strategic trade policy by giving priority to transforming domestic players into world-class transnational companies and allowing new entry to the home market only selectively. For instance, the national champion France Télécom linked up with its German counterpart DBP Telekom to be joined later by US Sprint, in the Global One venture, while the activities of other foreign companies in France were only marginal. In the case of the Netherlands, the government gave priority to supporting KPN/PTT Telecom in its bid to diversify into new product markets and enter new geographical markets (by participating in the Unisource and the AT&T-instigated Worldsource ventures) over encouraging new entry in the domestic market place.

Although all three countries belong to the European Community, this does not necessarily mean that the impact of European integration on telecommunications policy making in France, the UK and the Netherlands was the same. Being the first mover on the liberalisation/privatisation path in Europe in the mid-1980s, the British government was clearly ahead of the schemes set out by the European Commission for open terminal and services markets and open access to networks, starting from the early 1990s. Besides providing the European Union with real-life experiences of liberalised and privatised markets that are regulated to some extent, the British government collaborated with the Commission in pushing the other member states to open up their domestic telecommunications markets and make administrative structures more transparent.

The French and the Dutch governments, concerned about the one-sidedness of the British approach, also attempted to persuade the European Commission to follow their particular telecommunications trajectories. Compared to the radical pro-market initiatives suggested and implemented by the British, they pushed for a more modest pace of liberalisation and privatisation, taking into account the time needed for the national operator and the domestic market place to adjust to real competition. Especially between 1985–9, the Dutch model of combining an exclusive concession for the provision of basic services with the liberalising of terminal and services markets and corporatising the PTT administration, as suggested by the Steenbergen Committee, acted as a propelling force to establish one integrated European telecommunications market. Later, when the implementation of the European Commission's Green Paper had gained momentum, the Dutch were overtaken by the insistence of the European Community on abolishing cross-national trade barriers and attacking the exclusive rights of the national operators. In the case of France and the Netherlands, the European Commission acted both as an agent pushing for change and the speeding up of the adjustments process, and also as a serious constraint on the development of certain alternatives to the exclusion of others. The reform of the established PTT-controlled regime in the Netherlands and France had to meet the generic provisions of the Community's anti-trust policy as well as the sector-specific requirements on liberalising particular markets, separating regulatory and operational activities and ensuring open network provision. For instance, the French were more or less forced by the Commission to liberalise voice telephony and network provision. It took almost three years of tough negotiation before the Commission finally endorsed the proposed Atlas/Global One joint venture, including the French and the German PTO working together with the American operator Sprint in global service provision.

7.4 Concluding remarks

An explanation for this partial convergence of policy ideas and practices across Western Europe could be found in the increasing international dependencies between the USA, Japan and Europe in an emerging world economy, restricting the freedom to pursue a strictly national policy. The international conditions restricting the scope for go-it-alone strategies, or more precisely the American hegemony in telecommunications and computing have to be taken into account by policy makers in Western Europe and Japan in the formation and implementation of adequate adjustment policies. The deregulation of the American telecommunications industry from the 1960s onwards encouraged the leading telecommunications company AT&T and the leading computer supplier IBM to compete for market power and control in the developing market of telematics, integrating dataprocessing, computing, office automation, and telecommunications. Although it had to dispose of its local communications branches, AT&T was allowed to enter the markets of the dataprocessing industry and international services at home and the sale of equipment abroad. In terms of economies of scale, technological capabilities and experience with competition from new entrants in the domestic telecommunications services market (MCI and US Sprint), in equipment (GTE and Northern Telecom) and in the computing market (DEC, GE), AT&T and IBM were clearly ahead of their European equivalents. In the mid-1980s, the European telematics market was still organised at the member state level and divided among relatively small-scale telecommunications operators and the national champions of the computing industry, which were protected from foreign entry. Both the European telecommunications and computing industries, highly dependent upon state intervention through statutory monopolies, state enterprises, subsidies and public procurement, were very concerned about a potential entry of the American giants, AT&T and IBM.

The established network of European stakeholders, made up of national governments, PTTs and their domestic equipment suppliers, were well aware of their delicate position *vis-à-vis* their powerful American counterparts and of the devastating consequences of a once-and-for-all liberalisation strategy. Therefore, they opted for a restrictive and gradual opening up of the protected national market and the Common Market. This would give the major stakeholders in the dominant coalition in Western Europe time to formulate an adequate adjustment strategy. The fear of American commercial, organisational and technological superiority and the threat of American domination of the European telecommunications market has manifested itself in the UK, France, the European Community and, to a lesser extent, in the Netherlands. In the UK a link-up between BT and IBM to provide managed data services in the mid-1980s was not endorsed by the British government. Since the early 1990s, an ambitious BT has been aiming to penetrate the American market and to equal AT&T in size by linking up with its biggest domestic rival MCI. BT, however, did not manage to gain full control over MCI and lost out to WorldCom, who acquired the second American long-distance operator in the autumn of 1997. French telecommunications policy has been obsessed with the American challenge. Far-reaching government intervention and protection from foreign involvement in the reconstruction and modernisation of the domestic telecommunications industry was justified by at least three closely interrelated rationales. The first sector-specific argument refers to the emerging telematics sector as an infant industry, that required state leadership to foster and orchestrate technological and industrial development. The second more general argument was based on the leading principle in French industrial policy of reducing economic and technological dependence in the key industries and stimulating the autonomy of the French nation by supporting and building up a solid technological and industrial base domestically in the targeted sectors of electronics, manufacturing, nuclear industry and aerospace/aircraft manufacturing. The third argument refers to the propelling force of a foreign threat forcing the French to safeguard the nation's interests. France's telecommunications policy between 1975–93 to a large extent was a response to the challenge of IBM and AT&T in telematics (the Nora-Minc Report), the Japanese threat in electronics and semiconductors in the 1980s and the Anglo-Saxon bid for world hegemony (the global carriers BT/MCI and AT&T/Worldsource/Unisource), leaving France Télécom and the other European PTTs behind.

The attitude of the European Community reflects a combination of the French arguments, selectively supported by the British to support their domestic interests against American entry, and the free trade argument as laid down in the Treaty of Rome and reflected in the 1992 Internal Market Programme. In the early 1980s a common European Telecommunications policy was developed in response to the

American and Japanese domination in electronics, computing and telecommunications services. It was believed that co-ordination and collaboration of telecommunications policies at the EC level would generate the economies of scale and facilitate the rapid innovation required to keep up with the American and Japanese competitors. However, the creation of the Common Market also constrained the room for manœuvre of the individual member states in the formulation of an appropriate response by setting restrictions on industrial policy instruments, such as state subsidies, preferential treatment of national champions and statutory monopolies. The harmonisation of national commercial activities and market integration by the European Community fitted in perfectly with the American notions of trade liberalisation, justifying the search for access to the European market. For example, one of the big allies of the European institutions in the process of breaking down protective measures at national level and building up an Internal European Market, was American business, actively lobbying for international deregulation.

The Dutch also feared the entry of the large American firms in their protected home market, but the situation in the Netherlands was less conflicting than elsewhere. Although the Dutch subsidiary of IBM played quite an active role in the representation of the interests of the information industry, the relationship between domestic incumbent firms like PTT and Philips Telecom and the penetrating American firms was less adversarial in the Netherlands than in France, the UK or the European Community as a whole. For example, the strategic partnership with Philips in APT was the first strategic move of AT&T in Europe. In the early 1990s two more American-Dutch partnerships were established: the common effort of AT&T and PTT Telecom (together with the German and Danish PTTs) in modernising the telecommunications infrastructure in the Ukraine and the link up of Unisource (the joint initiative of the Swiss, Dutch and Swedish PTTs) and AT&T/Worldsource in 1994–5. This more pragmatic attitude of public and private stakeholders in the Netherlands towards foreign entry reflects the particular position of a small and relatively open economy.

Although the United Kingdom was clearly ahead of France and the Netherlands in the timing and degree of the implementation of telecommunications liberalisation, the way in which the preserved markets were demonopolised was quite similar: competition in the markets for terminal equipment and value-added services competition expanded only incrementally, not really touching upon the dominant stakes of British Telecom, PTT Telecom and France Télécom. After the new governance regime of market liberalisation and the corporatisation of the former PTT administration was installed, it was clear that the former PTTs still had substantial informational and bargaining powers at their disposal, which hampered the effective functioning of the newly established regulatory framework. Therefore the need was expressed for more precise and specific rules that would balance more evenly the interests of consumers, business users, PTT, and national and foreign service providers. The unequal distribution of power between a large and powerful regulated industry and a small and inexperienced domestic regulator became less unbalanced with the implementation of detailed rules regarding tariffs, interconnectivity, fair competition and quality of service levels. The domestic and international expansion of the former PTT administrations as emerging diversified and multinational conglomerates was going beyond the control of the national regulatory framework.

While national governments were moving towards detailed supervision and independent regulation at the national level, there was a comparable trend at EC level to match the powers of an internationalising telecommunications industry by stimulating more collaboration among the member states and supporting the growing importance of European institutions in telecommunications policy. The European Commission became actively involved in telecommunications regulatory reform through the increasing importance of its Competition Directorate in promoting general competition. It also initiated the creation of a loosely integrated federation of the member states' regulatory agencies to deal with the harmonisation and integration of legal provisions and rules concerning the opening up of various market segments, open network provision, standards and infrastructure development. The experiences with the liberalisation and privatisation of European telecommunications illustrate that effective regulatory reform not only presupposes deregulation (the abolition of rules), it also requires new detailed rules to guarantee the exclusive provision of dedicated services, equal access and fair competition. The retreat of the state in the domain of public utilities such as telecommunications, energy and railways, demanded more sophistication over the definition of the remaining core responsibilities. The new situation created by developments like liberalisation, privatisation and deregulation did not put an end to regulation but, paradoxically, required a process of regulatory reform in which the abolition of regulations was accompanied by the formation of new and sometimes more explicit or transparent rules. In such a new institutional environment the role of the government developed increasingly into that of trustworthy referee, putting the other roles of player/coach aside. Only recently this so-called 'paradox of regulatory reform' has been recognised in the UK, the Netherlands, France and the European Community (Majone 1989b, 1994a,b; Veljanovski 1989, 1991a).

Although the state is still involved as key shareholder and regulator in the new telecommunications setting, the former PTT administrations have clearly become more independent from the civil service. The imposed governmental controls on investment decisions, personnel policy, social service provision, pricing, profits, horizontal and vertical integration and cross-national partnerships and so forth have been abolished or substantially reduced in order to increase the new Public Telecommunications Operators' (PTO) corporate autonomy and flexibility. In the post-monopoly setting, characterised by market liberalisation, regulated provision of basic services and a more pluralist network, the PTTs have lost significant parts of their exclusive monopoly to new entrants like private carriers, service providers and foreign telecommunications companies. To protect their *de facto* core monopoly France Télécom, PTT Telecom and British Telecom have decided to follow an expansionist strategy of integrating both vertically and horizontally. In order to defend their dominant position at home, they have diversified into new profitable services like cellular communications (BT, PTT Telecom and FT), value-added networks (FT, BT, PTT Telecom), global network services (BT), cable systems (PTT Telecom), software and computing (FT), equipment manufacturing (FT and BT initially). The former Dutch, British and French PTTs have also integrated horizontally by building up cross-national partnerships with others to increase economies of scale and seek access to new geographical markets. The strategic alliances between BT and the US-based MCI, the Global One joint venture of the French and German PTOs with the US-based Sprint, and the Dutch-Swiss-Swedish-Spanish link up with AT&T in Unisource/Worldsource are illustrations of the emergence of multinational or even global carriers. Telecommunications has become a volatile market in which both domestic and international competition is intensifying, leading to price instabilities, excess capacities and a trend towards further concentration.

Although the Dutch, French and British PTOs have lost market shares, their bargaining power in the domestic equipment and services market is still substantial. The clear dominance of BT in nearly every part of the UK telecommunications market after almost a decade of drastic reform has eroded only very slowly. The position of PTT Telecom Netherlands and France Télécom *vis-à-vis* the competitors in their respective liberalised home markets is still unchallenged, resembling an offensively and defensively well-equipped Goliath against an emergent host of little Davids that have no common strategy or tactics. In France and the Netherlands the established interests of PTT Telecom and France Télécom were largely untouched: both in the traditional market segments and the newly developing markets like cellular communications, datatransport services, cable television and satellite services, market shares are still substantial. They have effectively delayed painful decisions that would allow for new entry by (potential) contenders or the creation of a level playing field. Instead they have opted for a step-by-step method of implementing fair competition. However, in emerging markets, where no hegemony by a powerful incumbent existed, competition has been introduced successfully. For example, workable competition has been effectively established in the UK cellular market. A duopoly of BT and (Racal) Vodafone was established in the mid-1980s when the mobile communications market was still in its infant stages. After the further liberalisation of the cellular market, the two incumbents faced competition from two newly licensed operators, One-2-One (Mercury) and Orange (Hutchinson/Microtel). The contest between two players of equal strength, followed by competition from new entrants, has triggered innovation and market efficiency in the domestic mobile market, making it internationally competitive and facilitating the overseas expansion of UK firms. While the British managed to promote and establish alternative operators that successfully rolled out their new network, attracted customers and reached workable interconnect agreement with the incumbent BT, the Dutch and the French, with their new operators such as Telfort, Enertel, Bouygues Télécom and Cégétel, are still in the pioneering stages.

8 Convergence and/or divergence in European telecommunications restructuring?

8.1 Introduction

In this study, we have argued that the individual policy response of each country is shaped by three independent variables to be found at different levels of analysis, namely the extra-national variable of structural forces that have an impact on the national telecommunications polity; a sectoral system variable including the composition of the domestic telecommunications industry of the three selected countries; and a national institutional variable accommodating extra-national developments with internal adjustments in the domestic telecommunications arena. The four *structural forces* of technological (r)evolution, newly emerging market patterns, international deregulation, and European integration, have effectively challenged the logic of a nationally oriented PTT system. An adaptation of the French, Dutch and British telecommunications sector to a new environment that is more dynamic, complex and uncertain, seems in order. The *sectoral system variable* refers to the structure of the domestic telecommunications sector, and the strategic choices made by dominant actors and ruling coalitions in the market place and the policy arena. This variable addresses the economic and political strength of the former PTT administration or the new Public Telecommunications Operator (PTO) *vis-à-vis* its favoured supplier(s), its customers (i.e. large and residential users), its (new) shareholders and last but not least, its (potential) competitors. It also includes the particular governance regime that shapes the transactions and the strategic interactions taking place between the stakeholders. The *national institutional variable* refers to the persistent routines, codes of conduct and rules that both constrain the policy-making process in the telecommunications domain and provide incentives to consider and act upon one particular set of strategies at the expense of other alternatives. National institutions set limitations on strategic behaviour and thereby constrain the feasibility of the various options available to the stakeholders in the domestic telecommunications field.

This research investigates whether one can detect convergent and/or divergent approaches to worldwide telecommunications restructuring, pursued by the three selected European countries. Here we will deal with evidence in the European telecommunications domain either for the proposition that 'the approach of various countries is becoming similar' or for the opposite, that 'country differences still matter in formulating appropriate policy responses'. In the concluding part of this comparative study, we will address the freedom to manœuvre that the Dutch, British and French states have in restructuring their domestic industry in the light of the new techno-economic and institutional contingencies in international telecommunications. Notwithstanding the structural forces pushing for the adoption of converging and similar adjustment strategies among the three countries, all three could also follow distinct strategies that might reflect different national policy preferences and further different domestic interests and strengths.

8.2 Structural forces forging convergence and the persistence of national diversity

From the late 1970s onwards, the nationally oriented telecommunications economies in Western Europe are challenged by:

1 the impact of technological developments;
2 the emergence of global and differentiated communications markets;
3 increasing international deregulation;
4 European integration.

These four structural forces (variable X1) press for a move away from the heavily protected and politicised monopoly framework of public service provision and universal access, towards an emerging new model in which there is more room for competition, as well as foreign market participation and less government intervention.

The first challenge to the established PTT system came from technological developments that led to alternative transmission technologies (satellites, cellular communications, cable networks) and the 'upgrading' of the core telecommunications infrastructure (high capacity broadband and software-based 'intelligent' networks) and to a plethora of value-added services. As a consequence, the traditional argument for a natural monopoly of the entire telecommunications industry was no longer valid and the reasons for active government intervention through a state-controlled PTT administration were eroded. Telecommunications has become more integrated with other business functions (like dataprocessing, office automation, industrial production, etc.) on the basis of advanced computer systems. As a consequence, the distinction between the public infrastructure and private business facilities has become blurred, especially where companies extended their information and engineering systems to privately developed telecommunications networks in order to meet their growing demand for communications services and the need to cut costs.

A second blow to the established system came from business and professional users who complained that their demand for low-cost, highly advanced and made-to-measure service provision could not be met by a monopolistic operator in a protected market. In particular international users effectively argued that the monopolistic control of the public telecommunications services and the restrictions on developing private alternative networks in Europe were frustrating the strong demand for a new range of 'global' one-stop-shopping and seamless services. Large business customers, such as airline companies, financial institutions and car manufacturers, started to construct their own computer-based networks, connecting their subsidiaries, suppliers and customers around the world, and manage their own datacommunications services. Other large international users relied upon specialised system integrators, including companies such as IBM, AT&T and commercially oriented European PTTs, to manage their corporate transactions and worldwide information flows.

A third reason for restructuring the national telecommunications polity in Europe was the challenge posed by the decision of the American government to deregulate its domestic equipment and services market. The US government had already started to question the characteristics of the public service and monopoly provision of the established telecommunications polity in the late 1960s, and was gradually moving towards a market-oriented industry that would enhance efficiency and stimulate innovation and international trade. The loosening and lifting of government regulations would allow large, vertically integrated companies like AT&T and IBM, relying upon a large home market and technology/market leadership, to become global telecommunications players. American business users in general, operating all over the world, would benefit from such a deregulated regime, which enabled competition between AT&T and IBM. Given US world leadership in both computing and telecommunications in the early 1980s, Japan and the European Union (EU) and its individual member states had to meet this challenge. At that time the EU consisted of a number of comparatively small, fragmented and sheltered markets, which were becoming a handicap to innovation, growth and international competitiveness. Therefore the pro-market programme of the US government and the (potential) entry of American business into the European market posed a severe threat to the governments of the EU member states.

The fourth structural force that pushed for an overhaul of the established PTT-centred system was the process of European integration, i.e. the gradual transfer of legislative powers from the national to the European level of decision making. Challenged by the structural forces of technological developments, market differentiation and international deregulation, the United Kingdom – and to some extent the Netherlands – embarked upon a three-fold policy in the early 1980s: to liberalise the market, to privatise (or corporatise) the PTT administration, and to (re)regulate market conditions and competitive behaviour. Other countries that were in the process of considering only marginal changes to the established regime (e.g. France), were surprised by the impact of the European Community's 1987 Green Paper and its subsequent regulations and directives aimed at liberalising the telecommunications market and questioning the PTTs' exclusive rights. From the late 1980s onwards, the freedom of the European Community's member states to protect their home market and further their own domestic interests was curtailed by an ever-expanding jurisprudence at Community level that took precedence over national legislation. For instance, the European institutions effectively contributed to a community-wide liberalisation of terminal equipment and value-added services, mobile and satellite communications, ensuring open access to public networks and structural separation between operating a network and regulating the industry. The most important measures were the opening up of network provision in 1996 and voice telephony in 1998: they are currently being implemented within the legislative frameworks of the member states. The impact of EC legislation on domestic telecommunications policy making is substantial: although directives leave a certain degree of freedom to make national law commensurate with EC legislation, EC regulations and directives are binding to each member state (in their entirety or in the results to be achieved, respectively).

All four structural developments, as discussed above, constrain European states in pursuing sovereign national strategies. The new market-based regime in European telecommunications can be characterised as more dynamic and differentiated than the traditional monopoly framework, offering new business opportunities for both the 'old' PTTs and their 'preferred' equipment manufacturers, and entry possibilities for new operators, service providers and foreign suppliers. The co-ordination of economic activities has changed from an administrative hierarchy, represented by the public operator and its monopoly, to a (more) competitive model that allows free entry in the terminal equipment and services market, and controlled entry to the areas of voice telephony and infrastructure provision. The PTT administrations were transformed from functional organisations to ones that have become more efficient and innovative. A new privatised and decentralised organisational form has allowed for the pursuit of market segmentation, product proliferation and price differentiation strategies (i.e. offering various products, serving different categories of users and introducing price flexibility), geographical expansion (through foreign investment and international strategic partnerships), strengthening customer relationships and account management, cutting costs and creating incentives for revenue enhancement and profits.

The public operator's interference in regulation and standards setting has become more transparent with a privatised and commercial PTT and the creation of independent regulatory agencies. Within the newly emerging institutional framework the character of government intervention changed from a truly 'interventionist' one, emphasising state monopolies, public ownership and planning, to a more detached regulatory role. European governments sought to introduce significant competition into the industry and delegate administrative responsibilities to relatively autonomous agencies, responsible for scrutinising fair competition, price controls, arbitration and dispute settlement, and for observing whether or not particular social objectives, such as universal service provision, are met. While there are still special regulatory requirements for market entry and strategic behaviour, European telecommunications policy has lost its immunity from the national and European anti-trust provisions and, as a consequence, the sector is gradually losing its unique framework and is becoming more like any other business activity. National public policies are now forced to include even transnational aspects of contemporary telecommunications; the entry of foreign firms to the domestic market, the (potential) behaviour of multinational business users, the policy-shaping role of the European Commission and other international institutions (such as the WTO/GATT), the strategies of overseas governments, transnationalisation through cross-border acquisition and strategic alliances, etc.

In the early 1980s, the UK was the first country in Europe to question the established monopoly regime and to embark upon a radical strategy of opening up the domestic market place, privatising the former General Post Office and establishing an independent regulatory structure separate from British Telecom and the Department of Trade and Industry, made up of the Office of Telecommunications (Oftel) and the Monopolies and Mergers Commission (MMC). In the late 1980s, the French and Dutch responsed in strikingly similar ways. They both safeguarded the monopoly of the PTT on networks and basic services and introduced competition in the peripheral market of equipment and value-added services and separated the regulatory and operational tasks of the PTT. But there were also differences between the two 'followers': while they adopted a more or less similar path of gradual market liberalisation, the emphasis given to privatisation and regulatory reform by the French and Dutch governments differed. The Dutch gave greater priority to a gradual preparation and implementation of privatisation of KPN/PTT Telecom (while keeping the telecommunications and postal services integrated within a holding structure) over establishing a transparent and independent regulatory structure. The French government, relatively early on, established a semi-independent regulatory structure through the Direction de Réglémentation in the early 1990s, the notion of corporatising and privatising France Télécom was highly controversial and could not be carried out immediately. The proposal to grant France Télécom corporate status and to privatise and sell off a minority of its shares was strongly opposed by a majority of its employees, fearful of losing their civil servants' status.

There also occurred a process of cross-societal emulation: the duopoly alternatives in network provision and cellular communications, as suggested and discussed in the Dutch and French case studies, clearly seem to have been inspired by the British experience. The strategy of the French government to create a quasi-vertically integrated telecommunications and electronics industry under state guidance (i.e. France Télécom) seems to have been inspired by the successful American Bell system before its 1984 divestiture. Another similar move made by the three domestic PTOs focused upon in this study, BT, KPN/PTT and France Télécom, was their link up through strategic alliances with American partners to ensure worldwide service provision. The Dutch KPN/PTT corporation joined forces with medium-sized PTOs from Sweden and Switzerland to form Unisource, and with the American giant AT&T, through the Worldsource venture. BT initially linked up with the second American long-distance operator, MCI, in the Concert joint venture, only to be left on its own in 1997 when MCI was acquired by WorldCom. France Télécom joined forces with the Deutsche Bundespost Telekom and the third American long-distance operator, Sprint, in the Global One alliance. Contrasting political decisions were also taken: for instance, while the Thatcher government pursued a radical privatisation and deregulation strategy aimed at enlarging the role of market forces throughout society, the Mitterand-Mauroy government entered on an ambitious path towards nationalisation to bring key industries under state leadership.

The United Kingdom was the first country in Europe to give political weight to the new 'demonopolisation' paradigm when it ferociously implemented a new governance regime in the domestic telecommunications sector, characterised by a privatised operator, a liberalised market (varying between controlled entry in network provision to free entry in services and equipment) and the creation of Oftel as an independent regulatory agency to ensure political objectives such as quality of service, tariff setting and fair competition. The driving forces behind this radical restructuring strategy were the Thatcher and Major governments with their pro-market and anti-government creed, and large business users, who demanded more advanced, efficient and customised communication facilities. The Netherlands and France were well behind the United Kingdom in adopting the new governance regime: more than ten years after the British government had privatised BT and established a market duopoly and the Oftel agency, the Dutch and French governments seriously started to consider network and voice competition and establish an independent regulator. It was not until 1994–5 that the Dutch floated the majority of KPN/PTT's assets and the French did not begin to privatise France Télécom until 1997. Even then they sold off only a first small tranche of the government's shares.

Despite the striking differences between the dates of introducing competition in the core markets of infrastructure provision and voice telephony between the United Kingdom on the one hand and the Netherlands and France on the other, the level of divergence in market liberalisation is significantly lower in the domain of end-user equipment, cellular networks, value-added services and global services provision. In these 'fringe markets' the picture is more varied: Britain was the first to liberalise mobile services and value-added networks, France was the first to have a relatively open market for peripheral equipment and tele-information services, and the Dutch PTT was the first to commit itself to a strategy of international partnerships by linking up with the Swedish, Spanish and Swiss PTTs in the cross-national alliance called Unisource. This, together with the AT&T-dominated Worldsource federation, provides global one-stop-shopping services for multinational businesses. In 1993, the Netherlands started to discuss and prepare a transitional duopoly in the voice telephony and infrastructure markets between the incumbent PTT, the potentially powerful Enertel/NS consortium of local cable operators, and alternative regional and national network operators. Although it was the first on the European continent to consider the two-stage implementation of full competition, the country lost its lead when these plans proved more difficult to implement than expected. Major disagreements between the constituents about strategy to be followed and the persistent impact of the four structural forces mentioned above, called into serious question the feasibility of putting into place an interim regime.

Britain's lead in the implementation of a market-oriented governance regime and its concomitant experiences with new structures is not a measure of its leadership in the European telecommunications industry. The results of the restructuring in Britain are mixed. BT was fully privatised, rationalised and transformed into a profit-driven company, serving the interests of its domestic and international customers and stockholders. As a result of the pro-market policies of the Conservative Thatcher/Major governments, new operators providing alternative network services have entered the UK market and effectively helped to make it one of the most liberal telecommunications markets in the world. Furthermore, a new administrative structure was established, consisting of industry-specific regulation carried out by the independent Oftel body and trans-sectoral competition policy through the Monopolies and Mergers Commission. Notwithstanding its unrivalled degree of innovation and structural reform, the framework of British telecommunications has several shortcomings. The incumbent BT, which was demerged (some of the policy makers originally wanted a US-style divestiture), clearly dominated the domestic telecommunications market. Furthermore, the reduction of BT's workforce by more than half has generated substantial social costs, the charges to residential users have gone up (while BT keeps making huge profits), the domestic manufacturing base has eroded, and there are complaints about the lack of political accountability and transparency of the newly established regulatory system. In the 1970s and 1980s, France carried out an impressive upgrading of its then backward network and effectively caught up with Germany and Britain. In some domains, it even overtook them: the French telecommunications network, for instance, is one of the most advanced in the world. The French government has established a solid and autonomous industrial and technological base with Alcatel and France Télécom now successfully expanding outside France.

The Dutch were right behind the British in opening up their telecommunications market, but they did not consider it feasible to implement fully fledged liberalisation and privatisation, given their conflict-ridden and radical character. After a broad investigation and intensive political discussion in preparing the new legislative framework of 1989, the Dutch pursued a gradual restructuring strategy aimed at achieving liberalisation of the market and the privatisation of KPN/PTT by 1998. Compared to the UK and France, where the telecommunications and postal branches were separated, the Dutch PTT remained integrated through a holding company that was partially privatised in 1994. In the course of 1998 the Dutch will follow the others in divesting KPN's telecommunications branch from its postal activities. Much like France, the Dutch were relatively slow in taking regulatory reform seriously: they sought gradually to establish a transparent administrative framework that would both facilitate and control competition in the core market segments and safeguard public interest. Initially, political support for the public monopoly model was strong and regulation was favourable to the PTT's interests. Clear priority was given to an even development, in which the PTT and government would be given sufficient time and space to restructure and build up the skills required to operate in a more commercial environment, and to accumulate regulatory knowledge and administrative capabilities. The interests of new (or potential) market players and business users were sacrificed in favour of those of the national champion and its patron, in order to facilitate the restructuring and international expansion of KPN/PTT and to prepare its eventual flotation. The Dutch government's meticulous plan to privatise only in the mid-term was aimed at allowing top management and staff sufficient time to adapt to competition and achieve the Treasury's objective of maximising the proceeds from the flotation.

The French followed a 'catching up' approach, aimed at strengthening the position of the domestic telecommunications industry both at home and abroad through the pursuit of interventionist industrial and trade policies. The strategy to close the technology gap with the market leaders relied upon a critical assessment of the long-term trends in the national economy, international trade, and a 'window of opportunity', temporarily created by technological innovation and the opening of new markets (Perez & Soete 1988). This scanning process was followed by a decision-making process concerning the priorities for economic development (i.e. the allocation of government subsidies), the timing of entry into new technology systems, and the formation of collaborative arrangements between the core firm(s) and related and supporting industries. The implementation of these state-led strategies was achieved through indicative planning, the (selective) use of entry barriers to the domestic market, credits provided by state-controlled financial institutions, and ambitious state-sponsored public works (i.e. network modernisation, Transpac, Minitel, Plan Calcul), carried out by privileged domestic companies. After a large-scale introduction in the domestic market, the French government vehemently pushed for the export of these technologies and the international expansion of these 'national champions' through the granting of export subsidies, trade diplomacy and strategic trade policy, and striking international strategic alliances.

This mercantilist strategy proved relatively successful with regard to the modernisation of the French public telecommunications infrastructure in the 1970s and the development and diffusion of new networks and Minitel services aimed at the consumer market. It also contributed to the effective restructuring of the domestic equipment industry by the combined nationalisation and consolidation of the various medium-sized suppliers into the envisaged 'national champion' CGE/Alcatel/CIT, which in the mid-1980s became a world player through the acquisition of the telecommunications branch of ITT. The French strategy proved less successful in the more dynamic markets of advanced services for large business users, provided at home or delivered around the globe. For instance, France's strong commitment to Minitel, due to the heavy investment outlays in network and equipment technologies in the 1980s and nationwide diffusion of Minitel services, might hamper the large-scale adoption of more advanced and globally available Internet services. Compared to the early internationalisers of BT and the Unisource participants, France Télécom (apart from some international activities in 'targeted' technological areas and former French colonies) was relatively late in establishing strategic links with international partners and thus joining the exclusive club of global carriers. One reason for the lack of commercial skills and strategic flexibility seems to be related to the semi-corporatised status of the telecommunications company, the strong resistance to change within France Télécom and the persistent patronage of the Treasury. Furthermore, because of the above-average performance levels in French telecommunications, neither economic nor political demands for a full-speed liberalisation were strong. The French government's plan to corporatise and privatise DGT/France Télécom in 1986–8 and 1993–4 was strongly opposed by its corps of engineers and regular staff, afraid of losing their civil service privileges and anticipating job cuts.

Although we have found a number of similarities between the countries investigated, there are notable differences where the timing and degree of implementation of structural reform measures is concerned. The British 'first-mover' strategy of carrying out privatisation and liberalisation as well as regulatory reform (i.e. combining deregulation with reregulation) in the early 1980s, was followed by a more patient and less radical approach on the part of the Dutch and the French. Besides the Thatcher ideology of private enterprise and free trade, the British response was influenced by a desire to strengthen the comparative advantage of the domestic services industry and attract economic activities from abroad, such as foreign investments and telecommunications traffic. The 'second-mover' strategies of the French and the Dutch, aimed at 'catching up' and 'following closely', respectively, included a more integrated approach to reinforcing the country's industrial/technological base by finding a 'trade-off' between economic objectives and socio-political concerns over safeguarding universal/public service obligations and maintaining employment levels. In France and the Netherlands policy makers attempted to combine a continued state-controlled monopoly on infrastructure provision and voice telephony with the promotion of competition in all other market segments. The quest for a compromise between these various objectives gave way to a lengthy and time-consuming process whereby liberalisation, privatisation and regulatory reform measures were implemented one at a time, to ensure social and political stability. While the British opted for a trajectory of 'creative destruction', emphasising the overhaul of existing technologies, markets and regulations, the Dutch and the French chose the path of socially acceptable competition, in which government would intervene selectively to protect the vital interests of national stakeholders and control domestic markets to accomplish socio-political goals other than open competition and free trade.

Although relatively similar to the British strategy, the policy responses of the Dutch and the French diverged completely with regard to the corporatisation/privatisation of the telecommunications administration, and to a lesser extent the establishment of a new regulatory structure. Whereas in France there existed little support for privatisation but there was enthusiasm for regulatory reform measures, the situation in the Netherlands was the reverse. There, privatisation of the Dutch KPN/PTT caused no major tensions, while the issue of establishing a more transparent regulatory structure independent of both the central government and market players, received only modest political backing.

8.3 Suggestions for further research

The last part of this chapter evaluates the analytical framework of this study and discusses suggestions for further research. Looking at the telecommunications polities of the three countries investigated, we can analyse the 'fit' between the French, Dutch and British telecommunications systems and their national institutional environments. First, there is no clear evidence of 'environmental determinism'. Although the techno-economic and international environment may exert strong stimuli to adjust on national telecommunications regimes, these do not necessarily lead to immediate and uniform policy responses. The policy space and decision-making freedom of the parties involved may be constrained, the critical processes of problem formulation, agenda setting, formation, enforcement and actual implementation of an appropriate policy response are still to a large extent shaped by domestic stakeholders and ruling coalitions. Techno-economic and international pressures both have a different impact upon domestic policy making in the various countries, because national institutions filter out certain issues, and private and public actors bring forth their own set of preferences and strategic alternatives. While the stakeholders in one country might propose the joint implementation of liberalisation, privatisation and regulatory reforms, stakeholders in another might reach a compromise and implement these measures sequentially or partially. As the three case studies have demonstrated, national decision makers were only loosely connected with their environment and – to a certain degree and temporarily – could insulate their domestic telecommunications from the effects of the structural forces of technological (r)evolution, shifting market supply and demand conditions, international deregulation and European integration. There was still substantial manœuvrability left for national governments and private stakeholders in France, the Netherlands and the UK to further their own particular interests and to make strategic choices about which particular path to follow.

Second, the feasibility of pursuing different trajectories becomes constrained over time. For instance, the ambitious nationalisation plans of the Mitterand-Mauroy government were no longer viable in a world economy increasingly characterised by economic interdependence and global capital flows. The timing of deregulation (or entry into growth markets in general) is highly relevant in this respect: rapid introduction of market competition and speedy reorganisation and privatisation of the former PTT administration can give a country a competitive edge, but at high social costs (industrial militancy, layoffs, sectoral uncertainty). Strategic variations and adaptive actions are subject to environmental selection, and, furthermore, some kind of fit between the internal organisation of a system and a particular selection environment to which the system must adapt itself is required in order to secure continuity. The national strategy chosen (i.e. to pioneer, follow or catch up) generally depends on the specific characteristics and capabilities of particular firms (such as the incumbent and new entrants) and other stakeholders in the industry, and the socio-political and economic goals as defined and advocated by the national government. There is no such thing as an ideal telecommunications system or national strategy. In his longitudinal study on the diffusion of the telephone, Flichy (1995:8) noticed clear differences between France and the UK, differences that still apply today: 'Whereas the French conceived of the telephone as an instrument of power, the English saw it more as a commercial tool.' While the British system was probably the most cost-effective and responsive to domestic demand and the French system the most technologically advanced, the Dutch were probably most effective in meeting the different socio-political demands. Through a gradual implementation of corporate and industrial restructuring and compensation for any losses by a combination of regional development, fiscal policy and collaborative export policy, the Dutch government managed to retain socio-economic cohesion.

Third, a social system such as the national telecommunications polity has to cope with changing environmental conditions and make adaptive shifts, while ensuring some degree of internal cohesion. Different environments place varying requirements on the social system: the rapidly changing and increasingly complex environment of present-day telecommunications, for instance, poses different demands on the organisation, compared to those of the stable and well-organised environment of the past. One could qualify UK telecommunications as an open and pluralist system that has accommodated itself to the turbulent environment relatively quickly. The UK system has established reciprocal linkages, tying it to those elements that surround and affect it, facilitating the pursuit of flexible adjustment strategies (e.g. the promotion of transnational solutions) and innovative policies (e.g. the creation of Oftel, duopolies, price controls and asymmetric regulation). The reverse of such an open and permeable system is erosion of the autonomy of, and domestic control over, the system, leaving it increasingly dependent on environmental dynamics (e.g. new entry and foreign direct investment). For instance, while the powers of the British government over BT have substantially diminished over the years, the roles of international partners and competitors, supranational institutions, such as the European Commission, the GATT and global financial investors (such as stockholders), have gained importance. Such an open system as UK telecommunications, that promotes first-mover strategies while running the risk of losing internal cohesion and co-ordination, could be characterised as the 'strength of weak ties' (Granovetter 1973).

The French telecommunications polity seems to reflect a closed and rational system that combines hierarchical control with self-sufficiency and hostility *vis-à-vis* its environment. Benefiting from its effective internal co-ordination mechanism, the French policy community was obsessed with catching up with its foreign competitors through high-risk leapfrogging strategies, while protecting the domestic base with coherent industrial policies (e.g. large R&D and public procurement programmes, the consolidation and/or nationalisation of leading firms). These policies proved extremely successful in the 1970s with the modernisation of its domestic telecommunications infrastructure. As the lengthy and extremely delicate process of privatising France Télécom and the slow diffusion of Internet applications due to the large installed base of outdated Minitels have shown, a pioneer fails to ignore his environment at his peril, as he thereby risks falling behind international competition. This situation also occurred at the end of the nineteenth century, when the French government made large-scale improvements to the state-run canal system and granted huge sailing ship subsidies, at a time when railways and steamships were taking over (Shonfield 1965). A relatively closed and rational system like the one found in France, based on strong domestic co-ordination and control, and to all intents and purposes ignoring the wider environment, could be characterised as the 'weakness of strong ties' .

The Dutch telecommunications polity seems to reflect a relatively open and natural system, that makes for domestic negotiations whilst keeping an eye on changes in the environment. The Dutch policy system, in which environmental shifts filter through, searches for a 'balanced' and modest adaptation to changing techno-economic and international conditions. This process is prepared and implemented carefully so as not to threaten the internal balance of power and the consensus reached in the negotiations in the course of the previous adjustment process. Thus the Netherlands has pursued a second-mover (or follower) strategy, but has lacked the decisive powers to adjust quickly, and therefore runs the risk of falling behind. Such a permeable and natural system as Dutch telecommunications, navigating between the open British approach and the French closed approach, could be characterised as the 'timidity of informal ties'.

This comparative study has predominantly focused on the national level of European states. Further research could focus on an examination of the precise roles of the European Commission, European Parliament and European Court of Justice, tracing the influence and power of non-European (i.e. American and Japanese) stakeholders in developing an EU-wide telecommunications policy. This would involve investigating issues such as the impact of techno-economic leadership and the benefits of foreign direct investment in Europe by American companies, a critical assessment of European strategic trade policies and a benchmarking study comparing the performance levels of the US and EU telecommunications industries. Another relevant research question is whether the emerging European federal regime, in which national regulators and the Commission work together, can learn something from the American policy process in which the Federal Communications Commission, the Anti-trust Division of the Department of Justice, the courts and state-level public utility commissions participate.

After investigating the responses of three distinct European countries, with varying domestic ruling coalitions and institutional structures, to the restructuring process in telecommunications, one could extend the explanatory power of the research framework a little further by looking at sectors with different but nonetheless related profiles. One might, for instance, look at the electricity utilities, postal services, public transport, water distribution, and at European countries comparable in size to the three investigated in this study, such as Belgium, Sweden, Germany, Finland, Switzerland and Italy. Such a broadening of the research could provide evidence on the likelihood of one particular country to follow the same path in restructuring its domestic airlines, energy, postal services, public transport and water distribution as it pursued in telecommunications. If this were the case, it would support the national diversity argument: that divergent strategic responses and policy styles of particular states to the restructuring process do take place in infrastructure industries. Or it might be that the transformation of European public utilities from nationally controlled monopolies into particular industry-based governance regimes (e.g. varying from international oligopoly to regional monopolies) takes place irrespective of national peculiarities and contingencies. If so, then this discovery would support the sectoral convergence thesis.

Such a research agenda could investigate whether dominant national responses exist in the infrastructure domain to structural techno-economic and global forces (i.e. a 'typically' Dutch, French, British or German approach) or whether there is evidence for dominant sectoral responses to the far-reaching shifts in technologies, markets and regulations across nations (i.e. reflecting the structural characteristics of telecommunications, energy, etc.). The institutional framework used to compare telecommunications policy could be refined further by selecting European countries with interesting sectoral and/or national institutional idiosyncrasies. For instance, future research could focus on the Danish and Finnish telecommunications sectors with an institutional legacy of different market characteristics (i.e. a system of regional monopolies and duopoly respectively); public utilities in countries with a federal two-tiered political system (Belgium, Germany); key sectors in national economies with a strong developmental character (Ireland, Spain, Portugal and Greece); countries with excellent or poor public sector performance (the Netherlands *vis-à-vis* Belgium); or compare labour-intensive with capital-intensive public utilities (postal services or railways with telecommunications or energy distribution).

In the late 1990s British Telecom, France Télécom and KPN/PTT Telecom face a business environment that is completely different from that of the early 1980s. Telecommunications is no longer a domestically controlled market with a pervasive role for the PTT administration, central government and national constituents. It is becoming an international oligopoly with the emergence of large diversified firms that link up with local and international partners to form global communications companies and provide a multitude of services. Operating at a variety of market levels, the strategic behaviour of the former PTT administrations, besides serving domestic customers and acting as a local constituency for their home governments, is contingent upon the role of their international business partners (whether AT&T, MCI or Sprint) effectively meeting the demands of large users for global services, and satisfying (future) shareholders. But as those international strategic alliances have shown (Telefonica leaving Unisource for Concert and WorldCom's acquisition of MCI), the relationships are fragile and commitments among the partners of the new global carriers are not yet fully binding.

One could further extend the chosen research model by adding a new variable, namely *company style* (X4) (see Figure 8.1). As a consequence of privatisation, liberalisation and regulatory reform measures, the dominant operator and the telecommunications services industry have become disentangled, thereby calling the relatively autonomous company-level of analysis into being. Firm-specific variables refer to the specific strategies and distinctive capabilities of commercialising and internationalising PTOs and their selection of the key stakeholders for their new business. Former PTTs have gained a substantial independence allowing them to develop their own skills (e.g. technical and marketing expertise, lobbying, partnerships with suppliers and users) and business portfolios (e.g. cost-leadership,

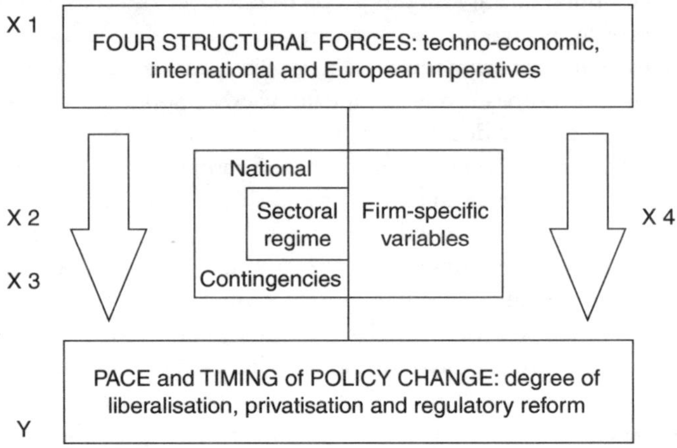

Figure 8.1 Extended research framework

differentiation, diversification), to set priorities on their own (e.g. profit maximisation, increasing market share and/or growth/market) and to choose their strategic partners (both domestic and international). Moreover, challenged by the emergence of new entry and international rivalry, the 'revitalised' telecommunications companies must position themselves strategically to anticipate competition, but they must also accommodate a broader category of stakeholders in their strategic and operational management. In addition to the traditional ones such as government, suppliers and the labour force, the stakeholders of today's PTOs also include stockholders, independent regulators, banks, leading-edge customers and, of course, (potential) competitors and strategic partners.

All these strategic choices might illuminate a particular (emerging) company style, that could cut across the carefully distinguished domestic-sectoral, national and international levels of analysis (as displayed in our research model). Although the discretionary authority of the former European PTTs has increased over the 1980s and the early 1990s, allowing for entrepreneurship, growth, profit maximisation and alliance formation, a substantial part of their routines, business operations and strategies are still shaped by persistent political and economic institutions at the national and (increasingly) EU level. Notwithstanding the impact of technological innovation and the internationalisation of markets and regulations, the (former) PTT administrations still dominate the domestic market. This holds true for the UK, the Netherlands and France. Governments' traditional concern about the public operator, national equipment champion and other important domestic stakeholders has not disappeared, although the relationships have become less overt and direct. The reasons for national government to safeguard the interest and protect the privileges of the PTT/PTO are to be found in fiscal policy (i.e. financial contributions to the Treasury, such as taxes, dividends and proceeds from the flotation), industrial policy (e.g. nationwide infrastructure, employment objectives, regional development) and technology policy (e.g.

electronics manufacturing and service provision as growth engines, and concerns about national independence/security). In a liberalised but still regulated telecommunications sector, governments still show substantial ambiguity between defining and regulating the domestic industry (i.e. licensing, policing the incumbent's pricing and market behaviour) and fostering the successful expansion of their privatised national champions into new product markets and geographical areas (by foreign acquisitions and corporate alliances). New entrants find it extremely difficult to enter this mercantilist web, willingly and unwillingly spun by the incumbent operators and their governments, in which the boundaries between the PTOs' regulated and non-regulated operations and their domestic and foreign activities are still far from transparent.

Telecommunications globalisation, in which governments and PTOs have become actively involved, seems to be characterised by strategic trade policy, in which market access and foreign investments are negotiated and trade deals are made between home and host governments, and between local and global business partners. The incumbent PTOs have succeeded in controlling their external environment by a mix of strategic and tactical moves aimed at maximising the dependency of third parties and minimising their own dependency upon others. Benefiting from their 'historic' assets, such as exclusive rights, a monopolistic market structure, their privileged relationships with key suppliers and leading-edge customers, and their strategic position in the national economy, the PTOs have effectively fended off threats and challenges to their dominant position in both the market place and the policy arena. They have also implemented programmes to enhance technical efficiency, innovation, diversification, and national and global integration, aimed at pre-empting competition and restricting market entry for outsiders. This research has shown that both history and nations play a significant role in the restructuring of firms in a globalising industry such as telecommunications. Despite the permeation of structural forces into domestic public monopoly regimes, the PTOs have managed to retain their solid base: their sunk investments, accumulated capabilities, inherited government privileges and access to new emerging markets of Internet-based services, allowing them to face a more competitive future with confidence, although they may be in for a busy time.

References

ABN AMRO Bank (1993), *European Telecom Services . . . A Call for Liberalisation and Growth*. Amsterdam: ABN AMRO.

AbvaKabo (1984), *Beleids – en Actieprogramma PTT 1984–88*. Zoetermeer: AbvaKabo.

Adams, W.J. and Stoffaes, C. (eds) (1986), *French Industrial Policy*. Washington: Brookings Institution.

Adviesgroep Rathenau (Advisory Committee Rathenau) (1979), *Maatschappelijke Gevolgen van de Micro-electronica. Rapport van de Adviesgroep Rathenau*. The Hague: Staatsuitgeverij.

Albert, M. (1991), *Capitalisme contre Capitalisme*. Paris: Editions du Seuil.

Aldrich, H. (1979), *Organizations and Environments*. Englewood-Cliffs: Prentice Hall.

Ambrosius, G. (1984), *Der Staat als Unternehmer. Öffentliche Wirtschaft und Kapitalismus seit dem 19. Jahrhundert*. Göttingen: Vandenhoeck & Ruprecht.

Amsden, A. (1989), *Asia's Next Giant. South Korea and Late Industrialisation*. New York: Oxford University Press.

Andeweg, R.B. (1994), 'Privatization in the Netherlands: The Results of a Decade.' In V. Wright (ed.), *Privatization in Western Europe. Pressures, Problems and Paradoxes*. London: Pinter, pp. 198–214.

—— (1989), 'Less Than Nothing? Hidden Privatisation of the Pseudo-Private Sector: The Dutch Case.' *West European Politics* 11 (4): 117–28.

Araskog, R.V. (1990), *De Strijd om ITT. Een Verslag uit de Directiekamer*. (Transl. of *The ITT Wars*). Utrecht NL: Veen.

Aronson, J.D. and Cowhey, P.F. (1988), *When Countries Talk: International Trade in Telecommunications Services*. Cambridge MA: Ballinger.

Arthur, B. (1996), 'Increasing Returns and the New World of Business.' *Harvard Business Review* July–August: 100–109.

—— (1988), 'Competing Technologies: An Overview.' In G. Dosi *et al.*, *Technical Change and Economic Theory*. London: Pinter, pp. 590–607.

Atkinson, M.A. and Coleman, W.D. (1989), 'Strong States and Weak States: Sectoral Policy Networks in Advanced Capitalist Economies.' *British Journal of Political Science* 19: 47–67.

Attali, J. and Stourdze, J. (1978), 'The Birth of the Telephone and Economic Crisis: The Slow Death of Monologue in French Society.' In I. de Sola Pool (ed.), *The Social Impact of the Telephone*. Cambridge MA: MIT Press, pp. 97–111.

Aurelle, B. (1986), *Les Télécommunications*. Paris: Editions La Découverte.

Baggott, R. (1989), 'Regulatory Reform in Britain: The Changing Face of Self-Regulation.' *Public Administration* 67: 435–54.

Baldwin, R. (ed.) (1995), *Regulation in Question. The Growing Agenda*. London: LSE/Merck, Sharp & Dohme.

Baldwin, R. and McCrudden, C. (1987), *Regulation and Public Law*. London: Weidenfeld & Nicolson.

Barreau, J. (1995), *La Reforme des PTT. Quel Avenir pour le Service Public*. Paris: Editions La Découverte.

Bartlett, C. and Ghoshal, S. (1991), *Managing Across Borders. The Transnational Solution*. Boston MA: Harvard Business School Press.

Bauer, M. (1988), 'The Politics of State-Directed Privatisation: The Case of France, 1986–88.' *West European Politics* 11: 49–60.

Bauer, M. and Cohen, E. (1981), *Qui Gouverne Les Groupes Industriels?* Paris: Editions du Seuil.

Beesley, M. (1981), *Liberalisation of the Use of British Telecommunications Network*. London: Department of Industry.

Beesley, M. and Littlechild, S. (1983), 'Privatization: Principles, Problems and Priorities.' *Lloyds Bank Review* 149 (July): 1–20.

Bennett, C.J. (1991), 'What is Policy Convergence and What Causes It?' *British Journal of Political Science* 21: 215–33.

Berenschot Consultancy (1985), *Status en Structuur van de PTT. Notitie ten behoeve van de Bijzondere Commissie PTT over Enkele Centrale Vragen naar Aanleiding van de Rapportage van de Commissie Steenbergen*. The Hague: Bureau Berenschot.

Berger, S. and Dore, R. (eds) (1996), *National Diversity and Global Capitalism*. Ithaca: Cornell University Press.

Bertho, C. (1981), *Télégraphes et Téléphones. De Valmy au Mircoprocesseur*. Pans: Le Livre de Poche.

Bessières, H. (1989), 'Bruxelles, Nouvelle Capitale des Télécoms.' *Télécoms Magazine* 29: 42–9.

Best, M. (1990), *The New Competition. Institutions of Industrial Restructuring*. Cambridge: Polity Press.

Blankart, C.B. and Knieps, G. (1989), 'What Can We Learn from Comparative Institutional Analysis? The Case of Telecommunications.' *Kyklos* 42: 579–98.

Bonnetblanc, G. (1985), *Les Télécommunications Françaises, Quel Statut, Pour Quelle Enterprise?* Paris: La Documentation Française.

Bradley, K. (1992), *Phone Wars. The Story of Mercury Communications*. London: Centure Business.

Breton, T. (1994), *Les Téléservices en France. Quels Marchés pour les Autoroutes de l'Information? Rapport au Ministre d'Etat, Ministre de l'Intérieur et de l'Aménagement du Territoire et au Ministre des Enterprises et du Dévelopement Economique*. Paris: La Documentation Française.

Brock, G.W. (1994), *Telecommunication Policy for the Information Age. From Monopoly to Competition*. Cambridge MA: Harvard University Press.

Bruce, R. (1981), 'US Regulatory Change Will Affect All Countries.' *Intermedia* 9 (6): 16–18.

BT (British Telecom) (1981–96), *Annual Reviews/Reports & Accounts*. London: British Telecom.

—— (1993), *Interconnection and Accounting Separation. BT's Response to Oftel's Consultative Document 'Interconnection and Accounting Separation'*. London: British Telecom.

—— (1991a), *The BT Commitment. Setting it Down in Black and White*. London: British Telecom.

—— (1991b), *Going Further Staying Closer. A Different Kind of Communications Company*. London: British Telecom.

—— (1991c), *Serving Telecommunications Customers. BT Plc's Response to the Government's Consultative Document 'Competition and Choice. Telecommunications Policy for the 1990s'*. London: British Telecom.

—— (1990), *Competitive Markets in Telecommunications. Serving Customers*. London: British Telecom.

BTG (Dutch Association of Large Telecommunications Users) (1993), *Commentaar van de BTG op de Hoofdlijnennotitie Wijziging Wet op de Telecommunicatievoorzieningen*. Bussum: BTG.

BTUC (British Telecom Unions Committee) (1991), *Quality and Choice. A Response to the Department of Trade & Industry's Consultative Paper 'Competion and Choice: Telecommunications Policy in the 1990s'*. London: BTUC.

—— (1990), *Telecommunications in the 1990s. A Submission to the Department of Trade & Industry on the Telecommunications Policy Review by the British Telecommunications Unions Committee*. London: BTUC.

—— (1984), *The Battle for British Telecom. The Story of the British Telecom Unions' Fight against the Bill to Privatise BT*. London: BTUC.

—— (1983a), *The American Experience . . . A Report on the Dilemma of Telecommunications in the U.S.A.* London: BTUC.

—— (1983b), *How Selling Off British Telecom Will Harm Consumers, the Blind and Disabled, Poor and Low Income Groups, British Business and the Economy*. London: BTUC.

Buchanan, J.M. (1980), 'Rent Seeking and Profit Seeking.' In J.M. Buchanan, R.D. Tollison and G. Tullock (eds), *Toward a Theory of the Rent-Seeking Society*. College Station: Texas A&M University Press, pp. 3–15.

Bulmer, S., George, S. and Scott, A. (eds) (1992), *The United Kingdom and EC Membership Evaluated*. London: Pinter.

Bunel, J. and Saglio, J. (1984), 'Employers Associations in France.' In J.P. Windmuller and A. Gladstone (eds), *Employers Associations and Industrial Relations. A Comparative Study*. Oxford: Clarendon Press, pp. 232–63.

Burns, T. and Stalker, G.M. (1961), *The Management of Innovation (1994)*. London/Oxford: Tavistock Publications/Oxford University Press.

Burton, J. (1983), *Picking Losers . . .? The Political Economy of Industrial Policy*. London: Institute of Economic Affairs.

BZW (1993), *KPN Koninklijke PTT Nederland. Company Review*. Amsterdam: Barclays de Zoete Wedd.

CA (Consumers' Association) (1991), *Competition and Choice – Telecommunications Policy for the 1990s*. London: Consumers' Association.

Campbell, J.L. and Lindberg, L.N. (1991), 'The Evolution of Governance Regimes.' In J.L. Campbell, J.R. Hollingsworth and L.N. Lindberg (eds), *Governance of the American Industry*. Cambridge: Cambridge University Press, pp. 319–55.

Campbell, J.L., Hollingsworth, J.R. and Lindberg, L.N. (eds) (1991), *Governance of the American Economy*. Cambridge: Cambridge University Press.

Capie, F. and Collins, M. (1992), *Have the Banks Failed Industry? An Historical Survey of Bank/Industry Relations in Britain, 1870–1990*. London: Institute of Economic Affairs.

CAPT (Commission for Advice on Postal and Telecommunications Policy) (1993a), *De WTV in de Steigers. Renovatie of Nieuwbouw*. The Hague: CAPT.

—— (1993b), *Telecommunicatiewetgeving in een Open Markt*. The Hague: CAPT.

Carter Committee (1977a), *Report of the Post Office Review Committee. Presented to Parliament by the Secretary of State for Industry by Command of Her Majesty*. Cmnd 6850. London: HMSO.

—— (1977b), *Appendix to the Report of the Post Office Review Committee. Presented to Parliament by the Secretary of State for Industry by Command of Her Majesty*. Cmnd 6954. London: HMSO.

Cawson, A. (ed.) (1985), *Organized Interests and the State. Studies in Meso-Corporatism*. London: Sage.

Cawson, A., Morgan, K., Webber, D., Holmes, P. and Stevens, A. (1990), *Hostile Brothers. Competition and Closure in the European Electronics Industry*. Oxford: Clarendon Press.

CCTA (Government Centre for Information Systems) (1994), *Information Superhighways: Opportunities for Public Sector Applications in the UK. A Government Consultative Report.* London: CCTA.

CEC (Commission of the European Communities) (1995), *Communication from the Commission. Green Paper on the Liberalisation of Telecommunications Infrastructure and Cable Television Networks. Part Two. A Common Approach to the Provision of Infrastructure for Telecommunications in the European Union.* Com (94)682. Brussels: Commission of the European Communities.

—— (1994), *Communication from the Commission. Green Paper on the Liberalisation of Telecommunications Infrastructure and Cable Television Networks. Part One. Principle and Timetable.* Com (94)440. Brussels: Commission of the European Communities.

—— (1993), *Communication to the Council and the European Parliament on the Consultation on the Review of the Situation in the Telecommunications Services Sector.* Com (93)159. Brussels: Commission of the European Communities.

—— (1992), *1992 Review of the Situation in the Telecommunications Services Sector. Communication by the Commission.* Sec (92)1048. Brussels: Commission of the European Communities.

—— (1988), *Towards a Competitive Community-Wide Telecommunications Market in 1992. Implementing the Green Paper on the Development of the Common Market in 1992.* Com (88)48. Brussels: Commission of the European Communities.

—— (1987), *Green Paper on the Development of the Common Market for Telecommunications Services and Equipment.* Com (87)290. Brussels: Commission of the European Communities.

—— (1985), *Completing the Internal Market: White Paper from the Commission to the European Council.* Com (85)310. Brussels: Commission of the European Communities.

—— (1984), *Communication from the Commission to the Council on Telecommunications.* Com (84)277. Brussels: Commission of the European Communities.

Cecchini, P. with Catinat, M. and Jacquemin, A. (1988), *1992 The European Challenge. The Benefits of a Single Market.* Aldershot: Wildwood House.

Chamoux, J.P. (1993), *Télécoms. La Fin des Privilèges.* Paris: Presses Universitaires de France.

Chandler, A.D. (1990), *Scale and Scope. The Dynamics of Industrial Capitalism.* Cambridge MA: Harvard University Press.

Chapman, C. (1990), *Selling the Family Silver. Has Privatization Worked?* London: Hutchinson.

Charkham, J. (1994), *Keeping Good Company. A Study of Corporate Governance in Five Countries.* Oxford: Oxford University Press.

Charon, J-M. (1987), 'Videotex: From Interaction to Communication.' *Media, Culture and Society* 9: 301–32.

Charon, J-M. and Cherki, E. (1984), 'Télétel, ou l'Abcès de Fixation.' In C. Ancelin and M. Marchand (eds), *Le Vidéotex. Contributions sur la Télématique.* Paris: Masson, pp. 11–21.

Chesnais, F. (1993), 'The French National System of Innovation.' In R. Nelson (ed.), *National Innovation Systems: A Comparative Analysis.* New York: Oxford University Press, pp. 192–229.

Child, J. (1981), 'Culture, Contingency and Capitalism in the Cross-National Study of Organizations.' *Research in Organisational Behaviour* 3: 303–56.

CIB (Centre for Information Policy) (1982), *Telecommunicatie in Nederland. Kansen en Bedreigingen. Reactie van het Centrum voor Informatiebeleid op het Rapport van de Commissie Swarttouw.* The Hague: CIB.

CIB-RCO (Centre for Information Policy/Council of Central Employers' Associations) (1987), *Concept-Commentaar inzake Rapport Zegveld (19-2-1987).* The Hague: CIB.

Clark, D.D. (1996), 'Adding Service Discrimination to the Internet.' *Telecommunications Policy* 20 (3): 169–81.

Cohen, E. (1992), *Le Colbertisme 'High Tech'. Economie des Télécom et du Grand Projet.* Paris: Hachette.

Cohen, E. and Bauer, M. (1985), *Les Grands Manoeuvres Industrielles.* Paris: Belfond.

Coleman, W.D. and Jacek, H.J. (eds) (1989), *Regionalism, Business Interests and Public Policy*. London: Sage.
Commissie Steenbergen (1985a), *Signalen voor Straks. Een Nieuwe Richting voor de PTT.* The Hague: Staatsuitgeverij.
—— (1985b), *Signalen voor Straks. Een Nieuwe Richting voor de PTT. Bijlagen.* The Hague: Staatsuitgeverij.
Commissie Swarttouw (1982), *Taak en Functie van de PTT in het Licht van de Informatieen Telecommunicatietechnologie.* The Hague: Staatsuitgeverij.
Commissie Zegveld (1986), *Op Weg naar Integratie. The Hague: Commissie Integratie Lokale Telecommunicatie-Infrastructuren.* The Hague: Staatsuitgeverij.
Consumentenbond (1993), *Telefoontarieven. Brief aan Ir. de Jong (Lid PvdA Tweede Kamerfractie), d.d. 17-3-1993.* The Hague: Consumers' Association.
Corby, M.E. (1979), *The Postal Business 1969–79. A Study in Public Sector Management.* London: Kogan Page.
Cornford, J. and Gillespie, A. (1993), 'Cable Systems, Telephony and Local Economic Development.' *Telecommuications Policy* 17: 589–602.
Cowhey, P.F. (1990), 'The International Telecommunications Regime: The Political Roots of Regimes for High Technology.' *International Organization* 44: 169–99.
Cowhey, P.F. and Aronson, J.D. (1993), *Managing the World Economy. The Consequences of Corporate Alliances.* New York: Council of Foreign Relations Press.
Crane, R.J. (1979), *The Politics of International Standards. France and the Color TV War.* Norwood NJ: Ablex.
Crouch, C. (1990), 'United Kingdom: The Rejection of Compromise.' In G. Baglioni and C. Crouch (eds), *European Industrial Relations: The Challenge of Flexibility.* London: Sage, pp. 326–55.
Crozier, M. and Friedberg, E. (1977), *L'Acteur et le Système. Les Contraintes de l'Action Collective.* Paris: Editions du Seuil.
Daalder, H. (1966), 'The Netherlands: Opposition in a Segmented Society.' In R.A. Dahl (ed.), *Political Oppositions in Western Democracies.* New Haven: Yale University Press, pp. 188–236.
Dang-N'Guyen, G. (1986), *A European Telecommunications Policy. Which Instruments for Which Prospects?* Brest (F): Ecole Nationale Supérieure des Télécommunications.
Darmon, J. (1985), *Le Grand Dérangement. La Guerre du Téléphone.* Rungis: JC Lattès.
David, P. (1986), 'Understanding the Economics of QWERTY: The Necessity of History.' In W.N. Parker (ed.), *Economic History and the Modern Economist.* Oxford: Basil Blackwell, pp. 30–49.
Davies, A. (1994), *Telecommunications and Politics. The Decentralised Alternative.* London: Pinter.
De Guers, A. (1987), *Casse Avenue de Ségur. La France dans la Guerre des Communications.* Paris: Editions Alain Moreau.
De Jong, H.W. (1992), 'Het Kartelparadijs: De Punten van Bezwaar.' *Economisch-Statistische Berichten* 23-9-1992: 921–7.
—— (1990), 'Nederland: Het Kartelparadijs van Europa?' *Economisch-Statistische Berichten* 14-3-1990: 244–8.
Delabarre, M. (1997), *Sur les Enjeux d'Avenir pour France Télécom. Rapport au Premier Ministre, au Ministre de l'Economie, des Finances et de l'Industrie, au Secrétaire d'Etat à l'Industrie.* Paris: Ministère de l'Economie des Finances et de l'Industrie.
Demarigny, F. (1996), 'Independent Administrative Authorities in France and the Case of the French Council for Competition.' In G. Majone (ed.), *Regulating Europe.* London: Routledge, pp. 157–79.

Derthick, M. and Quirk, P.J. (1985), *The Politics of Deregulation*. Washington: Brookings.

De Ru, H.J. (1992), 'De Lange Weg naar Privatisering in Nederland.' *TVVS* 1992/5: 111–16.

—— (1988), 'Die Nieuwe Wetgering inzake de PTT: Regulering Sui Generis voor een NV.' *Regelmaat* 1888/2: 115–20.

—— (1985), 'Public Enterprise in the Netherlands. A Tradition in Privatization.' *Annals of Public and Cooperate Economy* 56: 313–41.

De Swaan, A. (1988), *In Care of the State. Health Care, Education and Welfare in Europe and the USA in the Modern Era*. Cambridge: Polity Press.

De Zoete & Bevan (1985), *British Telecom. Post Privatisation*. London: De Zoete & Bevan.

—— (1984), *British Telecom*. London: De Zoete & Bevan.

DGPT (Direction Générale des Postes et Télécommunications) (1994a), *Rapport d'Activite 1994*. Paris: Ministères des Technologies de l'Information et de la Post.

—— (1994b), *Rapport d'Activite 1994. Annexes*. Paris: Ministères des Technologies de l'Information et de la Post.

DGT (1980–88), *Annual Reports*. Paris: Direction Générale des Télécommunications.

DiMaggio, P.J. and Powell, W.W. (1983), 'The Iron Cage Revisited: Institutional Isomorphism and Collective Rationality in Organizational Fields.' *American Sociological Review* 48: 147–60.

DoI (Department of Industry) (1982a), *A Programme for Advanced Information Technology*. The Report of the Alvey Committee. London: HMSO.

—— (1982b), *The Future of Telecommunications in Britain*. Cmnd 8610. London: HMSO.

—— (1978), *The Post Office. Presented to Parliament by the Secretary of State for Industry by Command of Her Majesty*. Cmnd 7292. London: HMSO.

Dommering, E.J. (1988), 'Gaat de PTT volgens Nieuwe Wetten Werken?' *Informatierecht AMI* 1988 (1): 3–9.

Douglas, M. (1986), *How Institutions Think*. Syracuse: Syracuse University Press.

Dreyfus, F-G., Morizet, J. and Peyrard, M. (eds) (1993), *France and EC Membership Evaluated*. London: Pinter.

DTI (Department of Trade and Industry) (1994), *Creating the Superhighways of the Future: Developing Broadband Communications in the UK*. London: HMSO.

—— (1991), *Competition and Choice: Telecommunications Policy for the 1990s*. Cmnd 1461. London: HMSO.

—— (1990), *Competition and Choice: Telecommunications Policy for the 1990s*. Cmnd 1303. London: HMSO.

Dumez, H. and Jeunemaitre, A. (1994), 'Privatization in France: 1983–1993.' In V. Wright (ed.), *Privatization in Western Europe. Pressures, Problems and Paradoxes*. Pinter, pp. 83–104.

Dyson, K.H.F. (1983), 'The Cultural, Ideological and Structural Context.' In J. Dyson and S. Wilks (eds) (1983), *Industrial Crisis. A Comparative Study of the State and Industry*. Oxford: Martin Robertson, pp. 26–66.

—— (1980), *The State Tradition in Western Europe. A Study of an Idea and an Institution*. Oxford: Martin Robertson.

Dyson, K. and Humphreys, P. (eds) (1990), *The Political Economy of Communications. International and European Dimensions*. London: Routledge.

—— (eds) (1986), *The Politics of the Communications Revolution in Western Europe*. London: Frank Cass.

ECTUA (European Telecommunications Users' Association) (1987), 'European Telecoms Users Credo.' *Transnational Data and Communications Report* October: 8.

Ergas, H. (1992), *'France Telecom: Has the Model Worked?'* A paper for a seminar organised by the Royal Norwegian Council for Scientific & Industrial Research, Oslo, 29 January 1992.

—— (1987), 'The Importance of Technology Policy.' In P. Dasgupta and P. Stoneman (eds), *Economic Policy and Technological Performance.* Cambridge: Cambridge University Press, pp. 51–96.

—— (1986), 'Does Technology Policy Matter?' *CEPS Paper No. 29*. Brussels: Centre for European Policy Studies.

—— (1984), *Why Do Some Countries Innovate More Than Others? CEPS Paper No. 5*. Brussels: Centre for European Policy Studies.

—— (1983), *Industrial Policy in France: The Case of Telecommunications.* Paper for a seminar on 'Industrial Policy and Structural Adjustment'. Naples, 21–22 April 1983.

ERT (European Roundtable of Industrialists) (1994), *Building the Information Highways to Re-engineer Europe. A Message from Industrial Users*. Brussels: European Roundtable of Industrialists.

—— (1986), *Clearing the Lines. A User's View on Business Communications in Europe.* Paris: European Roundtable of Industrialists.

—— (1985), *Changing Scales. A Review Prepared for the Roundtable of European Industrialists.* Brussels: European Roundtable of Industrialists.

Evans, P. (1995), *Embedded Autonomy. States and Industrial Transformation.* Princeton: Princeton University Press.

Feick, J. (1992), 'Comparing Comparative Policy Studies – A Path Towards Integration?' *Journal of Public Policy* 12: 257–85.

Fisher, F.M., McGowan, J.J. and Greenwood, J.E. (1983), *Folded, Spindled and Mutilated: Economic Analysis and U.S. vs. IBM.* Cambridge MA: MIT Press.

Flemming, J. (ed.) (1996), *The Report of the Commission on the Regulation of Utilities.* London: Hansard Society.

Flichy, P. (1995), *Dynamics of Modern Communication. The Shaping and Impact of New Communication Technologies*. London: Sage.

Foremann-Peck, J. and Müller, J. (eds) (1988), *European Telecommunication Organisations*. Baden-Baden: Nomos.

Foster, C.D. (1992), *Privatization, Public Ownership and the Regulation of Natural Monopoly.* Oxford: Basil Blackwell.

Fox, A. (ed.) (1986), *State, Finance and Industry. A Comparative Analysis of Post-War Trends in Six Advanced Industrial Economies.* Brighton: Wheatsheaf.

France Télécom (1989–96), *Annual Reports.* Paris: France Télécom.

Freeman, C. and Lundvall, B-A. (eds) (1988), *Small Countries Facing the Technological Revolution.* London: Pinter.

Freeman, G.P. (1985), 'National Styles and Policy Sectors: Explaining Structured Variation.' *Journal of Public Policy* 5: 467–96.

Fukuyama, F. (1995), *Trust: The Social Virtues and the Creation of Prosperity.* London: Hamish Hamilton.

Gamble, A. (1994), *The Free Economy and the Strong State. The Politics of Thatcherism* (2nd rev. edn.). Houndsmills: Macmillan.

Garnham. N. (1990), 'Telecommunications in the UK.' *Fabian Society Discussion Paper No. 1.* London: Fabian Society.

—— (1985), 'Telecommunications Policy in the United Kingdom.' *Media, Culture & Society* 7: 7–29.

Garnham, N. and Mansell, R. (1991), *Universal Service and Rate Restructuring in Telecommunications.* Paris: OECD.

Geelhoed, L.A. (1990), 'De Semi-Souvereine Staat.' *Socialisme en Democratie* 47 (2): 40–7.

—— (1989), 'De Agenda van de Begrotingswetgever in de Jaren Negentig.' *Beleid en Maatschappij* 16: 223–9.

Gerschenkron, A. (1962), *Economic Backwardness in Historical Perspective. A Book of Essays.* Cambridge MA: Belknap Press.

Gilpin, R. (1987), *The Political Economy of International Relations.* Princeton: Princeton University Press.

—— (1975), *U.S. Power and the Multinational Corporation. The Political Economy of Foreign Direct Investment.* New York: Basic Books.

Goetschy, J. and Rozenblatt, P. (1992), 'France: The Industrial Relations System at a Turning Point?' In A. Ferner and R. Hyman (eds), *Industrial Relations in the New Europe.* Oxford: Basil Blackwell, pp. 404–44.

Gombrich, E.H. (1979), 'Style.' In D.L. Sills (ed.), *International Encyclopedia of the Social Sciences.* Volume 15. New York: Macmillan/Free Press, pp. 352–61.

Gourevitch, P. (1986), *Politics in Hard Times. Comparative Responses to International Economic Crises.* Ithaca: Cornell University Press.

Grabher, G. (ed.) (1993), *The Embedded Firm. On the Socio-Economics of Industrial Networks.* London: Routledge.

Graham, C. and Prosser, T. (1991), *Privatizing Public Enterprises. Constitutions, the State, and Regulation in Comparative Perspective.* Oxford: Clarendon Press.

Grande, E. and Schneider, V. (1991), 'Reformstrategien und Staatliche Handlungskapazitäten. Eine Vergleichende Analyse Institutionellen Wandels in der Telekommunikation in Westeuropa.' *Politische Vierteljahresschrift* 32: 452–78.

Granovetter, M. (1992), 'Economic Institutions as Social Constructions: A Framework for Analysis.' *Acta Sociologica* 35: 3–11.

—— (1985), 'Economic Action and Social Structure: The Problem of Embeddedness.' *American Journal of Sociology* 91: 481–510.

—— (1973), The Strength of Weak Ties. *American Journal of Sociology* 78: 1360–80.

Grant, W. (1984), 'Large Firms and Public Policy in Britain.' *Journal of Public Policy* 4: 1–17.

Grant, W. with Sargent, J. (1991), *Business and Politics in Britain (1987).* Houndsmills: Macmillan.

Grant, W., Paterson, W. and Whitston, C. (1989), *Government and the Chemical Industry. A Comparative Study of Britain and West Germany.* Oxford: Clarendon Press.

Greenwood, J. and Jordan, G. (1993), 'The United Kingdom: A Changing Kaleidoscope.' In M.P.C.M. Van Schendelen (ed.), *National Public and Private EC Lobbying.* Aldershot: Dartmouth, pp. 65–89.

Greenwood, J., Grote, J.R. and Ronit, K. (1992), *Organised Interests and the European Community.* London: Sage.

Groenewegen, J. (1989), *Planning in een Markteconomie. Indicatieve Planning, Industriebeleid en de Rol van de Publieke Onderneming in Frankrijk in de Periode 1981–1986.* Delft: Eburon.

Groenewegen, J. and Beije, P.R. (1989), 'The French Communications Industry Defined and Analyzed Through the Social Fabric Matrix, the Filière Approach, and Network Analysis.' *Journal of Economic Issues* 23: 1059–74.

Hage, J. and Clignet, R. (1982), 'Coordination Styles and Economic Growth.' *The Annals of the American Academy of Political and Social Sciences* 459: 77–92.

Hagedoorn, J. and Schakenraad, J. (1992), 'Leading Companies and Networks of Strategic Alliances in Information Technologies.' *Research Policy* 21: 163–90.

Hall, M. (1987), *The City Revolution. Causes and Consequences.* Houndsmills: MacMillan.

Hall, P. (1993), 'Policy Paradigms, Social Learning, and the State. The Case of Economic Policymaking in Britain.' *Comparative Politics* 25: 275–96.

—— (ed.) (1989), *The Political Power of Economic Ideas: Keynesianism Across Nations*. Princeton: Princeton University Press.

—— (1986), *Governing the Economy: The Politics of State Intervention in Britain and France*. New York: Oxford University Press.

Hampden-Turner, C. and Trompenaars, F. (1995), *The Seven Cultures of Capitalism. Value Systems for Creating Wealth in the United States, Britain, Japan, Germany, France, Sweden, and the Netherlands*. London: Piatkus.

Hancher, L. (1996), 'The Regulatory Role of the European Union.' In H. Kassim and A. Menon (eds), *The European Union and National Industrial Policy*. London: Routledge, pp. 52–69.

Hancher, L. and Moran, M. (eds) (1989), *Capitalism, Culture and Economic Regulation*. Oxford: Clarendon Press.

Hannan, M.T. and Freeman, J. (1984), 'Structural Inertia and Organizational Change.' *American Sociological Review* 49: 149–64.

Harper, J.M. (1997), *Monopoly and Competition in British Telecommunications. The Past, the Present and the Future*. London: Pinter.

—— (1995), 'A 21st Century Structure for Telecommunications.' *Communications & Strategies* 17 (1): 55–76.

—— (1993), 'Shot in the Foot. UK Telecoms Policy.' In M. Christoffersen and A. Henten (eds), *Telecommunication. Limits to Deregulation*. Amsterdam: IOS Press, pp. 25–36.

Hart, J.A. (1992), *Rival Capitalists. International Competitiveness in the United States, Japan and Western Europe*. Ithaca: Cornell University Press.

Hawkings, F. (1987), 'Selling British Telecom.' *Policy Studies* 7 (3): 1–20.

Hayek, F. (1944), *The Road to Serfdom*. Chicago: University of Chicago Press.

Hayward, J. (1986), *The State and the Market Economy. Industrial Patriotism and Economic Intervention in France*. Brighton: Wheatsheaf.

HBS (Harvard Business School) (1989), *French Telecommunications in the 1980s (B). Case Study 9-389-037*. Boston MA: Harvard Business School Press.

—— (1988), *French Telecommunications in the 1980s (A). Case Study 9-388-160*. Boston MA: Harvard Business School Press.

Heald, D. (1988), 'The United Kingdom: Privatization and its Political Context.' West European Politics 11 (4): 31–48.

Heidenheimer, A.J., Heclo, H. and Adams, C.T. (1990), *Comparative Public Policy. The Politics of Social Choice in America, Europe and Japan* (3rd edn.) (1975). New York: St Martin's Press.

Hemerijck, A.C. (1994), 'Hardnekkigheid van Corporatistisch Beleid in Nederland.' *Beleid en Maatschappij* 21 (1/2): 23–47.

High Level Group on the Information Society ('Bangemann Group') (1994), *Europe and the Global Information Society*. Recommendations to the European Council. Brussels: Commission of the European Communities.

Hills, J. (1986), *Deregulating Telecoms: Competition and Control in the United States, Japan and Britain*. London: Pinter.

—— (1984), *Information Technology and Industrial Policy*. London: Croom Helm.

Hins, A.W. (1991), *Ontvangstvrijheid en Buitenlandse Omroep*. Deventer: Kluwer.

Hirschman, A.O. (1970), *Exit, Voice and Loyalty. Responses to Decline in Firms, Organizations and States*. Cambridge MA: Harvard University Press.

Hirst, P. and Thompson, G. (1996), *Globalization in Question. The International Economy and the Possibilities of Governance*. Cambridge: Polity Press.

HO/DoI (Home Office/Department of Industry) (1983), *The Development of Cable Systems and Services. Presented to Parliament by the Secretary of State for the Home Department and the Secretary of State for Industry*. Cmnd 8866. London: HMSO.

Hogwood, B. (1994), 'A Reform Beyond Compare? The Next Steps Restructuring of British Central Government.' *Journal of European Public Policy* 1 (1): 71–94.

Holcombe, A.N. (1911), *Public Ownership of Telephones on the Continent of Europe*. Boston: Houghton Mifflin.

Hollingsworth, J.R., Schmitter, P.C. and Streeck, W. (eds) (1994), *Governing Capitalist Economies. Performance and Control of Economic Sectors*. New York: Oxford University Press.

Hu, Y-S (1992), 'Global or Stateless Corporations Are National Firms with International Operations.' *California Management Review* Winter: 107–26.

Hulsink, W. (1994), 'From State Monopolies to Euro-Nationals and Global Alliances: The Case of the European Telecommunications Sector.' In J. van Dijck and J. Groenewegen (eds), *Changing Business Systems in Europe: An Institutional Approach*. Brussels: VUB Press, pp. 455–84.

Humphreys, P.J. (1990a), 'The Political Economy of Telecommunications in France: A Case Study of "Telematics".' In K. Dyson and P. Humphreys (eds), *The Political Economy of Communications. International and European Dimensions*. London: Routledge, pp. 198–228.

—— (1990b), 'The International Political Economy of the Communications Revolution: The Case for a Neo-Pluralist Approach.' *Government and Opposition* 25: 497–518.

Hutton, W. (1996), *The State We're In*. First published 1995. London: Vintage.

Hyman, H. (1989), 'Privatisation: The Facts.' In C. Veljanovski (ed.), *Privatisation & Competition. A Market Prospectus*. London: IEA, pp. 191–217.

Irwin, M.R. and Merende, M.J. (1989), 'Corporate Networks, Privatization and State Sovereignty. Pending Issues for the 1990s.' *Telecommunications Policy* 13: 329–35.

Jacquemin, A. and Wright, D. (eds) (1993), *The European Challenges Post-1992. Shaping Factors, Shaping Actors*. Aldershot: Edward Elgar.

Jänicke, M. (1990), *State Failure. The Impotence of Politics in Industrial Society*. Cambridge: Polity Press.

Jannès, H. (1970), *Le Dossier Secret du Téléphone*. Paris: Flammarion.

—— (1966), *Le Progrès Enchaîné. Téléphone - Logement*. Paris: Calmann-Lévy.

Johnson, C. (1982), *MITI and the Japanese Miracle. The Growth of Industrial Policy, 1925-1975*. Stanford: Stanford University Press.

Jordan, G. (1990), 'Sub-Governments, Policy Communities and Networks. Refilling the Old Bottles?' *Journal of Theoretical Politics* 2 (3): 319–38.

—— (1981), 'Iron Triangles, Woolly Corporatism and Elastic Nets: Images of the Policy Process.' *Journal of Public Policy* 1: 95–123.

Jordan, G. and Richardson, J. (1982), 'The British Style or the Logic of Negotiation?' In J. Richardson (ed.), *Policy Styles in Western Europe*. London: Allen & Unwin.

Katzenstein, P.J. (1985), *Small States in World Markets. Industrial Policy in Europe*. Ithaca: Cornell University Press.

—— (1978), 'Domestic Structures and Strategies of Foreign Economic Policy.' In P.J. Katzenstein (ed.), *Between Power and Plenty. Foreign Economic Policies of Advanced Industrial States*. Madison: University of Wisconsin Press, pp. 295–336.

Kay, J. (1984), 'The Privatization of British Telecommunications.' In D. Steel and D. Heald (eds), *Privatising Public Enterprises*. London: RIPA, pp. 77–85.

Keen, P.G.W. and Cummins, J.M. (1994), *Networks in Action. Business Choices and Telecommunications Decisions*. Belmont: Wadsworth.

Kelman, S. (1981), *Regulating America, Regulating Sweden: A Comparative Study of Occupational Safety and Health Policy*. Cambridge MA: MIT Press.

Keohane, R.O. (1984), *After Hegemony. Cooperation and Discord in the World Political Economy.* Princeton: Princeton University Press.

Kingdon, J.W. (1984), *Agendas, Alternatives and Public Policies*. Boston: Little, Brown and Company.

Kogut, B. (1991), 'Country Capabilities and the Permeability of Borders.' *Strategic Management Journal* 12: 33–47.

KPN Royal PTT Netherlands (1989–96), *Annual Reports*. Groningen: KPN.

Krasner, S.D. (1988), 'Sovereignty. An Institutional Perspective.' *Comparative Political Studies* 21: 66–94.

Labour Party (1983), *The New Hope for Britain. Labour's Manifesto.* London: Labour Party.

Landes, D.S. (1949), 'French Entrepreneurship and Industrial Growth in the Nineteenth Century.' *Journal of Economic History* 9: 45–61.

Lane, C. (1989), *Management and Labour in Europe. The Industrial Enterprise in Germany, Britain and France.* Aldershot: Edward Elgar.

Lane, J-E. and Ersson, S.O. (1991), *Politics and Society in Western Europe*. (2nd edn.) London: Sage.

Lash, S. and Urry, J. (1987), *The End of Organized Capitalism*. Cambridge: Polity Press.

Lasserre, B. (ed.) (1994), *Consultation Publique Organisée sur l'Evolution de la Réglementation Française des Télécommunications.* Paris: Ministère de la Poste, des Télécommunications et de l'Espace/DGPT.

Lehmbruch, G. (1984), 'Concertation and the Structure of Corporatist Networks'. In J.H. Goldthorpe (ed.), *Order and Conflict in Contemporary Capitalism.* Oxford: Clarendon Press, pp. 60–80.

—— (1982), 'Introduction: Neo-corporatism in Comparative Perspective.' In G. Lehmbruch and P.C. Schmitter (eds), *Patterns of Corporatist Policy-making*. London: Sage.

Lehmbruch, G. and Schmitter, P.C. (eds) (1982), *Patterns of Corporatist Policy-Making*. London: Sage.

Leray, A. (1994), *Les Télécoms en Questions. Privatisation ou Service Public?* Paris: Les Editions de l'Atelier.

Libois, L-J. (1983), *Genèse et Croissance des Télécommunications.* Paris: Masson.

Lijphart, A. (1984), *Democracies. Patterns of Majoritarian and Consensus Government in Twenty-One Countries*. New Haven: Yale University Press.

—— (1975), *The Politics of Accommodation. Pluralism and Democracy in the Netherlands*. Berkeley: University of California Press.

Lindberg, L.N. and Campbell, J.L. (1991), 'The State and the Organization of Economic Activity.' In J.L. Campbell, J.R. Hollingsworth and L.N. Lindberg (eds), *The Governance of the American Economy*. Cambridge: Cambridge University Press, pp. 356–95.

Lindberg, L.N., Campbell, J.L. and Hollingsworth, J.R. (1991), 'Economic Governance and the Analysis of Structural Change in the American Economy.' In J.L. Campbell, J.R. Hollingsworth and L.N. Lindberg (eds), *The Governance of the American Economy*. Cambridge: Cambridge University Press, pp. 3–34.

Lindblom, C.E. (1977), *Politics and Markets. The World's Political-Economic Systems.* New York: Basic Books.

Littlechild, S.C. (1983), *Regulation of British Telecommunications' Profitability.* London: Department of Industry.

—— (1981), 'Ten Steps to Denationalisation.' *Economic Affairs* 2 (1): 11–19.

—— (1978), *The Fallacy of the Mixed Economy.* London: Institute of Economic Affairs.

Lodge, G.C. (1987), 'Introduction: Ideology and Country Analysis.' In G.C. Lodge and E.F. Vogel (eds), *Ideology and National Competitiveness: An Analysis of Nine Countries.* Boston MA: Harvard Business School Press, pp. 1–28.

Longuet, G. (1988), *Télécoms. La Conquête de Nouveaux Espaces.* Paris: Dunod.

Lorenzi, J-H. and Le Boucher, E. (1979), *Mémoires Volées.* Paris: Ramsay.

Lukes, S. (1979), *Power: A Radical View (1974).* London: Macmillan.

Lyon-Caen, A. (1993), 'The French Case.' *Bulletin of Comparative Labour Relations (Special Issue 'Industrial Relations Developments in the Telecommunications Industry')* 25: 387–412.

McKendrick, G. (1987), 'The INTUG View on the Green Paper.' *Telecommunications Policy.* Volume II (December): 325–9.

MacKie-Mason, J., Shenker, S. and Varian, H.R. (1996), 'Service Architecture and Content Provision: The Network Provider as Editor.' *Telecommunications Policy* 20 (3): 203–17.

McKinsey & Company (1993), *Telecommunicatie in Nederland: Op weg Naar Wereldklasse.* Amsterdam: McKinsey & Company.

—— (1985), *Appendices. Signalen voor Straks. Een Nieuwe Richting voor de PTT.* Amsterdam: McKinsey & Company.

Majone, G. (1997), 'From the Positive to the Regulatory State: Causes and Consequences of Changes in the Mode of Governance.' *Journal of Public Policy* 17: 139–67.

—— (1994a), 'The Rise of the Regulatory State in Europe.' *West European Politics* 17(3): 77–101.

—— (1994b), 'Paradoxes of Privatization and Deregulation.' *Journal of European Public Policy* 1: 53–69.

—— (1992), 'Ideas, Interests and Policy Change.' *EUI Working paper SPS No. 92/21.* Florence.

—— (1991), 'Cross-National Sources of Regulatory Policy-Making in Europe and the United States.' *Journal of Public Policy* 11: 79–106.

—— (1989a), *Evidence, Argument, & Persuasion in the Policy Process.* New Haven: Yale University Press.

—— (1989b), 'Regulating Europe: Problems and Prospects.' In T. Ellwein *et al.* (eds), *Jahrbuch zur Staats- und Verwaltungswissenschaft* 3: 159–77.

Mansell, R. (1993), *The New Telecommunications. A Political Economy of Network Evolution.* London: Sage.

March, J.G. (1989), *Decisions and Organizations.* Oxford: Basil Blackwell.

March, J.G. and Olsen, J.P. (1989), *Rediscovering Institutions. The Organizational Basis of Politics.* New York: Free Press.

Mattelart, A. and Stourdze, Y. (1985), *Technology, Culture and Communication. A Report to the French Minister of Research and Industry.* (Transl. of *'Technologie, Culture et Communication. Réport pour la Ministère de la Récherche et de l'Industrie 1982).* Amsterdam: North-Holland.

Maurice, M., Sorge, A. and Warner, M. (1980), 'Societal Differences in Organizing Manufacturing Units: A Comparison of France, West Germany, and Great Britain.' *Organization Studies* 1 (1): 59–86.

Mercury (1993), *Interconnection and Accounting Separation. Submission by Mercury Communications.* London: Mercury Communications.

—— (1991), *Competition and Choice: Telecommunications Policy for the 1990s. The Mercury Reponse: Philosophy and Critical Issues.* London: Mercury Communications.

Mescheriakoff, A-S. (1991), *Droit des Services Publics.* Paris: Presses Universitaires de France.

Milner, H.V. and Yoffie, D.B. (1989), 'Between Free Trade and Protectionism: Strategic Trade Policy and a Theory of Corporate Trade Demands.' *International Organization* 43: 239–72.

Mission à la Réglementation (1988), *Rapport au Ministre.* Paris: Ministry of Post and Telecommunications.

MMC (Monopolies and Mergers Commission) (1986), *British Telecommunications Plc and Mitel Corporation. A Report on the Proposed Merger.* Cmnd 9715. London: HMSO.

Moon, J., Richardson, J.J. and Smart, P. (1986), 'The Privatisation of British Telecom: A Case Study of the Extended Process of Legislation.' *European Journal of Political Research* 14: 339–55.

Moore, J. (1986), 'Why Privatise (1983)?' In J. Kay, C. Mayer and D. Thompson (eds), *Privatisation and Regulation. The UK Experience.* Oxford: Clarendon Press, pp. 78–93.

Morgan, K., Harbor, B., Hobday, M., von Tunzelmann, N. and Walker, W. (1989), *The GEC-Siemens Bid for Plessey: The Wider European Issues. Working Paper No.2.* Brighton: CICT/SPRU.

MRI (Ministère de la Recherche et de l'Industrie) (1982), *Une Politique Industrielle pour La France. Actes des Journées de Travail des 15 et 16 Novembre 1982.* Paris: La Documentation Française.

MTIP (Ministère des Technologies de l'Information et de la Poste) (1995), *De Nouvelles Règles du Jeu pour les Télécommunications en France.* Paris: MTIP.

Mueller, D.C. (1989), *Public Choice II.* Cambridge: Cambridge University Press.

Mueller, F. (1994), 'Societal Effect, Organizational Effect and Globalization.' *Organization Studies* 15 (3): 407–28.

Müller, J. and Toker, S. (1994), 'Mobile Communications in Europe.' In C. Steinfield, J.M. Bauer and L. Caby (eds), *Telecommunications in Transition. Policies, Services and Technologies in the European Community.* Thousand Oaks: Sage, pp. 182–203.

Murphy, R. (1988), *Social Closure. The Theory of Monopolization and Exclusion.* Oxford: Clarendon.

Mytelka, L.K. (ed.) (1991), *Strategic Partnerships and the World Economy. States, Firms and International Competition.* London: Pinter.

NCC (National Consumer Council) (1993), *Paying the Price. A Consumer View of Water, Gas, Electricity and Telephone Regulation.* London: National Consumer Council.

—— (1991), *Telecommunications for UK Consumers. Response from the National Consumer Council to the Consultative Document 'Competition and Choice: Telecommunications Policy for the 1990s'.* London: National Consumer Council.

—— (1987), 'BT Since Privatisation – Is the Consumer Getting Better Service?' *Consumer Voice* Summer II: 1–2.

NCU (National Communications Union) (1992), *Telecoms Privatization: The British Experiment.* Geneva: PTTI.

—— (1985), *The Broad Strategy. A New Concept.* London: National Communications Union.

Nelson, R.R. (ed.) (1993), *National Innovation Systems. A Comparative Analysis.* New York: Oxford University Press.

—— (1984), *High-Technology Policies: A Five Nation Comparison.* Washington: American Enterprise Institute.

Nelson, R.R. and Winter, S.G. (1982), *An Evolutionary Theory of Economic Change.* Cambridge MA: Belknap Press.

Newman, K. (1986), *The Selling of British Telecom.* London: Holt, Rinehart and Winston.

Noam, E.M. (1992), *Telecommunications in Europe.* New York: Oxford University Press.

—— (1987), 'The Public Telecommunications Network: A Concept in Transition.' *Journal of Communication* 37: 30–48.

Nora, S. and Minc, A. (1978), *L'informatisation de la Société. Rapport à M. le Prèsident de la République.* Paris: La Documentation Française.

North, D.C. (1990), *Institutions, Institutional Change and Economic Performance.* Cambridge: Cambridge University Press.

NOTA (Netherlands Organisation for Technology Assessment) (1993), *Agenda for the Public Discussion on Telecommunications.* The Hague: NOTA.

Nouvion, M. (1982), *L'Automatisation des Télécommunications. La Mutation d'une Administration.* Lyon: Presses Universitaires de Lyon.

OECD (Organisation for Economic Co-operation and Development) (1997), *Communications Outlook 1997.* Paris: OECD.

—— (1996), *Information Infrastructure Convergence and Pricing: The Internet.* OCDE/GD(96)73. Paris: OECD.

—— (1995), *Communications Outlook 1995.* Paris: OECD.

—— (1992a), *Telecommunications and Broadcasting: Convergence or Collision?* ICCP 29. Paris: OECD.

—— (1992b), *Convergence Between Communications Technologies: Case Studies from North America and Western Europe.* ICCP 28. Paris: OECD.

Offe, C. (1981), 'The Attribution of Public Status to Interest Groups: Observations on the West German Case.' In S. Berger (ed.), *Organizing Interest in Western Europe. Pluralism, Corporatism and the Transformation of Politics.* Cambridge MA: Cambridge University Press, pp. 123–58.

Oftel (Office of Telecommunications) (1997), *Domestic Obligations in a Global Market.* London: Office of Telecommunications.

—— (1996a), *Promoting Competition in Services over Telecommunications Networks.* London: Office of Telecommunications.

—— (1996b), *Pricing of Telecommunications Services.* London: Office of Telecommunications.

—— (1995), *Effective Competition: Framework for Action.* London: Office of Telecommunications.

—— (1994a), *Interconnection and Accounting Separation: The Next Steps. Statement Issued by the Director General of Telecommunications.* London: Office of Telecommunications.

—— (1994b), *A Framework for Effective Competition. A Consultative Document on the Future of Interconnection and Related Issues.* London: Office of Telecommunications.

—— (1993), *Interconnection and Accounting Separation. Consultative Document Issued by the Director General of Telecommunications.* London: Office of Telecommunications.

—— (1992), *Policy on Separation and Interconnection. A Statement by the Director General of Telecommunications.* London: Office of Telecommunications.

—— (1991), *Competition and Choice: Telecommunications Policy for the 1990s. Statement from the Director General of Telecommunications.* London: Office of Telecommunications.

—— (1987), *British Telecom's Quality of Service. Statement of the Director General of Telecommunications.* London: Office of Telecommunications.

—— (1985a), *British Telecom's Procurement of Digital Exchanges. A Report by the Director General of Telecommunications.* London: Office of Telecommunications.

—— (1985b), *Determination of Terms and Conditions . . . under Section 7 of the Telecommunications Act 1984.* London: Office of Telecommunications.

Ohmae, K. (1990), *The Borderless World. Power and Strategy in the Interlinked Economy.* London: HarperCollins.

—— (1985), *Triad Power: The Coming Shape of Global Competition.* New York: Free Press.

Olsen, J.P. (1981), 'Integrated Organizational Participation in Government.' In P.C. Nystrom and W.H. Starbuck (eds), *Handbook of Organizational Design. Volume 2. Remodelling Organizations and Their Environments.* Oxford: Oxford University Press, pp. 492–516.

Olson, M. (1982), *The Rise and Decline of Nations. Economic Growth, Stagflation and Social Rigidities.* New Haven: Yale University Press.

OPT (Consultative Body on Postal and Telecommunications Policy) (1993), *Rapport van Bevindingen inzake Wetsvoorstel Mobiele Communicetie.* The Hague: OPT.
Ostrom, E. (1986), 'An Agenda for the Study of Institutions.' *Public Choice* 48: 3–25.
Ostry, S. (1990), *Governments & Corporations in a Shrinking World. Trade & Innovation Policies in the United States, Europe & Japan.* New York: Council of Foreign Relations Press.
Ostry, S. and Nelson, R. (1995), *Techno-Nationalism and Techno-Globalism. Conflict and Cooperation.* Washington: Brookings Institution.
Owen, B.M. and Braeutigam, R. (1978), *The Regulation Game. Strategic Use of the Administrative Proces.* Cambridge: Ballinger.
PA Consulting Group (1994), *Study of the International Competitiveness of the UK Telecommunications Infrastructure. Prepared for DTI.* London: PA Consulting Group.
—— (1993), *Mobicom Study. Importance of Wireless Communications in the Netherlands. Bottlenecks Preventing Mobile Take-Up and Possible Solutions.* Utrecht: PA Consulting Group.
—— (1991–2), *Onderzoek inzake Ontwikkelingen in de Telecommunicatie-Infrastructuur.* Phase 1 Report. Vol. 2, Baseline Papers. Utrecht: PA Consulting Group.
Parris, H. (1985), 'Public Enterprise in Great Britain.' *Annals of Public and Co-operative Economy* 56 (3): 393–410.
Perez, C. and Soete, L. (1988), 'Catching Up in Technology: Entry Barriers and Windows of Opportunity'. In G. Dosi *et al.* (eds), *Technical Change and Economic Theory.* London: Pinter. pp. 458–479.
PI (Privatisation International) (1997), *Privatisation Yearbook 1997.* London: Privatisation International.
Pigeat, H. and Virol, L. (1980), *Du Téléphone à la Télématique.* Paris: Commissariat Général de Plan.
Pinaud, C. (1985), *Entre Nous, Les Téléphones. Vers une Sociologie de la Télécommunication.* Paris: Insep.
Pitt, D.C. (1980), *The Telecommunications Function in the British Post Office. A Case Study of Bureaucratic Adaptation.* Westmead: Saxon House.
PO (Post Office) (1977), *Post Office Review Committee Report. Commentary by the Post Office.* London: BPO.
POEU (Post Office Engineering Union) (1985), *Making the Future Work. The Broad Strategy.* London: Post Office Engineering Union.
—— (1977), *Post Office Review Committee. Evidence Submitted by Post Office Engineering Union.* London: Post Office Engineering Union.
—— (1964), *The Telephone Ring. It's Time to Investigate.* London: Post Office Engineering Union.
Pool, I.de Sola (1990), *Technologies Without Boundaries. On Telecommunications in a Global Age.* Cambridge MA: Harvard University Press.
—— (1983), *Technologies of Freedom. On Free Speech in an Electronic Age.* Cambridge MA: Harvard University Press.
Porter, M.E. (1990), *The Competitive Advantage of Nations.* New York: Free Press.
—— (ed.) (1986), *Competition in Global Industries.* Boston MA: Harvard Business School Press.
Powell, W.W. (1990), 'Neither Market Nor Hierarchy: Network Forms of Organization.' *Research in Organizational Behaviour* 12: 295–336.
Powell, W.W. and DiMaggio, P.J. (eds) (1991), *The New Institutionalism in Organizational Analysis.* Chicago: University of Chicago Press.
Prahalad, C.K. and Doz, Y.L. (1987), *The Multinational Mission. Balancing Local Demands and Global Vision.* New York: Free Press.
Prévot (1989), *Rapport de Synthèse Remis par Hubert Prévot à Paul Quilès. A l'Issue de Débat Public sur l'Avenir du Service Public de la Poste et des Télécommunications.* Paris: Le Debat Public.

Prosser, T. (1986), *Nationalised Industries and Public Control.* London: Basil Blackwell.

PTT (Staatsbedrijf der Posterijen, Telegrafie en Telefonie) (1987), *Inspanningsverklaring Werkgelegenheid bij de Hoofddirectie Telecommunicatie.* The Hague: PTT.

—— (1985a), *Standpunt van de Bedrijfsleiding van PTT met Betrekking tot het Rapport van de Commissie Steenbergen.* The Hague: PTT.

—— (1985b), *PTT in Beweging.* The Hague: PTT.

—— (1980–8), *Annual Reports.* The Hague: PTT.

PTT Telecom (1993), *Lines to the Future.* The Hague: PTT.

RCO Council of Central Employers' Association (1994), *Nieuw Telecommunicatiebeleid. Brief aan de Minister van Verkeer en Waterstaat* (27-1-1994). The Hague: RCO.

Reagan, M.D. (1987), *Regulation. The Politics of Policy.* Boston: Little, Brown and Company.

Reich, R. (1991), *The Work of Nations. Preparing Ourselves for 21st-Century Capitalism.* New York: Vintage Books.

Renaud, J.L. (1990), 'The Role of the International Telecommunication Union: Conflict, Resolution and the Industrialized Countries.' In K. Dyson and P. Humphreys (eds), *The Political Economy of Communications. International and European Dimensions.* London: Routledge, pp. 33–57.

Richardson, J. (1994), 'The Politics and Practice of Privatisation in Britain.' In V. Wright (ed.), *Privatization in Western Europe. Pressures, Problems and Paradoxes.* London: Pinter, pp. 57–82.

—— (ed.) (1982), *Policy Styles in Western Europe.* London: Allen & Unwin.

Richardson, J. and Jordan, A.G. (1979), *Governing Under Pressure.* Oxford: Martin Robertson.

Richardson, J., Gustaffson, G. and Jordan, A.G. (1982), 'The Concept of Policy Style.' In J. Richardson (ed.), *Policy Styles in Western Europe.* London: Allen & Unwin. pp. 1–16.

Roobeek, A.J.M. (1990), *Beyond the Technology Race: An Analysis of Technology Policy in Seven Industrial Countries (1988).* Amsterdam: Elsevier Science Publishers.

—— (1988), 'Telecommunications: An Industry in Transition.' In H.W. De Jong (ed.), *The Structure of European Industry* (2nd edn.) Dordrecht: Kluwer, pp. 297–328.

Roobeek, A.J.M. and Broeders, J. (1993), 'Telecommunications: Global Restructuring at Full Speed.' In H.W. De Jong (ed.), *The Structure of European Industry* (3rd edn.) Dordrecht: Kluwer, pp. 273–306.

Rose, R. (1993), *Lesson-Drawing in Public Policy. A Guide to Learning Across Time and Space.* Chatham NJ: Chatham House.

Rugès, J-F. (1970), *Le Téléphone Pour Tous.* Paris: Editions du Seuil.

Ruigrok, W. and Van Tulder, R. (1995), *The Logic of International Restructuring.* London: Routledge.

Rutkowski, A.M. (1991), 'The ITU at the Cusp of Change.' *Telecommunication Policy* August: 286–97.

Sally, R. (1995), *States and Firms. Multinational Enterprises and Institutional Competition.* London: Routledge.

—— (1993), 'Alcatel's Relations with the French State: The Political Economy of a Multinational Enterprise.' *Communications & Strategies* 9 (1): 67–95.

Sauter, W. (1995), 'The Telecommunications Law of the European Union'. *European Law Journal* 1 (1): 92–111.

Savary, J. (1984), *French Multinationals* (Transl. of *Les Multinationales Françaises (1981)*) London/Geneva: Pinter/IRM.

Schattschneider, E.E. (1975), *The Semisovereign People. A Realist's View of Democracy in America (1960).* Hinsdale IL: Dryden Press.

Schiller, D. (1982), *Telematics and Government.* Norwood: Ablex.

Schmitter, P.C. (1979), 'Still the Century of Corporatism?' In P.C. Schmitter and G. Lehmbruch (eds), *Trends Toward Corporatist Intermediation*. Beverly Hills: Sage, pp. 7–52.

Schmitter, P.C. and Lehmbruch, G. (eds) (1979), *Trends Toward Corporatist Intermediation*. Beverly Hills: Sage.

Schneider, V. and Werle, R. (1990), 'International Regime or Corporate Actor? The European Community in Telecommunications Policy.' In K. Dyson and P. Humphreys (eds), *The Political Economy of Communications. International and European Dimensions*. London: Routledge, pp. 77–106.

Scott, C. (1994), 'The Development of a European Telecommunications Policy.' In C. Hicks (ed.), *Regulating Telecommunications – An International Assessment of Prospects and Strategy*. London: Centre for the Study of Regulated Industries, pp. 13–48.

Scott, W.R. (1995), *Institutions and Organizations*. Thousand Oaks: Sage.

Scrimgeour, Kemp-Gee & Co. (1984), *British Telecom*. London: Scrimgeour, Kemp-Gee & Co.

Segrestin, D. (1990), 'Recent Changes in France.' In G. Baglioni and C. Crouch (eds), *European Industrial Relations. The Challenge of Flexibility*. London: Sage, pp. 97–126.

Selznick, P. (1985), 'Focusing Organisational Research on Regulation.' In R. Noll (ed.), *Regulatory Policy and the Social Sciences*. Berkeley: University of California Press, pp. 363–7.

Servan-Schreiber, J-J. (1967), *Le Défi Américain*. Paris: Denoël.

Sharp, M. and Walker, W. (1991), 'Thatcherism and Technical Advance: Reform Without Progress. Part II: The Thatcher Legacy.' *Political Quarterly* 62 (3): 318–37.

Shonfield, A. (1965), *Modern Capitalism. The Changing Balance of Public and Private Power*. Oxford: Oxford University Press.

Slaa, P. (1993), 'Concurrentie op Mobiele Telecommunicatiediensten.' *Mediaforum* 5 (3): 26–30.

Smith, J. and Terry, M. (1993), 'The English Case.' *Bulletin of Comparative Labour Relations* (Special Issue on *Industrial Relations Developments in the Telecommunications Industry*) 25: 185–267.

Soeterbroek, F. and Walravens, A. (1985), *Privatisering in Nederland. Analyse, Kritiek en Alternatieven*. Zoetermeer: AbvaKabo.

Solomon, J. (1986), 'Telecommunications Evolution in the UK.' *Telecommunications Policy* September: 186–92.

Sorge, A. (1991), 'Strategic Fit and the Societal Effect: Interpreting Cross-National Comparisons of Technology, Organisation and Human Resources.' *Organization Studies* 12 (2): 161–90.

Spectrum Strategy Consultants (1996), *Development of the Information Society. An International Analysis*. London: Department of Trade and Industry.

STE (Society of Telecom Executives) (1989), *All on Your Own. The Case Against Personal Contracts*. Teddington: Society of Telecom Executives.

—— (1986), *Statement by STE on Future of British Telecom (BT)*. Teddington: Society of Telecom Executives.

Steel, D.R. and Heald, D.A. (1982), 'Privatising Public Enterprise: An Analysis of the Government's Case.' *Political Quarterly* 53: 333–49.

Steinfield, C., Bauer, J.M. and Caby, L. (eds) (1994), *Telecommunications in Transition. Policies, Services and Technologies in the European Community*. Thousand Oaks: Sage.

Stevens, C. (1990), 'Technoglobalism vs. Technonationalism: The Corporate Dilemma.' *Columbia Journal of World Business* Fall: 42–8.

Stopford, J. and Strange, S. with Henley, J. (1991), *Rival States, Rival Firms. Competition for World Market Shares*. Cambridge: Cambridge University Press.

Strange, S. (1992), 'States, Firms and Diplomacy.' *International Affairs* 68 (1): 1–15.

—— (1988), *States and Markets.* London: Pinter.

—— (1987), 'The Persistent Myth of Lost Hegemony.' *International Organization* 41: 551–74.

Streeck, W. (1993), 'The Social Dimension of the European Economy.' In D. Mayes *et al.*, *Public Interest and Market Pressures. Problems Posed by Europe 1992.* Houndsmills: St. Martin's Press, pp. 98–157.

Streeck, W. and Schmitter P.C. (1991), 'From National Corporatism to Transnational Pluralism: Organized Interests in the Single European Market.' *Politics & Society* 19: 133–64.

—— (eds) (1985), *Private Interest Government. Beyond Market and State.* London: Sage.

Sun, J-M. and Pelkmans, J. (1995), 'Regulatory Competition in the Single Market.' *Journal of Common Market Studies* 33: 67–89.

—— (1994), *Why Liberalisation Needs Centralistion. Subsidiarity and EU Telecoms.* CEPS Working Document No. 88. Brussels: CEPS.

Swann, D. (1988), *The Retreat of the State. Deregulation and Privatisation in the UK and US.* New York: Harvester Wheatsheaf.

Théry, G. (1994), *Les Autoroutes de l'Information. Rapport au Premier Ministre.* Paris: La Documentation Française.

Thompson, G. (1990), *The Political Economy of the New Right.* London: Pinter.

Thompson, J. (1967), *Organizations in Action.* New York: McGraw-Hill.

Tilly, C. (1984), *Big Structures, Large Processes, Huge Comparisons.* New York: Russell Sage Foundation.

TMA (Telecommunications Managers' Association) (1993), *Interconnection and Accounting Separation. The TMA Response to the OFTEL Consultative Document.* London: TMA.

—— (1991), *Competition and Choice: Telecommunications Policy for the 1990s. The TMA Response.* Orpington (Kent): TMA

TUA (Telecommunications Users' Association) (1991), *Review of the Government Consultative Document 'Competition and Choice': Telecommunications Policy for the 1990s.* London: TUA.

—— (1990), *Review of the Telecommunications Environment and Proposals Regarding Further Liberalisation of Regulations Governing the Carrying of Voice and Data Services.* London: TUA.

TUC (Trades Union Congress) (1985), *Stripping Our Assets. The City's Privatisation Killing.* London: TUC.

Unger, B. and van Waarden, F. (eds) (1995), *Convergence or Diversity? Internationalization and Economic Policy Response.* Aldershot: Avebury.

V&W (Ministry of Transport and Public Works) (1993), *Hoofdlijnennotitie Herziening Wet op de Telecommunicatie-Voorzieningen.* The Hague: Ministerie van Verkeer en Waterstaat.

Van Den Besselaar, P. and Visser, J. (1993), 'The Dutch Case.' *Bulletin of Comparative Labour Relations.* (Special Issue on *Industrial Developments in the Telecommunications Industry*) 825: 97–184.

Van Den Bosch, F.A.J. and De Man, A.P. (1994), 'Government's Impact on the Business Environment and Strategic Management.' *Journal of General Management* 19 (3): 50–9.

Van Den Brink, J.R.M. (1984), *Zoeken naar een 'Heilstaat'. Opbouw, Neergang en Perspectief van de Nederlandse Welvaartstaat.* Amsterdam: Elsevier.

Van Doorn, J.A.A. (1981), 'Corporatisme en Technocratie – Een Verwaarloosde Polariteit in de Nederlandse Politiek.' *Beleid en Maatschappij* 8: 134–49.

Van Putten, J. (1982), 'Policy Styles in the Netherlands: Negotiation and Conflict.' In J. Richardson (ed.), *Policy Styles in Western Europe.* London: Allen & Unwin, pp. 168–96.

Van Schendelen, M.P.C.M. (1993a), 'The Netherlands: Lobby it Yourself.' In M.P.C.M. Van Schendelen (ed.), *National Public and Private EC Lobbying.* Aldershot: Dartmouth, pp. 131–54.

—— (ed.) (1993b), *National Public and Private EC Lobbying*. Aldershot: Dartmouth.
—— (1991), *Dutch EC-Lobbying: Political Hooking on One's Own*. Paper presented at the ECPR Workshop on European Lobbying Towards 2000, Colchester (Essex) UK, 22–28 March 1991.
—— (1987), 'The Netherlands: From Low to High Politicisation.' In M.P.C.M. Van Schendelen and R.J. Jackson (eds), *The Politicisation of Business in Western Europe*. London: Croom Helm, pp. 59–83.
Van Schendelen M.P.C.M. and Jackson, R.J. (eds) (1987), *The Politicisation of Business in Western Europe*. London: Croom Helm.
Van Tulder, R. and G. Junne (1988), *European Multinationals In Core Technologies*. London: John Wiley & Sons.
VECAI (Association of Operators and Licensees of Cable Networks) (1990), *Toekomstvisie*. Dordrecht: VECAI.
Vedel, T. (1991), 'Les Filiales de l'Etat dans le Domaine des Télécommunications Depuis 1945: Des Colonies à la Déréglementation.' In C. Bertho-Lavenir (ed.), *L'Etat et les Télécommunications en France et à l'Etranger 1837-1987*. Genève: Librairie Droz, pp. 129–60.
—— (1987), 'Local Policies for Wiring in France.' In W.H. Dutton, J.G. Blumler and K.L. Kraemer (eds), *Wired Cities. Shaping the Future of Communications*. London: Cassell, pp. 255–78.
—— (1984), 'Les Ingénieurs des Télécommunications.' *Culture Technique* 12: 63–75.
Vedel, T. and Dutton, W.H. (1990), 'New Media Politics: Shaping Cable Television Policy in France.' *Media, Culture and Society* 12: 491–524.
Veljanovski, C. (ed.) (1991a), *Regulators and the Market. An Assessment of the Growth of Regulation in the UK*. London: Institute of Economic Affairs.
—— (1991b), 'The Regulation Game.' In C. Veljanovski (ed.), *Regulators and the Market. An Assessment of the Growth of Regulation in the UK*. London: Institute of Economic Affairs, pp. 3–28.
—— (ed.) (1989), *Privatisation & Competition. A Market Prospectus*. London: Institute of Economic Affairs.
—— (1987), *Selling the State. Privatisation in Britain*. London: Weidenfeld & Nicolson.
Vernon, R. (ed.) (1974), *Big Business and the State. Changing Relations in Western Europe*. London: Macmillan.
—— (1971), *Sovereignty at Bay. The Multinational Spread of U.S. Enterprises*. New York: Basic Books.
Vickers, J. and Yarrow, G. (1988), *Privatization: An Economic Analysis*. Cambridge MA: MIT Press.
—— (1986), 'Telecommunications' Liberalisation and the Privatisation of British Telecom.' In J. Kay, C. Mayer and D. Thompson (eds), *Privatisation and Regulation. The UK Experience*. Oxford: Clarendon Press, pp. 221–40.
Vietor, R.H.K. (1994), *Contrived Competition. Regulation and Deregulation in America*. Cambridge MA: Belknap Press.
—— (1989), *Strategic Management in the Regulatory Environment. Cases and Industry Notes*. Englewood Cliffs: Prentice Hall.
Visser, J. (1990), 'Continuity and Change in Dutch Industrial Relations.' In G. Baglioni and C. Crouch (eds), *European Industrial Relations. The Challenge of Flexibility*. London: Sage, pp. 199–242.
Visser, J. and Hemerijck, A. (1997), *'A Dutch Miracle.' Job Growth, Welfare Reform and Corporatism in the Netherlands*. Amsterdam: Amsterdam University Press.

Vogel, D. (1986), *National Styles of Regulation. Environmental Policy in Great Britain and the United States.* Ithaca: Cornell University Press.

Vogel, E.F. (1987), 'Conclusion'. In G.C. Lodge and E.F. Vogel (eds), *Ideology and National Competitiveness: An Analysis of Nine Countries.* Boston: Harvard Business School Press, pp. 301–23.

Volcansek, M. (1992a), 'Judges, Courts and Policy-Making in Western Europe.' *West European Politics* 15 (3): 1–8.

—— (1992b), 'The European Court of Justice: Supranational Policy-Making.' *West European Politics* 15 (3): 109–21.

Von Weizsäcker, C.C. (1986), 'Free Entry into Telecommunication.' In M.S. Snow (ed.), *Telecommunication Regulation and Deregulation in Industrialized Democracies.* Amsterdam: North-Holland, pp. 20–41.

Wade, R. (1990), *Governing the Market. Economic Theory and the Role of Government in East Asian Industrialization.* Princeton: Princeton University Press.

Walker, W. (1993), 'National Innovation Systems: Britain.' In R. Nelson (ed.), *National Innovation Systems. A Comparative Analysis.* Oxford: Oxford University Press, pp. 158–91.

Walker, W. and Sharp, M. (1991), 'Thatcherism and Technical Advance: Reform Without Progress. Part I: The Historical Background.' *Political Quarterly* 62 (2): 262–72.

Wassenberg, A.F.P. (1990), 'Games Within Games: On the Politics of Association and Dissociation in European Industrial Policy-Making.' In B. Marin (ed.), *Governance and Generalized Exchange. Self-Organizing Policy Networks in Action.* Frankfurt: Campus/Westview, pp. 256–88.

—— (1985), 'Organizational Instinct: On the Political Economy of Bargaining.' In A.H.G. Rinnooy-Kan (ed.), *New Challenges for Management Research.* Amsterdam: North-Holland, pp. 159–77.

—— (1982), 'Neo-Corporatism and the Quest for Control: The Cuckoo Game.' In G. Lehmbruch and P.C. Schmitter (eds), *Patterns of Corporatist Policy-Making.* London: Sage, pp. 83–108.

White Paper (1978), *The Post Office.* Cmnd 7292. London: HMSO.

Whitley, R. (1994a), 'Dominant Forms of Economic Organization in Market Economies.' *Organisation Studies* 15: 153–82.

—— (1994b), 'The Internationalisation of Firms and Markets: Its Significance and Institutional Structuring.' *Organization* 1 (1): 101–24.

—— (1992a), *Business Systems in East Asia. Firms, Markets and Societies.* London: Sage.

—— (1992b), 'Societies, Firms and Markets: The Social Structuring of Business Systems.' In R. Whitley (ed.), *European Business Systems. Firms and Markets in Their National Contexts.* London: Sage, pp. 5–45.

—— (ed.) (1992c), *European Business Systems. Firms and Markets in Their National Contexts.* London: Sage.

Wiener, M.J. (1981), *English Culture and the Decline of the Industrial Spirit, 1850-1980.* Cambridge: Cambridge University Press.

Williamson, O.E. (1993), 'Transaction Cost Economics and Organization Theory.' *Industrial and Corporate Change* 2: 107–56.

—— (1991), 'Comparative Economic Organization: The Analysis of Discrete Structural Alternatives.' *Administrative Science Quarterly* 36: 269–96.

—— (1985), *The Economic Institutions of Capitalism.* New York: Free Press.

—— (1975), *Markets and Hierarchies. Analysis and Antitrust Implications.* New York: Free Press.

Willis, D. and Grant, W. (1987), 'The United Kingdom: Still a Company State?' In M.P.C.M. Van Schendelen and R.J. Jackson (eds), *The Politicisation of Business in Western Europe*. London: Croom Helm.

Windmuller, J.P., De Galan, C. and Van Zweeden, A.F. (1987), *Arbeidsverhoudingen in Nederland*. Utrecht: Spectrum.

Wolters, M. and Coffey, P. (eds) (1990), *The Netherlands and EC Membership Evaluated*. London: Pinter.

Wood, J. (1994), The UK Telecommunications Environment Today. Springboard for the Future. London: FT.

Woodrow, B.W. (1991), 'Tilting Towards a Trade Regime. The ITU and the Uruquay Round Services Negotiations.' *Telecommunications Policy* August: 323–42.

Wright, V. (1987), *The Government and Politics of France* (2nd edn.) London: Hutchinson.

WRR (Scientific Council for Government Policy) (1993), *Shaping Factors for the Business Environment in the Netherlands after 1992. Report for the EC Cellule de Prospective*. Preliminary and Background Studies V 78. The Hague: WRR.

—— (1980), *Plaats en Toekomst van de Nederlandse Industrie*. The Hague: WRR.

Zukin, S. and DiMaggio, P. (eds) (1990), *Structures of Capital. The Social Organization of the Economy*. Cambridge: Cambridge University Press.

Zysman, J. (1994), 'How Institutions Create Historically Rooted Trajectories of Growth.' *Industrial and Corporate Change* 3 (1): 243–83.

—— (1983), *Governments, Markets and Growth. Financial Systems and the Politics of Industrial Change (1983)*. Ithaca: Cornell University Press.

—— (1977), *Political Strategies for Industrial Order. State, Market, and Industry in France*. Berkeley: University of California Press.

—— (1975), 'Between the Market and the State: Dilemmas of French Policy for the Electronics Industry.' *Research Policy* 3: 312–36.

Index

A-2000 (American-Dutch consortium) 204
ABN AMRO Bank 207
ADCs (access deficit contributions) 167
adjustment 49, 111; policies 4, 34, 103; politics of 119–26, 176–80, 236–41
Advisory Councils (Netherlands) 180, 182, 183, 184, 217–18, 219, 220; *see also* RATB
AFUTT (Association des Utilisateurs du Téléphones et des Télécommunications) 242
AGF 240, 265
Agritel 245
Air France 227, 235, 240, 241
Air Traffic Control System 123
Airbus 234
airline industry 43, 51
Airtouch 157
AIT (Association des Ingénieurs des Télécommunications) 242, 252
Albacom 157
Albert, M. 68
Alcatel 200, 225, 259, 260, 265, 270, 299, 300; Alsthom 227, 230, 262; CGE 239, 247, 249, 261–2, 294; CIT 8, 12, 78, 90, 242
alliances 18, 103, 152; R&D 25; services-information 81; strong 116; *see also* strategic alliances
Ameritech 95
Amsterdam 204; Stock Exchange 171, 208
ANF (Agence Nationale des Fréquences) 258, 275
anti-trust 4, 13, 26, 94, 96, 97, 175
APT (Philips/AT&T) 181
ART (Authorité de Régulation des Télécommunications) 258, 275
assets: productive 4–5; public, sale of 5, 84, 122; stripping 69; trading 113; undervalued 122
ASV (Swedish health and safety agency) 40
AT&T (American Telephone & Telegraph) 5, 12, 18, 66, 78, 90, 93, 146, 199, 200, 214, 247, 286; Bell Labs 241; break up 94, 97, 135; BT share sold 160; deregulated 287; Europeans joining forces with 248; joint ventures 97; licences 161; line leasing 145; long-distance services 95; permitted to enter dataprocessing industry 96; 'Project Atlantic' 260; Unisource 211, 213, 260, 274, 301; WorldSource 213, 269, 301; *see also* APT
Atlas 260, 269, 270
ATM switching technologies 242, 256
Attali, J. 244
Australia 60, 118, 147, 162
Austria 49, 58, 60
autonomy 4, 50, 63

BABT (British Approval Board for Telecommunications) 134, 145
Baker, Kenneth 137
Balladur, Édouard 240, 256, 272
Banco Nazionale del Lavoro 157
Banco Santander 157
bandwidths 84
Bangemann, M. 106, 107
Bank of England 116, 119
banks 61, 63, 91, 112; central 116, 119, 171; commercial 116; control over 46; investment 50, 116, 232; merchant 116
Banque Hervet 265
Barclays Bank 138
bargaining 21, 22, 23, 26, 62, 64; collective 4, 60; institutional 58; political 49; productive 58; role of 43; three dimensions of 73; tripartite 61
barriers 111; entry 31, 76, 98; innovation 11; *see also* trade barriers
Bauer, M. 230
Beesley, Michael 132, 133
Belgium 5, 20, 38, 68, 76, 205, 261, 269; cable penetration 106; corporatism 58, 60; dependence on world economy 48
beliefs: cultural 29, 67; ideological 68
Bell 94, 96, 241, 287; Atlantic 95, 97; CableMedia 148, 162; Canada 78, 141, 162, 286; South 95, 141, 197, 259; *see also* RBOCs
Benelux countries *see* Belgium; Luxembourg; Netherlands
Best, M. 44, 46
best practice 31
Blair, Tony 117, 125, 126

Blankart, C. B. 14
BNP (Banque National de Paris) 232, 236, 240
Bouyges 162, 239, 259
BP Mobile 269
Brazil 65, 66
Bretton Woods 3, 120
Britain *see* United Kingdom
British Aerospace 113, 146, 149
British Airways 122, 124
British Coal 123, 124
British Energy 123
British Gas 122
British Leyland 119, 122, 124
British Petroleum 123, 138
British Rail 124, 146
British Steel 119, 123
broadcasting 10, 11, 86, 87
Broeders, J. 90
BR-Telecommunications 148
Bruce, R. 96
Brussels 104
BSI (British Standards Institute) 134, 145
BSkyB 146, 157
BT (British Telecom) 8, 13, 103, 111, 112, 122–4 *passim*, 132–69, 260, 281; ban on entry to cable market 274; divisions 152–4, 166, 287; domestic and international expansion 298; drive to make more efficient 284; Dutch Railways link-up with 204; MCI and 269, 286; new service providers and 285; solid market presence outside Europe 286; Telefonica and 214; unionisation 126
BTG (British Technology Group) 92, 124, 196
BTUC (British Telecom Unions Committee) 135, 136, 143, 155
budgets 65, 82, 247; balancing 7; deficit 6, 113, 116, 177, 235; procurement 8
Bull 235, 239, 240, 252, 255, 260, 265, 268, 299; Honeywell 237, 238, 244, 249
bureaucracy 49, 63, 64; insulated 114; weak and incoherent 65
Burns, T. 44
Burton, J. 121

CA (Consumers' Association) 127, 142
cabinets 114, 177, 178, 179; left-wing 8
cable 11, 75, 78, 95, 97, 106
Cable Authority 147–8
Cable and Broadcasting Act (UK, 1984) 147
Cable Plan (France, 1981) 258
C&W (Cable and Wireless) 81, 111, 119, 123, 131, 259, 284; attempt to merge BT and 161; nationalised (1947) 126; solid market presence outside Europe 286; subsidiaries 138, 140, 143, 146, 149, 155, 161, 162, 285; success of 298
CWC (Cable and Wireless Communications) 148, 211
CAI-NKM 203
Campbell, J. L. 29
Canada 5, 60, 78, 150, 151, 157, 161, 286, 298
Canal Plus 213, 29
Cap Gemini 227
capital 65, 69, 115; allocations 26; free flow 72; intensity 7, 62; social 67; transnational 66; venture 264
capital markets 61; access to 5; competition 4; short-term loans through 116; well-developed 50
capitalism 25, 33, 68; global 44, 48; 'popular' 122
CAPs (competitive access providers) 147
CAPT (Dutch Commission for Advice on Postal and Telecommunications Policy) 196, 217–18
Caribbean 126, 162
cartels 19, 62, 76, 77, 79, 104, 247; international 5–10; 'paradise of Europe' 174
Carterphone Decision (1968) 94
Casema 197, 198, 200, 201, 203, 204, 213; demerging 222; successful bid for 268
CAT (Compagnie Auxiliere de Télécommunications) 264
CAT (French union) 253
CATV operators 148, 204, 274
Cawson, A. 23
CBI (Confederation of British Industries) 115–16
CCITT (International Consultative Committee for Telephone and Telegraph) 77
CDC (Caisse des Dépots et Consignations) 232, 241, 259
Cecchini, P. 101
CEGB (Central Electricity Generating Board) 123, 124
Cégétel 157, 260
Cellnet 149, 157
cellular radio 75, 78, 90
CEN-CENELEC 100
Central Europe 118, 209, 214
central planning 48
Centretel 269
CEPT (Conférence Européenne des Administrations des Postes et des Télécommunications) 77, 98, 100
CFDT (Confédération Française Démocratique du Travail) 231, 242
CGCT (Compagnie Générale de Constructions Téléphoniques) 90, 157, 225, 238, 242–4 *passim*, 259, 260, 261; Thomson-Houston 234
CGE (Compagnie Générale des Eaux) Alcatel 239, 247, 249, 261–2, 294
CGP (French Planning Commission) 228, 229, 232, 248, 249
CGT (Confédération Général du Travail) 230–1, 242, 265
chaebol 46
Chamoux, Jean-Pierre 271
Chandler, A. D. 33
Chapman, C. 124
Charkham, J. 45

Chataway, Christopher 128
chemical industry 62–3
Child, J. 67
China 46
Chirac, Jacques 235, 238, 295, 297, 298, 300
choice 29, 30; incremental 31; policy 32, 33, 43, 50; rational 28; shaped by the past 32; strategic 25, 32
Christian Democratic parties 4; Dutch 178, 184, 190
CIB (Dutch Centre for Information Policy) 183, 201
CIC Bank 241
CIGREF (Club Informatique des Grandes Enterprises Française) 92, 242
CII 234, 236, 244; Honeywell Bull 238, 244, 249
Cinevideo 213
CIT-Alcatel 8, 12, 78, 90, 242, 261
CI-Telecom 268
Citizen's Charter (UK) 125
Civil Service Commission (UK) 128
clientelism 65
Clignet, R. 23
CMA (Communication Managers' Association) 127, 155
CNCL (Commission Nationale de la Communication et des Libertés) 253, 271
CNES 227, 248, 265
CNET (Centre National d'Études des Télécommunications) 227, 241–2, 243, 246, 265, 294, 296, 299
CNT (Caisse Nationale des Télécommunications) 242, 245, 246
CNV (Dutch Christian Trade Unions Federation) 174, 179
coalitions 9, 113–14; distributional 59; rent-seeking 6, 76–84; ruling 48, 291–302; temporary 25
Codetel 242, 245
Cogecom (Compagnie Générale de Communication) 250, 255, 264, 265, 271
Cohen, E. 228, 230
COIFRA (French holding company) 265
COLT 147
Common Market 175, 181, 235; Telecommunications 13, 14
Communication Act (France, 1986) 271
Communism 68, 230, 234, 236, 237, 238, 240, 242, 294, 297
communitarianism 67, 68, 69
Compagnie Générale de Constructions Téléphoniques 90
Compagnie Générale des Eaux 239
Compagnie Nationale du Téléphone 245
Companies Act companies 134
comparative advantage 176
competition 3, 8, 11, 12, 64, 77, 96, 111, 112; controlled 81; cross-national policy 50; destruction 14; EC legislation 4; fair 68, 81, 82, 193, 194, 206; fierce 70, 105, 113; first steps to introduce 252–8; foreign 79, 88; furthering 117; general rules 25; healthy 57; insular 71; intense 90; introduction of 76; long-distance services markets 93; managed 131–9, 195; multinational 16; national policy 13; network 200–4; open 139–45; promoting 96, 98; public procurement market 104; reform of policies 48; stimulating 5, 101; strong policy 55; suppressing 32; worldwide 54; yardstick 4; *see also* international competition
Competition Act (Netherlands, 1956) 174, 175
Competition and Service (Utilities) Act (UK, 1992) 166
competitive advantage 56, 57, 67; international 96; search for 72; significant and sustained 91
competitiveness 11, 17, 57, 64, 96; differences in 66; handicapped 101; lack of 1; loss of 116; national 55, 56; policies to increase 49; *see also* international competitiveness
Compuserve 97
computers 9, 11, 70, 79; personal 257
Comsat 247
ComTel/IVS 216
Comvik/Tele 2 (Sweden) 162
Concert 92, 157, 160, 214, 269, 288
Concorde 234
consensus 39–40, 68, 75, 103, 192, 254; government, post-war 116; little scope for 113
Conservative party/government (UK) 111, 113, 114, 118–19, 291; Cable Programme (1982/83) 147; competition policy 117; election manifesto (1979) 131; free market and monetarist approach 120; Post Office privatisation 128; radical 121; relationship with CBI 115–16; widespread campaigns against privatisation 135–6; *see also* Heath; Macmillan; Major; Thatcher
consumer preferences 16, 72
contracts 24, 68; bilateral 4; procurement 18, 75, 101
control authoritative systems 45; credit and banks 46; family 46; organisational, strong 58; quest for 29; regulatory 40
convergence 11, 16–19, 21, 42, 71, 307–21; corporate 87; cross-national 15; pressures towards 44; service 87; technological 87
Corby, M. E. 129
core technologies/firms 53, 55
corporatism 35, 49, 58, 59; medium/strong/weak 60
corruption 65
costs 6, 9, 107; industrial growth 21; local telephone 7; marketing 90; R&D 12, 16, 20, 52, 72, 81, 90; sharing 80; sunk 5, 31; transaction 27, 29
Council on Post and Telecommunications Affairs (UK) 129, 131
Cowhey, P. F. 77

CPI (Consumer Price Index) 216
CPSA (Civil and Public Servants' Association) 127; *see also* NCU
credit 50, 61, 69, 116; control over 46
Crédit Agricole 249
Crédit Lyonnais 227, 232, 235, 236, 240
Crédit National 232
Créditel 245
créneau 233
Cresson, Édith 240
cross-subsidisation 6, 11, 27, 81, 187, 220, 253
CSA (Consul Supériure de l'Audiovisuel) 271, 272
CSF 229, 234, 247
culture: economic 54, 66–9, 116–17, 174–5, 232–4; 'institutional specificity of' 44; political 40
currency 113
customers: business 93; large 92, 112; multinational 92; residential 112; responsiveness 5, 10; services 89
customisation 85, 92
Czech Republic 214, 216

D66 (Dutch centre-left party) 178–9
Daewoo 229
Dansk Mobiltelefon 162
Dassault 238
datacommunications 79, 89, 93
dataprocessing 9, 10, 11, 91, 96; competition between telecommunications and 12
DBP (Deutsche Bundespost Telekom) 8, 105, 214, 255, 269, 270; AT&T's 'Project Atlantic' 260
DCS 1800 licence 199
decentralisation 80, 86, 116; collective bargaining 4; decision making 51; market forces 48; policy making 104; production 72
decision making 5, 16, 22, 23, 24, 42; centralised 104; collective 59; collegiate structures 45; decentralised 51; European level 13, 60; incremental and routinised 30; political 14, 58
defence 8, 18, 113, 227, 235
de-industrialisation 113
De Jong, H. W. 174
De Man, A. P. 56–7
demand 1, 56, 76; business community 7, 11, 81; changes in 51; corporate, rapid growth in 93; dynamic 80, 82; local 92; specialised 80
demergers 97, 222
demiurge 66
democratic regimes 38, 41, 53, 113
demonopolisation 2, 126–45, 180–98, 241–58
Den Uyl Cabinet (Netherlands, 1973–7) 176
Denmark 60, 76, 162, 269; *see also* Telecom Denmark
deregulation 3, 14, 18, 20, 51, 69, 82, 114; ambitious programmes 68; 'controlled' 105; financial services 117; gradual 111–12; implementation 115; impressive 84; international 12, 34, 76, 90, 93–9; labour markets 117; major programmes 83; national level 101; radical 4
Derthick, M. 30
De Swann, A. 2–3
Deutsche Telekom 265, 268, 269
DGT (Direction Générale des Télécommunications) 243–53, 258–65, 271, 272, 294–7 *passim*, 299, 300; *see also* France Télécom
diamond model 56, 57
differentiation 72, 75–6, 85
diffusion-oriented countries 52
digitalisation 11, 84
DiMaggio, P. J. 31
dirigiste countries 21
distribution: costs and benefits 21; information 11, 85; networks 86; power 15, 20, 44
divergence 107, 117, 307–21
diversification 7, 85, 89, 96, 97
diversity 308–15; national 15, 16, 19–22
divestiture 97
division of labour: geographical 73; inter-firm, internationallly concentrated 18, 48, 50, 69; international 117; transnational 235
DoI Department of Industry) 131, 135; White Paper (1982) 134
domestic responses 15–22
downsizing 69
DSM 171, 178
DTI (Department of Trade and Industry) 126, 132, 145, 146, 147, 152, 163, 164; 'Green Paper' (1990) 139–40, 141, 143, 144; White Paper (1991) 143, 144
duopolies 104, 114, 143–4, 148, 167, 195, 199, 206, 297; temporary 204; transitional 196
Duopoly Review (UK) 139, 143, 163, 167, 284
Dutch Rail 172, 178, 195, 196, 197, 198, 204
Dyson, K. H. F. 63, 66, 114

Eastern Europe 90, 118, 214
Eastern Telegraph 126
EC (European Community) 1–2, 17, 18, 59, 60, 76; Britain's relationship with 118; CAP (Common Agricultural Policy) 118; collaboration 236; divergence within 107; divided over strategy 105; fragmented markets 101; implementation of policies 180; institutions 13, 99–100; law 13, 100, 174; Leased Line Directive 290; legislation 194, 195, 218, 220; licensing 220; ONP directives 221; prohibitive rulings 175; regulations on national policy 280–1; re-regulation 4, 101; Single European Act (1987/1992) 70, 71, 101, 104; *see also* EU; *also under individual headings prefixed* 'European'
economic growth 26; little attention to 117; slowing down 59; stimulating 240–1; substantial 3

economies of scale 6, 16, 19, 20, 29, 85, 219, 254; minimum 90; promotion of 101; search for 72, 81; sufficient, failure to achieve 70–1
economies of scope 81, 254
ECTUA (European Telecommunications Users' Association) 92
EDCs (Economic Development Committees) 119
EDF (Electricité de France) 230, 241, 246, 253
EDI services 91, 211, 214
EDS 79, 80, 91, 92, 96, 146
EEA (Association of the Electronics, Telecommunications and Business Equipment Industries) 143
EEC (European Economic Community) *see* EC
efficiency 21, 25, 56, 59, 68, 96; allocative 26, 64; enhancing 112; gains 3, 98; operational 91; search for 29; stimulating 5
EGT (Enterprise Générale de Télécommunications) 260, 263
Electricité de France 227 *see also* EDF
electronic mail 78, 85, 88
electronics 18, 81; consumer 70; defence 113
Elf-Aquitaine 227, 230, 236, 240, 267
élites 65, 242, 252
employer organisations/associations 3, 8
employment 3, 58, 72; job security 7, 8, 79; policies 8
EMU (European Economic and Monetary Union) 69, 71, 175, 235, 241
Encompass Europe 212
Energis 146, 148
Enertel 197, 198, 204, 222, 269
entrants: new 15, 35, 82; prospective 27
entrepreneurship 3, 50, 54, 66, 68
entry 51, 86, 117; barriers to 31, 76, 98; competitive 14, 94; conditions 12; control of 4; foreign markets 81; restricted 79; stipulations 27
environmental issues: policy/regulations 41, 42; pressure groups 62
E-Plus Mobilfunk 162
equity 68; sharing 24; state 117
Ergas, H. 51, 52
Ericsson 79, 90, 151, 157, 181, 199, 200, 237, 247; *see also* SFTE; Thorn Ericsson
ERT (European Round Table) 60, 107
ESA/Ariane 234
Esprit Telecom 147
ETSI (European Telecommunications Standards Institute) 99, 102, 103
EU (European Union) 59, 60, 62, 98, 107, 235, 301; legislative requirements 295; liberalisation 199, 226; unemployment 177
Eucom 269
Eunetcom 269
Eureka programme 236
Eurodisney 267
Europe Research and Technology Community 101
European Commission 13, 14, 70, 71, 81, 102–6, 107, 118, 269, 270, 293; Action Programme (1984) 100–1; attempts to persuade 301; French regularly in trouble with 240; Green Paper (1987) 99, 101–2, 111, 272, 302; Internal Market 13, 70, 71, 101, 118, 175, 195, 218, 220, 237, 281; joint ventures approved by 160–1; liberalisation proposals 205, 260, 273; Merger Control Regulations 103, 175; Open Network Provision 260; package deals challenged by 292; player in its own right 290; policy 194, 274, 280; Taskforce on Information Technologies and Telecommunications 100
European Council of Ministers 14, 70, 102, 104, 106
European Court of Justice 71, 103, 104, 290
European Parliament 14, 71, 102, 104
Eurotel 261
Eutelsat 157, 269
Evans, P. 63, 65, 66
excess capacity 91, 93
exchange rates: fixed 3; high 113
exit 31, 51
exports: high prices 113; joint trade policy 21; niche 20; strong emphasis on 49; subsidies 53, 75
External Financing Limit 121, 126
externalities 6, 38

Fabian Society 139
FCR (France Câbles et Radio) 255, 264
Federa consortium 207
federalism 60
Ferranti 124
Fiat 262
fibre optics 84, 85, 89, 92
Fifth Republic (France) 237
filiales 264
filière 233, 248
financial resources 45, 61; key industries 79, 116
financial sector/institutions 61, 63, 113, 290; Conservative party links 116; deregulating 117
Finextel 242, 245
Finivest/Mediaset 157
Finland 60, 147
fiscal policy 53
five-year Plans (France) 237, 243, 245, 246, 247, 248, 252
FNV (Dutch Federation of Trade Unions) 173–4, 179
FO (Force Ouvrière) 231, 242, 253, 267
Fourth Republic (France) 237
France 1–2, 20, 38, 118, 162, 213, 216, 225–78; adjustment strategies 37; BT partnership 157; cable prices 107; central government 21; collaboration 116; comparative analysis 34, 279–306; convergence/and or divergence in restructuring 307–21; economy 68, 226;

France – *continued*
interest associations 92; mission-oriented 52; organisation of work and vocational training 44; policy style 40; protectionism 105, 106, 227; regime type 35, 49; state-controlled credit-based system 61; unbalanced corporatism 60
France Télécom 8, 204, 225–6, 239, 240, 241–2, 259, 294, 295, 298; AT&T's 'Project Atlantic' 260; functions of DGT split from 272; future competitiveness 257; government majority holding in 255–6; mission statement 254–5; participation in various media 274; performance levels 297; privatisation 262–70; reformed structure 258; supervision of 253
Francetel 242, 245
franchising 4, 25, 202, 216
Franco-German agreements/partnerships 235, 269, 270
free-riding problems 24
free trade 12, 16–17, 17–18, 19, 21; fostered 116; global 69, 72–3; market leaders more oriented towards 50; promotion of 52, 55, 117; selective adherence to 53; stimulating 114
FT&T (France Téléphone & Télégraph) 294
Fujitsu 151
Fukuyama, F. 67

gains: efficiency 3, 98; financial, short-term 69; industrial growth 21; redistributional 79
Garnham, N. 139, 151, 164–5, 233
GATT (General Agreement on Tariffs and Trade) 12, 99; *see also* WTO
Gaulle, Charles de 235, 236, 239
Gaz de France 227
GEC 8, 78, 90, 113, 149–51, 157; Telecom 127
GEIS (General Electric Information Services) 9, 79, 80, 81, 86, 91, 92, 96
Germany 14, 21, 151, 162, 181, 213, 216, 234, 261, 286; autonomous action by firms 62–3; banks 61, 63; BT partnership 157; chemical industry 62; collaboration 116; consensus and reactive attitude 39; corporatism 58, 60; diffusion-oriented 52; economic power 118; economy 68; *Länder* 62, 63, 105; organisation of work and vocational training 44; protectionism 105; *see also* DBP
Gerschenkron, A. 50
Giscard d'Estaing, Valéry 245, 247
global forces 15–22
Global One 92, 260, 268, 269, 270, 286, 288
globalisation 12, 15, 16, 18, 34, 51, 75–6, 88–93; high-tech industries 72; increasing 42, 52
'glocalisation' 73
GM 79, 80, 91, 92, 96, 146
GMC Talkland 162
GNS 153, 160
'golden shares' 124, 208
Gourevitch, P. 18, 21, 35, 37, 53–73, 295
governance 25, 39, 68, 99; comparing regimes 32–6; established regime 75; ideal-typical structures 24; regimes vary across countries 45; self 114; transformations 27–32
government/state intervention 2, 3, 25, 49, 82, 105, 111, 114, 119, 121; far-reaching 34, 115; moderate 21; sector characterised by high degree of 5–6; strong public power epitomised by 228
GPT (GEC Plessey Telecommunications) 151
Grandes Écoles 228, 230, 243, 245
Grands Corps 228, 252, 253, 283
Grant, W. 62
Greece 105, 106, 107, 162, 269
GSM provision 149, 162, 197, 198, 200, 205, 206, 209
GTE 90, 94, 161

HACA (Haute Autorité de la Communication Audiovisuelle) 258, 270–1
Hage, J. 23
Hall, Lord 128
Hall, P. 30
Hampden-Turner, C. 67
Harper, J. M. 151
Havas 239, 249, 265
Hayek, F. von 122
HDTP (Dutch regulatory and supervisory body) 191, 200, 217, 219, 221, 289
health 69; and safety 40
Heath, Edward 120, 121, 128
hegemony 12, 17, 73, 77
Heidenheimer, A. J. 38
hierarchies 45, 81
High Level Group on the Information Society 106, 107
Hills, J. 98
Hirschman, A. O. 51
Hirst, P. 72, 73
HKT (Hong Kong Telephone Company) 162, 286
Hoechst 62
Holcombe, A. N. 77
holding companies 61
Honeywell 237, 238, 244, 249
Hong Kong 46, 118, 126; Telecom 162; Telecommunications 285; *see also* HKT
Hoogoven 171
Hotchkiss-Brandt 234
Hu, Y.-S. 52
Hull Communications 147
Hungary 214
'husbandry' 65, 66
Hutchison 149

IBCN (Integrated Broadband Communication Network) 84–5
IBM 9, 12, 18, 53, 66, 79–81 *passim, 86, 91–3 passim*, 165, 240, 247; anti-trust suit against

94, 96; base to rival 227; competing against 248; diversification 97; joint venture with BT 152
ICL (International Computers Limited) 151, 152
ideas 30, 38, 67; cross-cultural exchange 48
ideology 29, 49, 50, 84; economic 54, 66–9, 116–17, 174–5
IG Farben 62
IMF (International Monetary Fund) 120
imports 49, 113
incentives 28, 29, 35; constraints on 19; financial 66
Independent Broadcasting Authority 148
India 65, 269
individualism 67, 68
Indonesia 209, 214
Indosuez 239
industrial relations 44, 58, 115
industrialisation 46, 50, 113
Industry Act (UK, 1980) 124
industry associations 62, 63
inefficiency 7, 10, 84, 113
'infant industry' 50, 55, 197, 206
inflation 3, 113
Infonet Services Corporation 269
infrastructure 10, 11, 51, 65, 85, 86, 102; alternative 84; backbone 88; control over 88, 89; core 87; decentralised 80; digital 106; fragmented 105; future 100–1; institutional 50; lagging development 106; liberalisation of 107; monopolistic providers 7, 14, 103; network 106; strategically important 8; sunk costs 5
ING Bank 206
Inmarsat 157, 269
innovation 29, 30, 54, 55, 57, 89; barriers to 11; blocking 14; goods rlated to 51; improved 83; investment in 69; lacking in flexibility 113; little attention to 117; market 117; policies 8, 20, 52, 103; promotion of 10; slowness in bringing to market 101; stimulating 3, 5, 18, 105; technological 1, 112, 117
INS (International Network Services) 152
'insiders' 16, 72
Institute of Directors 115
institutional forms 24; foreign 31
Intelmatique 268
Intelsat 157, 269
Intercable 216
interconnectivity 77, 86, 87
interdependence 16, 18, 25, 42, 43; economic and political 71–2; seen as triangular diplomacy 73
interest intermediation 58–63, 115–16, 173–2, 230–2; organisation of 53
Interim Act (Netherlands, 1966) 198, 204
Interkabel Polska 216
international calls 6, 9
international competition 18, 43, 70, 71, 73, 79, 97; firms shielded from 113; hampered 101; little attention to 117
international competitiveness 20, 53, 55, 84; lack of 117
internationalisation 10, 48, 54, 62, 71; increased 99
Internet 84, 87–8, 89, 257, 267; *see also* ISPs
investment 57; foreign direct 16, 17, 43, 49, 50, 64, 71, 97, 112; inward 20; national differences 43; R&D 69, 116; return on 69; shared 9; transaction-specific 8
Ionica 298
IRC (Industrial Reconstruction Corporation) 119, 120, 121
Ireland 60, 106; Telecom Eireann 214
ISDN (Integrated Services Digital Network) 84, 85, 102, 104, 148, 250, 267, 283
ISO (International Standards Organisation) 99
ISPs (Internet service providers) 88
ISR (international simple retail) licence 161
Istel 146
IT (information technology) 12, 65, 66, 85, 86, 87, 101, 151, 250
Italtel 90, 97, 181
Italy 60, 63, 105, 106, 107, 118, 205, 261; BT partnership 157; *see also* Italtel; Telecom Italia
ITC (UK Independent Television Commission) 148
ITJ Japan 268
ITT (International Telephone and Telegraph Corporation) 90, 237, 238, 242, 247, 296; Alcatel 200, 262; CGE 261; Dialcom 157; NSEM 181, 199; World Communications 94
ITU (International Telecommunications Union) 9–10, 19, 77, 98, 99
Ivory Coast 268

Jannès, H. 244
Japan 1, 5, 12, 13, 18, 20, 57, 90, 98, 100, 113; central government 21; consensus inspired by normative values 39–40; deregulation 76, 98; developmental state 65; economic power 118; economy 68; fast industrialisation 46; fierce competition from 70; government bureaucracy 64, 65; high-tech threat from 101; increasing competition from 3; interdependence 16, 72; *kaisha* 46; lean production 48; mercantilism 69; need to harmonise regulations with Europe and America 62; orientation 52; penetration of markets 162; protectionism 96, 97, 99; socio-economic change and innovation 51; subsidiaries 117; unbalanced corporatism 60; worldwide liberalisation 93; *see also* Fujitsu; ITJ; KDD; Matsushita
Jeumont Schneider 260
Johnson, C. 64–5
joint ventures 16, 24, 72, 90, 97, 146–7, 151, 157, 203, 260; approved by FCC 160–1; far-reaching 152, 197–8; international 264

Jordan, G. 42
Joseph, Sir Keith 132
Jospin, Lionel 240, 241, 257, 267
JOVE initiative 152, 165
Juppé, Alain 240

Kabelmedia 216
KDD Japan 213, 214, 269
Kelman, S. 40, 41
KEMA (Dutch certification agency) 200
key industries 79, 116, 245, 247
Kingston upon Hull 133, 166
Knieps, G. 14
Kogut, B. 43
Kok, Wim 178, 179
KPN/PTT (Netherlands) 178, 179, 197, 198, 204, 218, 221–2, 286, 301; exclusive concession reduced 199; Kabel/Casema 198, 203, 213; privatisation 207–16; privileges and obligations 193; protecting 195; valuation and flotation 196
KPN Royal PTT Netherlands 191, 207
Krasner, S. D. 32

labour 8, 44, 45, 58, 59, 60, 69; access to resources 45; centralised 3; concertation without 60; disciplined 50; discretion of 61; free flow 72; intensity of 7; shedding 90, 300
labour markets 3; deregulating 117; policy 44; reduced controls 115
Labour party/government (UK) 111, 113, 114, 116; Carter Committee (1976) 129–30, 131, 292; industrial policy 119; interventionist policies 117, 119; nationalisation 126; oposition to Thatcher's policies 135; perceived failure 121; policy experts 151; strategy to break 122; trade unions and 115, 127, 136, 138, 146; *see also* Blair
Lagardère 229
laissez-faire 3, 27, 119
Landes, D. S. 231
Lane, C. 43, 46
Lasserre, Bruno 273
LAT (living apart together) relationship 91
law EC 13, 100, 174; private 4; public to civil 5; supremacy of 64
leadership: government 116; strategic 51; technological 69; trial-and-error 31
Lebanon 269
legislature 38
Lehmbruch, G. 60
Liban Cell 269
Liberal Conservatives (Netherlands) 178, 179, 190, 202
Liberal-Conservative governments (France) 229, 238, 240, 252–4 *passim*, 298
Liberal Democrats (Netherlands) 178, 179
liberalism 35, 49, 63, 68
Libertel 162, 199, 206
licences 95, 107, 115, 198–200 *passim*; EC-wide 220; GSM 199, 206; ISR 161
Licensing Act (Netherlands, 1986) 204
Lijphart, A. 38, 172
Lille 260
Lindberg, L. N. 29
Lindblom, C. E. 17
Littlechild, S. C. 3, 137, 164
LMT 242, 247
loans: longer-term 50; short-term 116
Lodge, G. C. 67, 68
London 113, 116–18 *passim*, 132, 138, 146, 147, 284, 290; Underground 123
Longuet, Gérard 252, 255, 273
Lubbers, Ruud 177, 179, 184
Lucent 97
Lucky Goldstar 66
Lukes, S. 23
Luxembourg 58, 68, 106
Lyons 260

Maastricht 71, 105, 106, 241
McCaw Cellular 157, 160
McDonnell Douglas 157
McKinsey & Co 185, 194–5
Macmillan, Harold 124
macro-economics 8, 21, 53, 54, 60, 61, 114; Keynesian 3, 119, 120
Majone, G. 3, 31
Major, John 117, 118, 123, 124
Major Operations programme (Netherlands) 177–8
management 44, 45, 69; buy-outs 5; company-grown 46; economic 114; involvement of banking community in 116; paternalistic 46
Mannesman 162, 260; Arcor 213
Mansell, R. 233
manufacturing sector 113, 116
Marconi 126
market share 16, 17
marketing 90
markets 3, 5, 24, 44; access to 17, 18, 52, 69, 97, 99, 161; chemical 62; controls 4, 27; 'core' 82, 102; decisions left to 116; differentiation of 12, 34; dominance in 97; emerging 90; fast growing 93; foreign 17, 81; fragmented 70, 101; 'fringe' 81, 105; heavily regulated 17; imperfections 26; incompatible 100; global/international/world 11, 16, 17, 48, 53, 63, 66, 72, 75, 90, 98, 111; key 14, 16, 104; long-distance 93, 94; mobile communications 205–7; national 90, 92; markets open and competitive 51; organisation of 45; politicised 8, 21; promising 78; regional 16; restructuring 29; saturating 105; segments 6, 12; self-regulated 68; state-owned enterprises 66; stimulating 114; terminal equipment 103; too small 13; value-added 96; voice services 104; *see also* capital markets; deregulation; entry; globalisation; protectionism

Marshall Plan (1947) 228
Martin Dawes 162, 205, 268
Matra 90, 238, 239, 249, 299
Matra-Hachette 227
Matsushita 227
Maurice, M. 44
Mauroy, Pierre 235, 236, 238, 251, 294, 297
Maxwell 146
MCI 12, 81, 86, 94, 95, 96, 97, 152, 161, 167, 286; far-reaching agreement with Europeans 269; *see also* Concert
Mercury 81, 126, 133, 136–48 *passim*, 152, 155, 165, 166, 281, 284; Communications Ltd 161, 285; One-2–One 149, 162; solid market presence outside Europe 286; success of 298
mergers 52, 72, 90, 97; cross-border 71; *see also* European Commission (Merger Control Regulations); MMC
Merlin 153
Merrill Lynch 91
Mexico 268
MFS 97, 147, 204
MHP (Dutch Federation of White-collar Employees) 174
micro-economics 28
micro-electronics 11, 80; advanced 87; R&D 124; socio-economic consequences of 181
Microtel 149
microwave technologies 92, 93
Middle East 162
'midwifery' 65, 66
Millicom 149
Minc, Alain 87, 247, 248, 249, 250
Minitel 249, 250–1, 256, 257, 265; US Community Link Associates 268
Mission à la Réglementation 271–2
mission-oriented countries 52
Mitel Corporation 150, 157, 160
MITI (Japanese Ministry of International Trade and Industry) 64
Mitterrand, François 235, 236, 238, 251, 271, 294, 295, 297, 298
MMC (Monopolies and Mergers Commission) 117, 125, 150, 163–4, 289
mobile communications 205–7
MobilRom 269
Mobistar 269
Mobutu, S. S. 65
Moldova 269
Monnet, Jean 236
monopolies 1, 13, 22, 24, 35; abuse of power 26; challenging/questioning 11, 18; divestiture of 12; overmanning 3; privatising 4; representative/ representational 25, 59, 60; restriction of 112; virtual 114, 142; *see also* MMC; PTTs
monopsony 78–9
Moore, John 121–2
most-favoured nation principle 18
Motorola 90, 240
Mueller, F. 43
multimedia 106; applications 89; conglomerates 11, 76; emerging market 87
multinationals 9, 12, 17, 52, 53, 75; dilemma for 14; expanding reach 48; favourable conditions to 70; inward investment 20; rise of 16; *see also* PTTs; transnationals
multiplicity of forms 33, 34

National Bus Company 124
National Grid 123, 146
National Power 123
national security 6, 8, 26, 51, 69, 117; strong commitment to 113
nationalisation 3, 21
NATO (North Atlantic Treaty Organisation) 69, 245
NCC (National Consumer Council) 127, 142, 166
NCR Handelsblad 220
NCU (National Communications Union) 127, 136, 155
NCW (Dutch Christian Employers' Union) 173
NEB (National Enterprise Board) 120, 124
NEC 90, 240
NEDC (National Economic Development Council) 119, 121
Nelson, R. R. 55
neo-liberalism 4, 83, 135, 253, 290
Netherlands 1–2, 8, 21, 38, 162, 170–224, 261, 262, 269; adjustment strategies 37; BT partnership 157; cable penetration 106; comparative analysis 34, 279–306; consensus and reactive attitude 39; convergence/and or divergence in restructuring 307–21; dependence on world economy 48; economy 68, 171; interest associations 92; largest cable operator in 268; regime type 35, 49; size of economy 20; strong corporatism 60; *see also* Casema; KPN; Postal Girobank; Postal Savings Bank; Postbank
networks 6, 24, 52, 102, 104, 111; access to 14, 106; alternative 11; analogue 84; 'bypass' 92, 94, 105; cable 85; computer 84; corporate-industrial 55; distribution 86; engineering and interoperability 8; expansion 80; fibre-optic 89; interconnected 86; interdependent 25; interorganisational 51; large-scale private user 81; leased-facilities 87; obligation 46; sectoral 42; social relations 28; supplier 20; supportive 15; *see also* INS; ONP; PCNs; VANs
New Institutionalism 27
New Right governments 82–3
New Zealand 38
NHS (National Health Service) 119
NIE (New Institutional Economics) 27, 28
NKT/PTT Contest (Dutch certification agency) 200
NMA (Dutch anti-trust authority) 175

Noam, E. M. 6, 79, 93
Nora, Simon 87, 247, 248, 249, 250
Nordic-Tel 162
North, D. C. 31–2
Northern Telecom 78, 79, 90, 151, 160
Norway 60
nouvelles cathédrales 234
NRAs (National Regulatory Authorities) 107
NRDC (National Research and Development Company) 124
NS (Dutch railway company) 172, 178, 195, 196, 197, 198, 204
NUON/Telekabel 204
Nynex 95, 97, 141, 162; CableComms 148

OECD (Organisation for Economic Co-operation and Development) 18, 62, 257
Offer (Office of Electricity Regulation) 125
Ofgas (Office of Gas Supply) 125
OFT (Office of Fair Trading) 117, 125, 134, 164
Oftel (Office of Telecommunications) 111, 125, 134, 136, 140, 142, 143, 145, 146, 148, 150, 155, 281–3, 285; BT accountability 166–7; DGT (Director General of Communications) 163–4, 289; statutory duty 168; strong impetus to effective duopoly 165
Ofwat (Office of Water) 125
Ohmae, K. 16, 72–3
Olivetti 97, 106, 181
ONP (Open Network Provision) 104, 107, 197, 198, 220, 221
openness 1, 14, 104, 111
OPT (Dutch consultative body) 206, 218
OPTA (Dutch autonomous authority) 221, 222
Orange 147, 149
Orbitel 162
organisation studies 43–8
OSHA (American health and safety agency) 40
outsiders 77
ownership: conditions of 25; foreign 4; private 5; public 3, 5; state 61; stipulations of 27

PABXs (Private Automatic Branch Exchanges) 97, 130, 150, 157, 260
Pacific Basin *see* Australia; Hong Kong; Taiwan
Pacific Link Communications 162
Pacific Rim 160, 162, 213; *see also* Japan; Singapore; South Korea
Pacific Telesis 97
PAFE (Programme d'Action par la Filière Electronique) 249
PAL system 234
Panafon 162, 269
pantouflage 230
Paribas 230, 238, 239
Paris 244, 260; stock exchange 232, 238, 239
partnerships 63; international 16, 17, 72, 157, 162, 227; social 21, 49, 115; strategic 43, 73, 81
path-dependent behaviour 32, 296
PCF (French Communist party) 230
PCNs (Personal Communications Networks) 140, 143, 144, 149
performance 4, 7, 55, 58; across countries 44; determination of 28; differences in 66; national, superior 46; negative influence on 69; poor 11, 113
Philips 8, 53, 78, 90, 97, 106, 171, 199, 200; CII and 244; *see also* APT
Plan Calcul 227, 243
Plan de Composants 227
Plan de Rattrapage Téléphonique (1970) 246
Plan Filière Electronique 252
Plessey 8, 78, 90, 113, 127, 149–51, 157
pluralism 37, 46, 58, 59, 115; disjointed 60
POEU (Post Office Engineering Union) 126, 128, 130, 131, 136–7, 146; *see also* NCU
Poland 216, 269
policy making 9, 25, 28, 31, 40–1, 52, 113–14; changes in 22–3; decentralisation of 104; EC 104; incentives and constraints 35
political economy 48–53
politics 41; comparative 38; confrontational 114; domestic 15, 16; economic adjustment 119–26, 176–80, 236–41; stable 49
polytechniciens 244
Pompidou, Georges 245
Pool, I. de Sola 86, 87
Porter, M. E. 54, 56–7, 67
Portugal 106, 161
positioning 54, 69–73, 117–19, 175–6, 234–6
Post Office Acts: (Netherlands, 1954) 190; (UK, 1969) 127–8
Post Office (UK) 111, 126–30, 132, 138, 149, 297; White Paper (1987) 131
Postal Girobank (Netherlands) 180
Postal Savings Bank (Netherlands) 180
postal services *see* PTTs
Postbank (Netherlands) 178, 179, 213
POTS market 250
POUNC (Post Office Users' National Council) 126, 128, 130, 131, 136
Powell, W. W. 31
power: abuse of 26; dependencies 30, 38; distribution 15, 20, 44; EC institutions 13; economic 118; executive 38, 113; legislative 104; national economies losing 72; political 114; quest for 29; union 3, 114
PowerGen 123
pressure groups 58; environmental 62
Prévot/Quilès Report (1989) 253–4
prices 3, 24, 104, 106–7; caps 117, 137, 164, 216; competitive 68; controls 4; equipment 8; export 113; non-discriminatory 88; setting 6, 27; share 116
problem solving 30, 39; (re)active approach 40
procurement 19, 51, 53, 104; budgets 8; contracts 18, 75, 101; EC regime 100; policies 8, 98

product life cycles 72, 81, 90
production 16, 24, 26, 73; decentralised 72; factors of 55; innovating 29; lean 48; selective 49; shortening life-cycles 12; steel 66; upgrading 69
production profiles 53, 54–8, 112–13, 171–2, 226–7
productivity 89; improved 83
profit: retained 116; short-run, maximising 113
property rights 24, 25, 29, 45; defining and enforcing 51; emphasis on 68
protectionism 6, 12, 16, 49, 52, 65, 90, 91; adverse effect on 96; avoided 72; 'infant industry' 55; key industries 79; only way to combat 97; rejection of 73; *see also* France; Germany; Greece; Italy; Japan; Spain
PSBR (Public Sector Borrowing Rate) 121, 126, 134
PTT Telecom (Netherlands) 193–205, 207, 216–20; *see also* KPN/PTT
PTTs (postal, telegraph and telephone monopolies) 2, 5–10, 14, 19, 75–110; British 111–69; Dutch 171–2, 177–224; French 225–78; *see also* BT; DBP; France Télécom; KPN; PTT Telecom; Telefonica
public expenditure: controlling 116; cutbacks/curbing 3, 4, 16; reducing 117
'public goods' 51, 52
public policy studies 38–43
PUK (Pechiney-Ugine-Kuhlman) 238, 240

quality 56
'quasi-trading blocs' 17
Quirk, P. J. 30

R&D (research and development) 96, 120, 241, 261, 265; alliances 25; comprehensive framework 13; cooperative 51; coordination of 296; costs 12, 16, 20, 52, 72, 81, 90; duplicated 101; expenditure 113, 171; funding75, 256; international cooperation 42; investment in 69, 116; joint 8, 53, 100, 243; long-term collaboration 79; micro-electronics 124; miltary 227; subsidising 66, 91; support for 62, 227
Rabobank 207
Racal 146, 148, 149, 162, 281, 284
Radio Monte Carlo 229
RAM Mobile Data 268
RATB (Dutch Advisory Council on Post and Telecommunications) 187, 189, 191
Rathenau (Dutch Advisory Committee) 181–2
Rattrapage Téléphonique, Le Plan de (1970) 243–7
RBOCs (Regional Bell Operating Companies) 12, 94, 95, 96, 97, 287
RCA 95; Global Communications 94
RCO (Dutch Council of Central Employers' Association) 196, 202
reactive approach 39, 40
Reagan, Ronald 83
recession 3, 14, 114
redistribution 6, 7, 79, 80, 81, 105
Reed-Elsevier 171
'regionalisation' 73
regulations 12, 31, 59; adversarial 114; agency role 26–7; anti-trust 26; asymmetric 117; burdensome 17; bureaucratic 114; competition 102; convergence of 71; EC 13, 118; environmental 41, 42, 62; flexible 9, 79; loosening and lifting 18; public 27; self 115; social 26; structural developments in 14; supranational 10
regulatory reform 111–278 *passim*, 288–91
Renault 227, 235, 236, 240, 241, 249
rent-seeking coalitions 6, 76–84
Résaux Câbles 216
resource allocation 3, 24, 51, 72–3
restructuring 2–5, 10, 13, 29, 32, 35, 81, 105, 111; comparative analysis 15–22; convergence and/or divergence 307–21; domestic 48; external 63; industrial 57, 69; internal 63; international 37–74; radical 113
Reuters 80
revenues 5, 6; division of 9; redistributed 7; sharing 10
Revised Telecommunications Bill (Netherlands, 1998) 198
Rhône-Poulenc 238, 240, 267
Richardson, J. 39
rivalry 56, 93, 101; strategic 71; transnational 53
Rolls Royce 124
Rolm 97
Romania 269
Rome, Treaty of (1957) 4, 13, 71, 100, 103, 118
Roobeek, A. J. M. 90, 91
Rose, R. 31, 32
Roulet, Marcel 272
Royal Dutch Shell 52, 171
Royal Mail 123, 132
RPI (retail price index) 164–5, 247, 272
Ruigrok, W. 52, 55, 73
rules 25, 27, 32, 34, 64, 78; competition 103, 106; devising 41; enforcement 40, 41; free trade 19; implementation 40, 41; interdependent 28; making 40; persistent 1
Rutkowski, A. M. 98–9
RVPT (Dutch supervisory body) 191

Saint-Gobain 230, 238, 239
salaries 8, 79, 89
Samsung 66
satellites 11, 75, 78, 84, 86, 89, 92, 97
Sauter, W. 100
SBS (Satellite Business System) 97, 247
Scandinavian countries 58, 68, 105; *see also* Denmark; Finland; Norway; Sweden
Schattsneider, E. E. 22
Schiller, D. 92, 98
Schiphol Telematics 212

Schmitter, P. C. 60
Schneider 244
Schuman, Robert 236
Scott, C. 100
Scottish Hydro-Electric 146
ScottishPower 146
SECAM standard 234
Securicor 149
security *see* national security
SEMA Group 268
semiconductors 70, 100
service providers 10, 86; multiple 87; private 11; *see also* ISPs
services industries 113
SFR (Société Française du Radiotéléphone) 157, 162, 257, 259, 260, 274
SFTE 242, 243
Sharp, M. 120–1
Shonfield, A. 233
Siemens 8, 53, 90, 106, 151, 181, 270
Singapore 57; Telecom 141, 213, 269
Single Market Programme 175
Siris 213, 260
SITA (Société Internationale de Télécommunications Aéronotiques) 92, 214
SLECs (Sociétés Locales d'Exploitation du Cable) 259
Slovenia (Slovtel) 269
'smokestack' industries 113
SNA architecture 152
SNCF (French railways) 157, 239, 253, 260, 274
social benefits 101
Social Democrats 7, 179, 180, 184, 190
'social-market economy' model 68
social security 69
socialism 115, 242; French governments 229, 237–40 *passim*, 249, 251–3 *passim*, 271, 294, 297–8
Société Anonyme status 258
Société Générale 230, 232, 236, 239, 261, 267
Société Nationale 252–3
Sociétés de Financement 242
sociology 27, 28
Socotel 242
Sofrecom 264
software 70, 90; advanced technologies 85
Sorge, A. 43, 44, 46
Sotelec 242
South Africa 162, 269
South Korea 46, 50, 57, 65, 66, 229
Soviet Union (former) 48, 118
Space 262
Spain 76, 105, 106, 118, 261, 286; BT partnership 157; *see also* Telefonica
specialisation 46, 85; international trade 18; niche 49
Sprint 86, 96, 145, 160, 213, 256, 260, 265, 268–70 *passim*, 286, 301
SPT Telecom 214
stagflation 59
stagnation 4
Stalker, G. M. 44
standards/standardisation 10, 19, 42, 51, 78, 89; idiosyncratic 11, 90, 91, 99; national, diverging 101; stimulating 102; transparent 86; *see also* CEN-CENELEC; ETSI; ISO
state: organisation 113–15, 172, 227–30; role 29, 63–6; 'strong' 117; structure 53; *see also* government/state intervention
state-owned enterprises 4, 66
statism 35, 49
STC (Standard Telephones and Cables) 127, 128, 149, 151, 157, 160
STE (Society of Telecom Executives) 127, 155
Steenbergen Committee (Netherlands) 185–90, 302
stock markets 45, 61, 69, 116
Stopford, J. 73
Stourdze, J. 244
Strange, S. 73
strategic alliances 12, 66, 82; build up of 194; cross-national 13, 213; establishing 72; international 16, 92, 214, 255, 256
strategies 34, 44, 51, 81, 12; adjustment 37, 61, 111, 112; cartel-oriented 23; catch-up 50, 64, 65, 66; common 14, 100; competitive 55; diversification 7, 85, 96; financial 116; firm 56; grand design 21; high-risk 238; liberal 114; long-term 85; market-oriented 23; national 2, 37–74; niche 20; search 32; state-oriented 23; two-fold 53
Streeck, W. 60
structural adjustment 37, 112
structural forces 10–15, 34; challenging the European regime 75–110
subcontracting 25
subsidiaries 117, 132, 138, 161, 162, 198, 200, 237, 238, 241, 255, 260; important 265; loss making 160; semi-independent 263
subsidies 114; export 53, 75; R&D 66, 91
Suez Bank 238, 239
sunrise industries 113
sunset industries 50, 113
Surfnet 212
Swann, D. 4
Swarttouw Committee (Netherlands) 182–5
Sweden 5, 21, 150, 162, 181, 242, 286; BT partnership 157; consensus inspired by normative values 39–40; diffusion-oriented 52; policy making 40–1; regime type 49; strong corporatism 60; *see also* Telia; Televerket
SWIFT (Society for Worldwide Interbank Financial Telecommunications) 92
Swiss Telecom 214
switching equipment 8, 79, 81, 84, 88; digital 90; second or third suppliers 91; selling abroad 96

Switzerland 38, 49, 60, 68, 211, 286; *see also* Swiss Telecom
SWPs (Sectoral Working Parties) 120
Syncordia 153, 157, 160, 286

T&T Act (Netherlands, 1904) 200
Taiwan 46
Talkline 205
tariffs 10, 77, 81, 85, 86; common 103; customised 92; drastic reduction 93; flat-rate 87; intra-Community 104; long-distance, expensive 11; low 105; *see also* GATT
taxes 101; corporate rulings 115; high, avoidance of 17
TDF (Télédiffusion de France) 241, 248, 255, 258, 264
techno-globalism 15, 52, 53
techno-nationalism 15, 52, 53
technological change 16
technological revolution 11, 34, 75, 84–8
technology: policy 50, 51, 52; transfer 51, 55
Telecential 216
Telecom de Argentina 268
Telecom Denmark 157, 214
Telecom Finland 147
Telecom Italia 211, 260
Telecom Portugal 161
Télécom Système Mobile 264
Telecommunications Acts: (France, 1988) 258 (Netherlands, 1989) 193, 198, 293; (UK, (1981/84) 133–4, 136, 137
Telefication 200
Telefonica 161, 181, 214, 261
telegraph services *see* PTT
telematics 93, 96, 247–52
Télématique, Le Plan (1976–80) 247–52
Telenor 157
Telenordia 157
telephone services *see* PTT
Télésystèmes 264
Télétel 250–1, 256, 268
Télétel Vitesse Rapide 268
Telettra 262
Televerket/Telia 213
TeleWest 148
telex 9, 84, 85, 91
Telfort 157, 198, 204, 207, 222
Telia 147, 213, 214
Telkom 214
Telkomsel 216
TelMex 268
Telstra 147
TEMA (Telecommunications Engineering and Manufacturing Association) 131
tendering 4, 103
terminal equipment 84, 103
TFI (first French TV channel) 239
TGVs (French high-speed trains) 234
Thatcher, Margaret 43, 83, 111, 114–24 *passim*, 131–2, 135, 136, 147, 161, 281, 284, 290, 291, 297, 300
Théry, Gérard 256–7
Thompson, G. 72, 73
Thomson 225, 227, 237, 240, 243, 249, 252, 255, 260, 294, 299; Brandt 238; CGE agreement 261; combined with Schneider in CII 244; Consumer Electronics 268; CSF 229, 234, 247; SA 268; SGS 265
Thorn Ericsson 149
Thorn-EMI 157
Tilly, C. 33
TMA (Telecommunications Managers Association) 92, 127, 141–2, 166, 294
TND (Dutch regulatory unit) 221
Torch Ltd 146–7
Tories *see* Conservative party/government
Total 240
Toyota 227
trade: international/world 11, 18, 19, 52, 71, 81, 112; liberalisation 50, 69, 99; national differences 43; strategic policies 53; *see also* free trade; trade barriers; trade policy
trade barriers 16, 19, 55, 70, 71, 72, 73; abolishing 101, 105
trade policy 69, 73; joint export 21; managed 55
training 51, 55; investment in 69; vocational 44
transactions 8, 23, 24, 25, 29
transformation: governance 27–32, 33, 34; structural 49
transmission: alternative 10, 86; leased 92; technologies 84
transnationals (transnational companies, *see also* multinationals) 15, 16, 18, 19, 52, 92; demands of 70; giant 96; large 62; power games 72; rivalry 53; suppliers of switching equipment 79
Transopen 212
Transpac 249–50, 255, 264, 270
Travel-net 212
Trentes Années Glorieuses 237
Trompenaars, F. 67
TUA (Telecommunications Users Association) 127, 142
TUC (Trades Union Congress) 115, 120, 127
Tymnet 157

UAP 227, 240, 249, 265
UCW (Union of Communications Workers) 127, 130, 155
Ukraine 209
unemployment 3, 113, 177, 240–1
Unilever 52, 171
unions 8, 58, 62; fall in membership 115; rent-seeking coalitions 6, 81; substantial power 3; *see also* AIT; BTUC; CAT; CFDT; CGT; CMA; CNV; CPSA; FNV; FO; MHP; NCU; NCW; POEU; STE; TUC; UCW
Unisource 92, 209, 260, 286, 288; At&T 211, 213, 274, 301; Telematic systems and Services 214

United Kingdom 1–2, 5, 38, 57, 100, 111–69; adjustment strategies 37; cable prices 107; chemical industry 62; comparative analysis 34, 279–306; convergence/and or divergence in restructuring 307–21; deregulation 4, 12, 76, 84, 98, 107; duopoly 104, 114, 143–4, 148, 167, 297; economy 68; environmental policy making 41; government factions and main companies 63; interest associations 92; liberalisation 105; market-based financial system 61; mission-oriented 52; organisation of work and vocational training 44; policy style 40, 43; regime type 35, 49; size of economy 20; weak corporatism 60; worldwide liberalisation 93; *see also* BT; Mercury; Racal; Vodafone
United Nations 9
United States 1, 13, 18, 20, 48, 57, 100; British collaboration with 118; Communications Act (1934) 94, 97; Computer Inquiries (1971/1980/ 1986) 95; decision-making 14; Department of Justice 94, 97; deregulation 76, 84, 93–9, 107; disjointed pluralism 60; economic power 118; economy 68; environmental policy making/regulations 41, 42; FCC (Federal Communications Commission) 94, 97, 160, 289; fierce competition from 70; health and safety policy 40; hegemony 12, 17; high-tech threat from 101; interdependence 16, 72; liberalisation 93, 96; market-based financial system 61; mission-oriented 52; Modified Final Judgement 12, 95, 96; private firms 5; PUCs (Public Utility Commissions) 94; radical/ambitious deregulation and 4, 12; regime type 35, 49; regulatory policy 30, 43, 114; self-assertive values 41; socio-economic change and innovation 51; Specialised Common Carrier Decision (1971) 94–5; Star Wars 235; subsidiaries 117; Telecommunications Act (1996) 97; trade liberalisation 69; traffic routing to 145; vertical integration 78
universalising 33, 34
Uplink 146
US West 95, 148, 149, 204, 268
Usinor Sacilor 238, 240, 241
USSR *see* Soviet Union (former)
utilities 4, 115, 117; transformed 5
UUNet 97

value-added services 85, 97, 153, 211, 252, 253; developing 91; 'fringe' markets liberalised 81–2; high-quality products 113; market for 96; *see also* VANs
Van Agt cabinets (Netherlands) 177
Van den Bosch, F. A. J. 56–7
VANs (value-added network services) 89, 96, 132, 133, 138, 146, 152; licences and access charges 260
Van Tulder, R. 52, 55, 73
variables: dependent 35; extra-national 34; national 35, 53–73; sectoral 35
variation-finding comparisons 33–4
VEBA 162, 259, 260
VECAI 202, 203
vertical integration 78, 79, 88, 288
Vesatel initiative (1991) 213
VIAG Interkom 157
Vickers, J. 122
videoconferencing 84, 88, 89
videotechnologies 11
Videotex NL 268
Videotron 148, 162
Viditel 211
Vision Networks 216
VNO (Federation of Dutch Industries) 173
Vodam/Vodac 162
Vodafone 81, 147, 149, 162, 206, 259, 281, 284; success of 298
Vogel, D. 41, 68

Wade, R. 65
wages 3, 101
Walker, W. 120–1
Warner, M. 44
Water Authorities (England and Wales) 123
Western Electric 94, 95, 96
Western Union 94
White Papers, Netherlands: (1984/85) 184–5, 188, 190, 201; (1988) 202
Whitley, R. 43, 45, 46
Williamson, O. E. 23
WIND 269
Windmuller, J. P. 173
WorldCom 97, 147, 161, 204
Worldpartners 213
WorldSource 213, 269, 301
Wright, V. 227
WTO (World Trade Organisation) 17, 18–19, 62, 69, 99, 119

Yarrow, G. 122
Yorkshire electricity 147

Zaire 65
Zegveld Committee (1986) 201–2
Zysman, J. 61, 232